The Nanotechnology Driven Agriculture

Nanotechnology is believed to accelerate our fight to sustain and enhance crop productivity for the ever-increasing world population. It has been reckoned as one of the safest and most cost-efficient techniques to boost crop productivity in the future. The ever-increasing state-of-the-art availability of various nanomaterials has allowed us to pursue their beneficial properties in agronomy. Nanotechnology will help minimize the dependency on chemical fertilizers and existing crop production techniques that have already been exploited to their maximum potential. Therefore, in the present context, it could well be realized that agriculture will be driven by nanotechnology in the future.

This book focuses on the application of nanotechnology for enhancing crop production through the application of nanofertilizers or nanocomposites. Several avenues of nanotechnology are beneficial in improving crop productivity in a sustainable manner, which has been presented in a comprehensive way. The book also delves into the mechanistic view of nanoparticle functioning and its role in stress alleviation. In addition, the book presents some recent insights into the application of nanotechnology for post-harvest management, stress tolerance and usage as nanobiosensors.

Broadly, the book will encompass the following advances in the field, distinguishing it from other published volumes. The salient features include:

- Role of nanoparticles in improving abiotic stress tolerance in plants.
- Role of nanoparticles in protection against pathogens and pests.
- Mechanism of nanoparticle-induced plant responses.
- Synthesis and modification of nanoparticles to enhance their biological efficacy.
- Prospects of nanofertilizers, nanoformulations, nanopesticides, etc., and their beneficial attributes.

This book, therefore, presents this emerging topic and the most recent innovations in this field for postgraduate students, researchers and faculty members working in the fields of plant science, microbiology, biotechnology, agricultural sciences, etc.

The Nanotechnology Driven Agriculture

The Future Ahead

Edited by
Swarnendu Roy
and Akbar Hossain

CRC Press
Taylor & Francis Group
Boca Raton London New York

CRC Press is an imprint of the
Taylor & Francis Group, an **informa** business

Cover image credit: Swarnendu Roy

First edition published 2024
by CRC Press
2385 NW Executive Center Drive, Suite 320, Boca Raton FL 33431

and by CRC Press
4 Park Square, Milton Park, Abingdon, Oxon, OX14 4RN

CRC Press is an imprint of Taylor & Francis Group, LLC

ISBN: 9781032453217 (hbk)
ISBN: 9781032453248 (pbk)
ISBN: 9781003376446 (ebk)

DOI: 10.1201/9781003376446

Typeset in Times New Roman
by Deanta Global Publishing Services, Chennai, India

Contents

Preface

Increasing crop productivity to feed the ever-increasing world population has been a significant challenge in modern agriculture. Agriculture productivity has reached its theoretical maximum, and the increasing impacts of several environmental stresses significantly affect the same. At the same time, the use of chemical fertilizers and pesticides has been known to have dire consequences. Given these circumstances, there is an emergent need to look towards alternative strategies or techniques that can help in enhancing crop productivity without harming the environment.

In the past few decades, the potential of nanotechnology has been realized in many sectors, including electronics, drug development, food sciences, environmental safety, agriculture, etc. Nanotechnology has been a boon for humanity in every aspect and is believed to be the next big thing that could drive the much-needed second wave of a green revolution. In the agriculture sector, the potential of nanotechnology for developing nanofertilizers and nanopesticides has been explored. The ever-increasing state-of-the-art availability of various nanomaterials has allowed us to venture into their beneficial properties in agronomy. Therefore, in the present context, it could be well realized that nanotechnology will drive future agriculture. However, there is still a long way to go; therefore, we urgently need to shift our focus towards nanotechnology to boost our agricultural productivity. This book will attempt to illuminate this emerging topic and cover the most recent innovations in this field.

Editor Biographies

Swarnendu Roy, PhD, is presently working as an assistant professor at the University of North Bengal, Siliguri, India. He has more than 13 years of teaching experience. He has published more than 40 articles in reputed journals with high-impact factors, such as *Water Research, Scientific Reports* and many others. He has recently edited two books published by Springer and has also published several book chapters in edited volumes to date. Presently, he is actively engaged in research topics encompassing stress management in plants by engineered nanoparticles and the synthesis of starch-based biofilms and edible coatings. He is presently supervising four research scholars. He received International Travel Support (ITS) from DST, Govt. of India in 2019 to present his research paper at the International Conference on Integrative Plant Physiology in Sitges, Spain. He is a life member of prestigious societies such as the Indian Science Congress Association and the Association of Food Scientists and Technologists (India).

Akbar Hossain, PhD, is working as a senior agronomist at the Bangladesh Wheat and Maize Research Institute. He has edited/is editing several books in his research areas, which include plant stress physiology, maize production and uses, input use efficiency for food and environmental security and plant defense mechanisms. He has also authored more than 200 journal articles related to agronomy, plant breeding, plant physiology, nanotechnology etc. He is serving as a Research Supervisor of the M.Sc. students in Agronomy and Soil Science and Irrigation Engineering at Hajee Mohammad Danesh Science and Technology University in Dinajpur, Bangladesh. He is also very interested in the prospects of nanotechnology in agriculture in the modern era of climate change.

Contributors

NPM Abeysinghe
Department of Export Agriculture, Faculty of Agricultural Sciences, Sabaragamuwa University of Sri Lanka

Muhammad Ahmad
Department of Agronomy, University of Agriculture Faisalabad
Punjab Pakistan

Zahoor Ahmad
University of Central Punjab Constituent Punjab College Yazman Road Bahawalpur
Punjab Pakistan

Sarfraz Ahmed
Department of Botany, University of Okara
Punjab Pakistan

N Akter
Plant Breeding Division, Bangladesh Agricultural Research Institute (BARI)
Gazipur, Bangladesh

Muhammad Ashar Ayub
Institute of Agro-Industry and Environment, The Islamia University of Bahawalpur, 63100
Punjab, Pakistan

Amir Aziz
Department of Soil and Environment Sciences, College of Agriculture, University of Sargodha
Sargodha, Pakistan

Prasanta Kumar Bandyopadhyay
Department of Agricultural Chemistry and Soil Science, Bidhan Chandra Krishi Viswavidyalaya
Mohanpur, Nadia, West Bengal, India

Shehar Bano
Department of Plant Biotechnology, National University of Sciences and Technology (NUST)
Pakistan

Anamika Barman
Division of Agronomy, ICAR, Indian Agricultural Research Institute
New Delhi, India

Celaleddin Barutçular
Department of Field Crops, Faculty of Agriculture, Cukurova University Adana
Turkey

Saika Bashir
Proteomics Laboratory, Division of Plant Biotechnology, Faculty of Horticulture, Sher-e-Kashmir University of Agricultural Sciences & Technology of Kashmir
J&K, India

Aneesa Batool
Proteomics Laboratory, Division of Plant Biotechnology, Faculty of Horticulture, Sher-e-Kashmir University of Agricultural Sciences and Technology of Kashmir
J&K, India
and
Department of Chemistry, Bhagwant University of Ajmer
Rajasthan, India
and
Department of Chemistry, Govt. College for Women, Cluster University Kashmir
Srinagar, India

Sujit Kumar Behera
Department of Nematology, OUAT
Bhubaneswar; Odisha, India

Anurag Bera
Department of Agronomy, Institute of Agricultural Sciences, Banaras Hindu University
Varanasi, Uttar Pradesh, India

Zahoor Ahmad Bhat
Division of Plant Pathology, Sher-e-Kashmir University of Agricultural Sciences & Technology of Kashmir
J&K, India

Ramireddy Boppella
Department of Chemistry, Yonsei University, 50 Yonsei-ro Seodaemun-gu, Seoul, Republic of Korea
and
Department of Mechanical Engineering, Colorado State University, Fort Collins
Colorado, United States

Swapnadip Bose
Department of Agronomy, Bidhan Chandra Krishi Viswavidyalaya
Mohanpur, Nadia, West Bengal, India

Muhammad Adnan Bukhari
Department of Agronomy, Faculty of Agriculture and Environmental Sciences, The Islamia University of Bahawalpur
Punjab, Pakistan

Aniruddha Chatterjee
Centre for Advanced Materials Research and Technology, Department of Plastic and Polymer Engineering, Maharashtra Institute of Technology
Aurangabad, Maharashtra, India

Arkadev Chattopadhyay
Department of Nematology, Assam Agriculture University
Jorhat, Assam, India

Manoranjan Das
Faculty of Agriculture, Department of Nematology, OUAT
Bhubaneswar; Odisha, India

Sanjoy K Debsharma
Plant Breeding Division, Bangladesh Rice Research Institute (BRRI)
Gazipur, Bangladesh

Madhurima Dey
Department of Agronomy, Bidhan Chandra Krishi Viswavidyalaya
Mohanpur, Nadia, West Bengal, India

Ayman El Sabagh
Department of Field Crops, Faculty of Agriculture, Siirt University
Siirt, Turkey
and
Department of Agronomy, Faculty of Agriculture, Kafrelsheikh University
Kafr Elsheikh, Egypt

Faria Fatima
Department of Agriculture, IIAST, Integral University
Kursi road, Lucknow, Uttar Pradesh, India

GAH Galahitigama
Department of Export Agriculture, Faculty of Agricultural Sciences, Sabaragamuwa University of Sri Lanka

Rajeev K. Gupta
Punjab Agricultural University
Punjab, India

T Halder
Plant Physiology division, Bangladesh Rice Research Institute (BRRI)
Gazipur, Bangladesh

Akbar Hossain
Division of Agronomy, Bangladesh Wheat and Maize Research Institute
Dinajpur, Bangladesh

Khurshed Hussain
Division of Vegetable Sciences, Sher-e-Kashmir University of Agricultural Sciences & Technology of Kashmir
J&K, India

Samuel J. Ippolito
Centre for Advanced Materials and Industrial Chemistry (CAMIC), School of Science, RMIT University
Melbourne, Victoria, Australia
and
School of Engineering, RMIT University
Melbourne, Victoria, Australia

Muhammad Aamir Iqbal
Department of Agronomy, University of Poonch Rawalakot
Azad Kashmir, Pakistan

Avdhut Jadhav
Functional Materials Laboratory, Department of Mechanical Engineering, Maharashtra Institute of Technology
Aurangabad, Maharashtra, India

Jawaria Jameel
Department of Botany, Sub campus Rahim Yar Khan, The Islamia University of Bahawalpur
Punjab Pakistan

Rupak Jena
Nematology, ICAR-National Rice Research Institute
Cuttack, India

Shravanti Joshi
Functional Materials Laboratory, Department of Mechanical Engineering, Maharashtra Institute of Technology
Aurangabad, Maharashtra, India

Anu Kalia
Punjab Agricultural University
Punjab, India

Pragnya Khamari
Department of Chemistry, National Institute of Technology Jamshedpur
Adityapur, Jamshedpur, Jharkhand, India

Fiza Rashida Khan
Department of Chemistry, National Institute of Technology Jamshedpur
Adityapur, Jamshedpur, Jharkhand, India

H Khatun
Plant Breeding Division, Bangladesh Rice Research Institute (BRRI)
Gazipur, Bangladesh

M Khatun
Plant Breeding Division, Bangladesh Rice Research Institute (BRRI)
Gazipur, Bangladesh

Kunal Kumar
Functional Materials Laboratory, Department of Mechanical Engineering, Maharashtra Institute of Technology
Aurangabad, Maharashtra, India

Ravinder Kumar
ICAR-Central Potato Research Institute
Shimla, India

Rishav Kumar
Department of Agronomy, Dr. Rajendra Prasad Central Agricultural University, Pusa
Bihar, India

Milan Kumar Lal
ICAR-Central Potato Research Institute
Shimla, India

Subrata Mahanta
Department of Chemistry, National Institute of Technology Jamshedpur
Adityapur, Jamshedpur, Jharkhand, India

Richa Prasad Mahato
Division of Microbial Technology, CSIR-National Botanical Research Institute
Rana Pratap Marg, Lucknow, India

Sagar Maitra
Department of Agronomy, Centurion University of Technology & Management
Paralakhemundi, Odisha, India

Firdose A. Malik
College of Temperature Sericulture, Sher-e-Kashmir University of Agricultural Sciences and Technology of Kashmir
J&K, India

Satyabrata Mangaraj
Odisha University of Agriculture and Technology
Bhubaneswar, India

Solange I. Rivera Manrique
Grupo de Investigación sobre Aplicaciones Nano y Bio tecnologías para la Sostenibilidad (ANBioTS-LASAE), Departamento de Ingenierías Química, Electrónica y Biomédica, División de Ciencias e Ingenierías, Campus León, Universidad de Guanajuato, Guanajuato C.P. 37150
México

Alonso Méndez-López
Departamento de Botánica. Universidad Autónoma Agraria Antonio Narro. Calzada Antonio Narro 1923
Buenavista, Saltillo, Coahuila. C.P. 25350. México

Katyayani Mishra
Department of Biotechnology, Era University
Lucknow, Uttar Pradesh, India

Debojyoti Moulick
Department of Environmental Science, University of Kalyani
Nadia, West Bengal, India

Hira Nawaz
Plant Protection Department Faculty of Agriculture, Cukurova University Adana
Turkey

Harsita Nayak
Institute of Agricultural Sciences, Banaras Hindu University
Varanasi, India

Momina Nazir
Department of Chemistry, Govt. College for Women, Cluster University Kashmir
Srinagar, India

Rabindra Kumar Paikaray
Odisha University of Agriculture and Technology
Bhubaneswar, India

Mukesh Patra
Centurion University
Parlakhemundi, Gajapati, Odisha, India

Binaya Kumar Pattnaik
Symbiosis Institute of Geoinformatics (SIG), Symbiosis International (Deemed University) (SIU)
Pune, India

Hermes Pérez-Hernández
Departamento de Botánica. Universidad Autónoma Agraria Antonio Narro. Calzada Antonio Narro 1923
Buenavista, Saltillo, Coahuila. C.P. 25350. México

Ratneswar Poddar
Department of Agronomy, Bidhan Chandra Krishi Viswavidyalaya
Mohanpur, Nadia, West Bengal, India

Jyotirmayee Pradhan
Department of Chemistry, National Institute of Technology Jamshedpur
Adityapur, Jamshedpur, Jharkhand, India

Shriram Ratan Pradhan
Odisha University of Agriculture and Technology
Bhubaneswar, India

Biswajit Pramanick
Department of Agronomy, Dr. Rajendra Prasad Central Agricultural University
Pusa, Bihar, India

MA Rahman
Plant Breeding Division, Bangladesh Rice Research Institute (BRRI)
Gazipur, Bangladesh

Vishnu D. Rajput
Academy of Biology and Biotechnology, Southern Federal University
Stachki 194/1, Rostov-on-Don, Russia

Anuj Ranjan
Academy of Biology and Biotechnology, Southern Federal University
Stachki 194/1, Rostov-on-Don, Russia

Subhash Rathod
Functional Materials Laboratory, Department of Mechanical Engineering, Maharashtra Institute of Technology
Aurangabad, Maharashtra, India

Disna Ratnasekera
Department of Agricultural Biology, Faculty of Agriculture, University of Ruhuna
Sri Lanka

Amber Raza
Department of Botany, The Govt. Sadiq College Women University Bahawalpur Punjab

Popy R Roy
Seed Regulation and Quality Control Wing, Seed Certification Agency (SCA), Ministry of Agriculture, Government of the Peoples Republic of Bangladesh
Gazipur, Bangladesh

Aryadeep Roychoudhury
Discipline of Life Sciences, School of Sciences, Indira Gandhi National Open University
Maidan Garhi, New Delhi, India
Pakistan

Ylias M. Sabri
Centre for Advanced Materials and Industrial Chemistry (CAMIC), School of Science, RMIT University
Melbourne, Victoria, Australia
and
School of Engineering, RMIT University
Melbourne, Victoria, Australia

Priyanka Saha
Division of Agronomy, ICAR- Indian Agricultural Research Institute
New Delhi, India

Subhadip Saha
Department of Agricultural Chemistry and Soil Science, Bidhan Chandra Krishi Viswavidyalaya, Mohanpur, Nadia, West Bengal, India

Chandan Sahu
Gangadhar Meher University
AmrutaVihar, Sambalpur, Odisha, India

Salinda Sandamal
State Key Laboratory of Systematic and Evolutionary Botany, Institute of Botany, Chinese Academy of Sciences
Beijing, China
and
University of Chinese Academy of Sciences
Beijing, China

Sundas Sana
Department of Botany, Sub campus Rahim Yar Khan, The Islamia University of Bahawalpur
Punjab Pakistan

Mahima Misti Sarkar
Plant Biochemistry Laboratory, Department of Botany, University of North Bengal
Raja Rammohunpur, West Bengal, India

Alma Serafín-Muñoz
Departamento de Ingeniería Ambiental, Divisi6n de Ingenierías, Universidad de Guanajuato
Campus, Guanajuato 36000, México

AKM Shalahuddin
Plant Breeding Division, Bangladesh Rice Research Institute (BRRI)
Gazipur, Bangladesh

Nafisa Siddiqui
Department of Chemistry, National Institute of Technology Jamshedpur
Adityapur, Jamshedpur, Jharkhand, India

Ankur Singh
Post Graduate Department of Biotechnology, St. Xavier's College (Autonomous), 30, Mother Teresa Sarani
Kolkata, West Bengal, India

Pallavi Singh
Department of Microbiology, Kanya Gurukul Campus, Gurukul Kangri University
Haridwar, India

Shiv Vendra Singh
Departemnt of Agronomy, Graphic Era Hill University, Dehradun
Uttarakhand, India

Parvaze A. Sofi
Division of Genetics and Plant Breeding, Sher-e-Kashmir University of Agricultural Sciences and Technology of Kashmir
J&K, India

Suchi Srivastava
Centre of Advanced Study in Botany, Banaras Hindu University
Varanasi, UP, India-221005

Jebi Sudan
Proteomics Laboratory, Division of Plant Biotechnology, Faculty of Horticulture, Sher-e-Kashmir University of Agricultural Sciences & Technology of Kashmir
J&K, India

Manorama V. Sunkara
Department of Polymers and Functional Materials, CSIR-Indian Institute of Chemical Technology
Uppal Road, Tarnaka, Hyderabad, Telagana, India

T Suthajini
Department of Biosystems Technology, Faculty of Technology, Eastern University
Sri Lanka

MA Syed
Plant Breeding Division, Bangladesh Rice Research Institute (BRRI)
Gazipur, Bangladesh

S Temuhi
Department of Nematology, Assam Agriculture University, Jorhat
Assam, India

Asanka Tennakoon
Department of Biosystems Technology, Faculty of Technology, Eastern University
Sri Lanka

Isuri Thakshila
Department of Biotechnology, Faculty of Agriculture and Plantation Management, Wayamba University of Sri Lanka
Makandura, Gonawila (NWP), Sri Lanka

PWM Tharindi
Department of Export Agriculture, Faculty of Agricultural Sciences, Sabaragamuwa University of Sri Lanka

Rahul Kumar Tiwari
ICAR-Central Potato Research Institute
Shimla, India

Bhagwan Toksha
Centre for Advanced Materials Research and Technology, Department of Plastic and Polymer Engineering, Maharashtra Institute of Technology
Aurangabad, Maharashtra, India

Edgar Vázquez-Núñez
Grupo de Investigación sobre Aplicaciones Nano y Bio tecnologías para la Sostenibilidad (ANBioTS-LASAE), Departamento de Ingenierías Química, Electrónica y Biomédica, División de Ciencias e Ingenierías, Campus León, Universidad de Guanajuato
Guanajuato C.P. 37150, México

Ileana Vera-Reyes
Department of Plastics in Agriculture, Centro de Investigación en Química Aplicada
Saltillo, Coahuila, Mexico

Ejaz Ahmad Waraich
Department of Agronomy, University of Agriculture Faisalabad
Punjab Pakistan

Gulam Nabi Yatoo
Department of Chemistry National Institute of Technology
Hazratbal Srinagar, India

Manaal Zahera
Department of Biotechnology, Era University
Lucknow, Uttar Pradesh, India

Sajad Majeed Zargar
Proteomics Laboratory, Division of Plant Biotechnology, Faculty of Horticulture, Sher-e-Kashmir University of Agricultural Sciences & Technology of Kashmir
J&K, India

1 Recent Developments and Applications of Nanocomposites in Agriculture

Properties, Benefits and Prospects

Edgar Vázquez-Núñez, Ileana Vera-Reyes, Alma Serafín-Muñoz, Hermes Pérez-Hernández, Solange I. Rivera Manrique and Alonso Méndez-López*

1.1 INTRODUCTION

The world population is estimated to be 9.6 billion people by 2050; food demand will increase by 70% to meet global requirements (Guha et al., 2020). As a result, global concerns are emerging that overuse of natural resources, such as global climate change, will affect food production around the world. Global warming has increased global temperature, and atmospheric CO_2 levels, increasing dry spells in some places and precipitation in others (Seddon et al., 2020). Climate change is a major challenge to the sustainability of agriculture due to the depletion of natural resources (Srivastav et al., 2021), leading to lower agricultural productivity and threatening food security as well as food and feed safety (Farooq et al., 2019).

To meet the growing demand for food, agriculture has been enhanced and expanded to achieve an increase in production. Sustainable agriculture is key to global food security in the face of climate, social and commercial uncertainties. Sustainable agriculture uses integrated practices of plant and animal farming systems to improve farmers' incomes with environmental integrity and social sustainability. Among the emerging strategies to deal with this scenario, nanotechnology represents an important alternative to solve these problems, including solutions for soil and water treatment, designing nanofertilizers and nanopesticides to reduce the amount of conventional fertilizers and pesticides used, while increasing food production and quality (Younis et al., 2021).

* Corresponding author.

DOI: 10.1201/9781003376446-1

The principle behind agronanomaterials is based on their success in controlled release systems, defined as the osmosis-regulated transport of active ingredients from a reservoir and transformation to a specific target region, maintaining the concentration of the active ingredient during a long time (Vejan et al., 2021). In this sense, nanocomposites (NCs) represent an alternative for slow and durable delivery of agrochemicals, ensuring increased availability of nutrients to plants (Dhiman et al., 2022); NCs are composites where one of the phases is on the nanometer scale.

Smart NCs are designed to selectively release nano-sized nutrients, i.e. a) particles or emulsions, delivered as direct nutrients (Vejan et al., 2021), b) encapsulated nanofertilizers specifically designed for controlled release in a variety of ways: Slow-release capsule system for a longer duration, fast release on contact with the target surface and conditioned release system, based on pH/temperature/humidity, release based on magnetic/ultrasonic pulse (Guha et al., 2020); c) nutrient delivery through complex NCs, integrated into the matrix of organic polymers of biological or chemical origin, serving as a carrier (Kalia et al., 2020). The NCs-related field has become relevant in recent years and in-depth research has been developed to reach the final stage of application and validation in field conditions, generating knowledge important to consider in the future.

1.2 METHODS TO SYNTHESIZE NANOCOMPOSITES

Nanocomposites can be classified according to the type of reinforcement and matrix materials present in them. According to the type of matrix material, NCs are divided into three classes: Polymer matrix NCs, ceramic matrix NCs and metal matrix NCs (Bhatia et al., 2023).

1.2.1 Synthesis of Polymer-based NCs

Polymer-based NCs are frequently used for the delivery of agrochemicals. In general, the most used method to synthesize this type of nanocomposite is the sol-gel technique, involving steps like hydrolysis and polycondensation, gelation, ageing, drying densification and crystallization (Bokov et al., 2021). Among the most common are the silica-based hybrid nanomaterials and zeolites (Guha et al., 2020).

1.2.2 Synthesis of Ceramic-based NCs

The most common methods are the conventional powder method, polymer precursor route, spray pyrolysis, vapour techniques i.e., chemical vapour deposition (CVD) and physical vapour deposition (PVD), and chemical methods, including sol-gel process, precipitation, and colloidal techniques, and the synthesis based on templates (Parvej et al., 2022). Several factors affect the sol-gel technique, i.e., type of solvent, pH, precursor, the ratio of metal/water during the synthesis, etc. allowing the variability of the synthesis to control the structural and chemical properties of the material. When talking about the synthesis of CNTs-ceramic NCs, it is remarkable to mention the hot press method, CVD or spray pyrolysis, catalytic decomposition and solvothermal process (Barbaros et al., 2022).

1.2.3 Synthesis of Metal-based NCs

The most used techniques are spray pyrolysis, rapid solidification, PVD, CVD, liquid metal infiltration, chemical methods i.e., colloidal and sol-gel processes and electrodeposition (Tariq et al., 2021). Few reports are found regarding the application of the rapid solidification technique, possibly due to the very specific thermal conditions producing a material with multiphase characteristics, which results in difficulty to control and maintain. It has been also difficult to obtain NCs with fine particles due to their agglomeration and non-homogeneous distribution (Pawlik et al., 2020). For this case, the use of ultrasound treatment has helped importantly to improve the contact between the matrix and the nanoparticles (Mohanty et al., 2020). In this context, Table 1.1 describes the most common techniques to synthesize NCs, remarking on their advantages and drawbacks.

TABLE 1.1
Methods to synthesize NCs and their advantages and drawbacks

Methods	Advantages	Drawbacks	References
Powder process	Simplest process available	Limited yield and formation rate, it is needed high temperature, multiple phases formed in the final material	Rane et al., 2018; Ranjan & Bajpai, 2021
Sol-gel process (ceramic NCs)	Simple process, high chemical homogeneity, products with high purity, preferably used for alkoxides	Low content of voids, therefore low superficial area	Mezan et al., 2019; Sharma et al., 2020; Sheng et al., 2019
Spray pyrolysis	Used for ultrafine and homogeneous powders in multi-component systems, high reproducibility in size and purity	High costs	Ardekani et al., 2018; Gharbi et al., 2021
Liquid infiltration	Obtention of materials with different stiffness and enhanced wear resistance; rapid solidification; process relatively easy to scale	Use of high values for temperature; segregation of the reinforcements; formation of non-desirable by-products	Rane et al., 2018; Shi et al., 2022
Rapid solidification process (RSP)	Effective and simple to carry out	Non-homogeneous appearance in NCs; applicable only for metal-metal NCs	Bharti et al., 2020; Bhatia et al., 2023
High-energy ball milling	Uniform distribution		Fathy et al., 2019; Kamrani et al., 2018

1.3 NANOCOMPOSITES FOR CONTROLLED RELEASE OF MICRO- AND MACRO-NUTRIENTS IN SOIL AND THEIR IMPACT ON AGRICULTURE

Nitrogen (N), phosphorus (P), potassium (K), magnesium (Mg), sulfur (S) and calcium (Ca) are essential macro-nutrients for plant development; meanwhile, micro-nutrients are trace elements necessary for higher plants such as boron (B), chlorine (Cl), copper (Cu), iron (Fe), manganese (Mn), molybdenum (Mo) and zinc (Zn), required in small quantities important for different metabolic processes (Grusak, 2001). However, conventional fertilizers that provide these nutrients are associated with pollution due to leaching and inequality between release rates and crop requirements. Therefore, using fertilizers with a controlled or slow release (CRF) stands out among the recent solutions studied to mitigate these impacts.

Nanoencapsulation reduces the rate of dissolution of the agrochemical and allows slow, sustained release of actives, improving macro- and micro-nutrients availability (Jakhar et al., 2022). Bio-based polymeric nanocarriers (BPNC) are the most studied materials, such as alginate, cellulose, chitosan, lignin and poly(lactic acid) (PLA), due to their biodegradability. However, BPNC has some disadvantages in mechanical properties. Therefore, they are reinforced with two or more inorganic or natural materials resulting in a multi-component system called composite (Elemike et al., 2022). Table 1.2 shows different composites used in the synthesis of CRF.

The release control of fertilizer depends on polymer-based coatings properties, i.e., the nature of the polymer, composition, concentration in the coating solution, degradability, solubility, thickness, porosity, techniques used for the coating, etc. and soil properties (soil type, pH, temperature, moisture, microbial activities, etc.) (Fertahi et al., 2021). Additionally, some biopolymers improve water-retention and water holding capacity (WHC) of soils due to the polymer chains forming strong hydrogen bonds between water molecules; hydrogel polymers are an approach to improving the efficiency of water and fertilizer release (Kim et al., 2022).

The nutrient release from CRF takes place in three phases, i) lag phase, the material absorbs water and starts to swell, the water must establish a flux of water entering and the solute leaving; ii) the second phase, the fertilizer inside is dissolved, the difference among concentrations makes saturated solution, which allows the fertilizer to through the polymer coating or the swollen hydrogel network; iii) decay phase, the fertilizer has been dissolved, resulting in a reduction of the rate concentration (Lawrencia et al., 2021). The mechanism for releasing in CRF is swelling and diffusion, and therefore, different mathematic models have been used to describe the mass transfer associated (Table 1.2).

TABLE 1.2
Biopolymer nanocomposites for controlled release of micro- and macro-nutrients

Material	Fertilizer	Release	Time release	Reference
Cellulose nanocrystals hydrogels/ reinforced core-shell gelatin/ alginate hydrogel	Hydrogels were swollen in NPK aqueous solution (10 mg mL^{-1})	Diffusion	After 72 h, released the NPK, showed a good water absorption and retention capacity after 48 h	do Nascimento et al., 2022
Carboxymethyl cellulose CMC/ NPK/zeolite	NPK (16:16:16); zeolite with variations (0.01, 0.1, 0.5, 1 wt.% to CMC)	N and P nutrients were released by pseudo-second order kinetic models; K release followed the Korsmeyer-Peppas kinetic models	After 24 hours the percentage release of N, P and K was 10.94, 16.08, 0.13%	Wahyuni et al., 2022
Regenerated cellulose (RC)/MAP	Monoammonium phosphate (MAP) fertilizer/5% RC	Diffusion	Release rates was 12 h in water and 19–20 days in soil with coating thickness	
Poly (4-vinylpyridine) (P4VP) and carboxymethyl cellulose (CMC)	Urea; $CaCO_3$	Urea release through Super case II transport mechanism. Calcium nitrate, the release mechanism is via the quasi Fickian	CMC/P4VP was decomposed by microorganisms in 50% at day five	Mohamed et al., 2022

(Continued)

TABLE 1.2 (CONTINUED)
Biopolymer nanocomposites for controlled release of micro- and macro-nutrients

Material	Fertilizer	Release	Time release	Reference
Cellulose nanofibrils/ KNO_3	KNO_3	Diffusion and biodegradation	The nutrient release was 8 h in water and over 80 days in soil. The biodegradation efficiencies were up to 75% in 120 days	França et al., 2022
Sodium alginate (SA)/acrylic acid (AA)/acrylamide (AM)/ ammonium persulfate (APS)/ microcrystalline cellulose (MCC)	Urea	Schott's pseudo-second order swelling kinetic model	Nitrogen complete release after 20 days. Enhanced soil water-retention capacity after the addition of 1 % of the polymer	El Idrissi et al., 2022
Cellulose (sugarcane bagasse) / (AHG)/$(NH_4)_2HPO_4/CO(NH_2)_2$	Potassium (KCl)	Korsmeyer-Peppas model	Potassium release was observed in three stages 41.2, 58.1 and 67.2% potassium within 5–10 min (stage I), followed by 3.58%, 2.2% and 1.7% after 6–7 h, and lastly release of 15.65%, 8.9% and 5.6%	Messa et al., 2022a
Poly(lactic acid) (PLA) composites				
Poly(lactic acid) PLA/ poly(butylene adipate co-terephthalate) (PBAT) (Ecoflex®)/ bentonite clay	NPK (4-14-8)	Semi-empirical models of Higuchi and Korsmeyer-Peppas	Ecov/6Bent/3NPK system release an 81.85% after 120 h	da Silva et al., 2022
Biochar/PLA	Ammonium sulfate as N source	Diffusion/swelling	BCRNF-PLA 10% showed the longest N-releasing time (25 days), followed by BCRNF-PLA 6% (19 days), and BCRNF-PLA 3% (four days)	Rubel & Wei, 2022

(Continued)

TABLE 1.2 (CONTINUED)
Biopolymer nanocomposites for controlled release of micro- and macro-nutrients

Material	Fertilizer	Release	Time release	Reference
Polylactic acid (PLA)/carboxymethyl cellulose sodium salt (CMCNa)/whey-polysaccharide-based hydrogel	KNO_3/Urea	Fickian or non-Fickian diffusion	KNO_3 was released by 85% in 96 h. The soil amended with hydrogel increased by 30% water retention	Durpekova et al., 2022
Chitosan				
Chitosan-kaolinite	NPK	P release followed the Korsmayer-Pepas model	P release rate from coated NPK granules reached 14% after 6 h in water; 30% in soil	Eddarai et al., 2022
Chitosan/graphene oxide	Potassium nitrate	NS*	Released 80 % of content after four days	T. Li et al., 2019a

*NS: Not specified

1.4 ALLEVIATION OF BIOTIC AND ABIOTIC STRESS IN AGRICULTURE BY USING NCS

When the plant perceives some adverse environmental condition, it responds with biochemical, physiological and transcriptomic modifications to mitigate stress (Esgario et al., 2020; González-García et al., 2021). These conditions mainly promote the inhibition of their growth derived from the induction of reactive oxygen species (ROS), which are then used as signalling molecules to activate the tolerance mechanisms to stress.

1.4.1 Enhancement of Plant Growth

Nanofertilizers can be absorbed both through the roots and leaves of plants, so the route of exposure and the mode of application influence the behaviour, bioavailability and absorption in crops (El-Ramady et al., 2018). For example, in potato plants, NPK nanofertilizers played an important role in increasing productivity and tuber quality (Abd El-Azeim et al., 2020). ZnO and TiO NCs improved plant height, number of leaves, fresh and dry weight, as well as stem thickness of eggplant, chile and tomato (Younes et al., 2020). In wheat, the NCs of Fe_3O_4 at doses of 200 and 500 $mg{\cdot}L^{-1}$ improved plant growth, photosynthesis and respiration, in addition to increasing the content of photosynthetic pigments in the leaves; this increased ascorbate peroxidase activity and reduced malondialdehyde (MDA) content in roots and leaves (Feng et al., 2022).

1.4.2 Pathogen Control and Disease Management

Nanocomposite matrices not only provide support for nanoparticles but can also enhance antimicrobial performance to broaden their potential applications (Sidhu et al., 2020). Another benefit of the use of NCs compared to nanoparticles is the reduction in the toxicity of some nanoparticles, such as Ag, Cu and Zn oxides. (Ali et al., 2020).

The mechanisms of antifungal activity in NCs include the generation of ROS, which damages the biomolecules of pathogens leading to cell death (Alghuthaymi et al., 2021); adhesion to the cell membrane, causing damage to the membrane structure, membrane permeability, and transport (Ahmed et al., 2021); penetration inside the cell, causing denaturation of proteins and degradation of the DNA or mRNA (Al-Dhabaan et al., 2017). For their part, Ahmed et al. (2022) reported that chitosan and iron nanocomposite reduced the incidence of bacterial leaf blight disease caused by the bacterial pathogen *Xanthomonas oryzae* pv. *oryzae* by modulating antioxidant enzymes (superoxide dismutase, peroxidase, and ascorbate peroxidase) and an increase in the transcription level of defence-related genes (OsPR, OsSOD and OsAPX). Table 1.3 presents a summary of some nanocomposites that have been evaluated against pathogens in different crops.

TABLE 1.3
Nanocomposites to improve plant growth, control pathogens and disease management

Nanocomposites	Plants or substrate	Experimental conditions	Effects	References
Chitosan-silicon (Cs–Si NCs)	Maize (*Zea mays* L.)	Laboratory and field	NCs promoted growth and yield in maize by foliar application. With the seed treatment, the seedlings exhibited a higher vigour index (SVI) up to 3.7 times higher compared to SiO_2	Kumaraswamy et al., 2021
Biogenic synthesized zinc oxide nanofertilizer	Wheat (*Triticum aestivum* L.)	Field	Nanozinc oxide at a concentration of 80 ppm increased plant height, seeds per spike, 100 seed weight, yield and biomass accumulation compared to control and zinc nitrate treatment	Sheoran et al., 2021
Combination of NPK nanofertilizers with commercial nanoparticles of iron, zinc, and CeO_2 (NP)	Cabbage (*Brassica oleraceae* L.)	Field	Nanofertilizers improved cabbage circumference size compared to the control. Increased head girth and height with NPK + Fe nanofertilizer. The average head weight of the cabbage increased three times more than the control when the cabbage was fertilized with NPK + CeO_2 NP. The chlorophyll content of leaves increased	Abdulhameed et al., 2021
Nanocomposites	Plants or substrate	Studied pathogen	Effects	References
Graphene oxide (GO)-Fe3O4	Grapevine (*Vitis vinifera*)	*Plasmopara viticola*	Excellent protective and antifungal activity by effectively suppressing sporangia germination and inhibiting downy mildew development	Wang et al., 2017

(Continued)

TABLE 1.3 (CONTINUED)
Nanocomposites to improve plant growth, control pathogens and disease management

Nanocomposites	Plants or substrate	Experimental conditions	Effects	References
Chitosan-Zn-Cu	Strawberry (*Fragaria* spp.)	*Botrytis cinerea*	Inhibited the growth of *Botrytis cinerea* and reduced the number of rotten fruits	Al-Dhabaan et al., 2017
Chitosan-Cu	Cotton (*Gossypium hirsutum*) Chickpea (*Cicer arietinum*)	*Rhizoctonia solani* *Fusarium oxysporum*	It provided effective disease control, and it was shown that Chitosan-Cu NCs can suppress cotton seedling disease caused by *R. solani* in vivo. In chickpea, these NCs showed a reduction of the disease of 46.67% in the disease caused by *F. oxysporium*	Abd-Elsalam et al., 2018; Kaur et al., 2018
Chitosan-Mg	Rice (*Oryza sativa* L.).	*Acidovorax oryzae* *Rhizoctonia solani*	It significantly inhibited the growth of pathogens	Ahmed et al., 2021
Chitosan-Ag	Chickpea (*Cicer arietinum*) Peach (*Prunus persica*)	*Fusarium oxysporum* *Pseudomonas syringae*	In chickpea, it showed good antifungal efficacy and inhibited the pathogen by 33.33%; likewise, it reduced the bacterial canker disease of peach trees	Kaur et al., 2020; Shahryari et al., 2020
Chitosan-Fe	Rice (*Oryza sativa* L.)	*Xanthomonas oryzae*	Significantly reduced the incidence of bacterial leaf blight disease by up to 67.1%	Ahmed et al., 2022
Chitosan-Clay	Citrus crops	*Penicillium digitatum*	Complete inhibition of the pathogen was observed due to damage and collapse of the hyphae	Youssef & Hashim, 2020

1.4.3 Pollutant Stress

Nanometric products such as chitosan (Ch), biochar (BC), silica, nanoclay and nano-zeolite, among others, combined with nanoparticles of copper (Cu), silver (Ag), silicon oxide (SiO), iron ($CoFe_2O_4$, Fe_3O_4, c-Fe_2O_3, nZVI, among others) and titanium (TiO_2) form effective NCs to alleviate pollution stress in plants (Guha et al., 2020).

Maksoud et al. (2022) prepared and tested the Cs/Ag/$MnMgFe_2O_4$ nanocomposite in cabbage plants (*Brassica oleracea* var. *capitata*) grown in hydroponics under cadmium (Cd) stress. The results revealed an increase in the photosynthetic rate, stomatal conductance and transpiration rate compared to the control treatment. In addition, the cadmium content in the leaves and roots decreased. Rizwan et al. (2019) evaluated the combination of NMs, Fe and Zinc in wheat plants grown in pots under greenhouse conditions. The authors revealed that in wheat plants at high concentrations of 100 mg L^{-1} and 20 mg L^{-1} of Fe and Zn NPs, biomass, plant length and grain weight increased, compared to the control treatment.

1.4.4 Salinity Stress

Salinity stress alters nutrient uptake, decreases water uptake and indirect drought stress due to osmotic stress, which is followed by ionic toxicity due to the accumulation of Na^+ and Cl^- ions in the cell cytoplasm (Ludwiczak et al., 2021). Laboratory, greenhouse and field studies have shown it with different crops. For example, in stevia plants (*Stevia rebaudiana* Bertoni) grown in pots conditioned with NaCl at concentrations of 0, 50 and 100 mM, compared to the control, the nanocomposite Se and chitosan (Cs-Se) at a concentration of 20 mg L^{-1}, increased the plant growth, photosynthetic activity, antioxidants and oil content when plants were exposed to the highest concentration of salinity. Therefore, the authors suggested that the Cs-Se nanocomposite is promising for alleviating salinity stress in stevia plants (Sheikhalipour et al., 2021).

Potted *Mentha spicata* L. plants treated with chitosan and salicylic acid applied at 1% w/v (foliar spray) improved photosynthetic rate and protein content compared to the control. It is suggested that the nanocomposite can be an alternative for enhancing production in regions of the world with saline soil problems (Hassanpouraghdam et al., 2022).

1.4.5 Drought Stress

Although studies have shown significant effects in alleviating drought stress with simple nanomaterials, NCs can improve water stress efficiency. Nassaj-Bokharaei et al. (2021) evaluated the combination of natural char nanoparticles (NCNPs) with hydrogel (NCNPs/hydrogel) on tomato plants (*Lycopersicon esculentum* Mill.). The authors evaluated three levels of concentration (0, 0.3 and 0.6 % (w/w)) with three levels of WHC of 50%, 75% and 85%, severe stress, medium and without stress, respectively. The NCs at levels of 0.6% and 75% WHC reduced the effects of water deficit stress on tomato growth and development. Also, an increase in soil microbial biomass was observed. The results suggested that the nanocomposite (NCNPs/hydrogel) is promising for maintaining soil moisture, plant nutrients and soil microbial activity, with positive consequences for tomato development. Kenawy et al. (2022) evaluated two levels of water stress and the cellulose and activated carbon (SANC) nanocomposite on *Vicia faba* L. plants grown in pots with natural soil (60 days of study). The results indicated that compared to the control treatment, the SANC nanocomposite alleviated water stress by boosting the growth and physiological and biochemical characteristics of broad beans, which contributed to better growth of stressed broad beans. Table 1.4 also shows other examples of the alleviation of pollutants, salinity and drought stress.

TABLE 1.4
Nanocomposites used to alleviate stress from pollutants, salinity and drought

Nanocomposites	Plants or substrate	Effects	References
Combination of melatonin (Mel) with gold nanoparticles (Mel-AuNP)	Rice (*Oryza sativa* L.)	The concentration of 200 µM Mel-AuNP improved photosynthetic and enzymatic activity, increased melatonin content and mitigated oxidative stress by reducing the Cd content in the plant	Jiang et al., 2021
Chitosan with SiO_2 and TiO_2 nanomaterials	Irrigation water	NCs adsorbed between 290 and 300 mg gr of Cd. Furthermore, inhibited the growth of Escherichia coli and Staphylococcus aureus	Banivaheb et al., 2021
Biochar NCs and the magnesium and manganese oxides (BNC-MgO + BNC-MnO; 12.5 g kg^{-1} of the soil of both compounds)	Safflower (*Carthamus tinctorius* L.)	Increased the content of elements in the plant tissue, photosynthetic pigments, water in the leaves and reduced the generation of ROS, under saline conditions, compared to the control	Ghassemi-Golezani et al., 2021; Rane et al., 2018; Ranjan & Bajpai, 2021
BNC-MgO + BNC-MnO (12.5 g kg^{-1} of the soil of both non-materials)	Safflower (*Carthamus tinctorius* L.)	The NCs was the treatment that most reduced saline toxicity in safflower due to a reduction in the absorption of sodium and osmotic stress	Farhangi-Abriz & Ghassemi-Golezani, 2021
Surface-functionalized CNDs (sugar-terminated CNDs (glucose and trehalose))	Soja (*Vigna radiata* L.)	The functionalized CDs in plants under salinity stress improved seed germination, growth, pigment content and ionic and osmotic balance	Sarkar et al., 2022

1.5 REMOTION OF AGROCHEMICAL COMPOUNDS BY USING NCS

Agrochemicals used in agriculture often fail to reach the target sites and exhibit efficacy errors i.e., disproportionate dissolution rate removal or leaching due to wind, rain and sunlight (Okey-Onyesolu et al., 2021). Runoff and leaching of agrochemicals lead to low input use efficiency to maintain high agricultural productivity. To maximize the benefits of agrochemicals and reduce their potential toxicity, NCs offer important advantages to be applied in the field.

1.5.1 Polymer Matrix NCs

The wide field of study of available monomers, oligomers and chemicals leads to the design of polymeric matrices with adjusted physicochemical properties such as molecular structure, chemical stability, transparency and conductivity, among others (Wei et al., 2020). Within this field, several biopolymeric nanocomposites can be adsorbent materials with environmentally friendly and sustainable characteristics that are effective in removing harmful pesticide residues from wastewater.

Among the biopolymers, polysaccharides have emerged as sustainable alternatives as a direct consequence of some intrinsic advantages, such as their non-toxicity, biocompatibility and biodegradability, as well as their wide availability and low production costs. Cellulose, alginate, chitosan and guar gum are some of the many polysaccharides that can be used (Idumah et al., 2019). For example, several studies reported that chitosan-ZnO nanoparticles effectively adsorb permethrin in solution (Dehaghi et al., 2014). Another study focused on chitosan-modified CuO surface indicated that organophosphorus pesticides such as malathion and dichlorvos were effectively removed (Badawy et al., 2018). Similarly, silver nanoparticles coated with chitosan were effective in removing atrazine from the medium (Saifuddin et al., 2011).

Another study was based on the design of effective hybrid biosorbent materials for paraquat water remediation (Baigorria & Fraceto, 2022). In this study, chitosan matrix adsorbent hydrogel nanocomposites with clays were synthesized using simple methodologies. Extensive and diverse porosity was observed among the studied beads, with interconnected micron-sized macropores. Aqueous adsorption tests showed excellent results for nanobeads. The addition of nanoclays to chitosan biopolymeric matrices was critical for effective pesticide adsorption. Optimal removal conditions were found at neutral pH, 25°C and the lowest adsorbent concentration. Reuse and regeneration/reuse tests showed the possibility of maintaining acceptable paraquat removal capacity during the first two reuse cycles of the materials without paraquat desorption.

In the field of water treatment with agrochemicals harmful to health, nanocomposite membranes have received increasing attention due to environmental concerns. The basic point is to strategically manipulate their structural and physicochemical properties (hydrophilicity, porosity and thermal stability) (Wei et al., 2020). For example, the hyperbranched, multi-walled carbon nanotubes/poly (vinylidene fluoride) MWNTs/PVDF nanocomposite membrane with good surface hydrophilicity and pore structure showed high water transport (Zhao et al., 2012). Under the same context, isocyanate-treated graphene oxide/polysulfone ultrafiltration membranes (iGO/PSF) exhibited optimal antifouling properties due to their higher surface smoothness, higher hydrophilicity and more negative zeta potential (Zhao et al., 2013). Another study demonstrated a non-solvent-induced phase separation approach to modify polyethersulfone membranes with different concentrations of TiO_2 nanotubes (TNTs). TiO_2-modified polymer nanocomposite membranes resulted in higher organic matter rejection (53.9%) and water flux (82 L/m^2 h at 40 psi) than the unmodified membrane.

1.5.2 Ceramic Matrix NCs

The presence of agrochemicals is frequently detected in water bodies and soils, as they have been used extensively as pre-emergence control of broadleaf weeds and annual grasses in fields. Their harmfulness and carcinogenic potential are well documented and are of increasing concern because they can be hazardous at low concentrations (Rasmussen et al., 2015).

There are several studies about ceramic nanocomposites in the field of agrochemistry. For instance, a study was focused on zeolite-based magnetic NCs, formed by a dispersion of Fe magnetic metal nanoparticles (between 5 and 25 nm) embedded in a ceramic matrix based on amorphous silica and alumina. This work reported the concentration of agrochemicals below 10 mol/L (Zhang & Goss, 2022); they suggested that an adsorbent, viable for practical applications, whose task is to remove simazine from water, should work well in the simazine concentration range from around 9–90 mol/L down to below 0.25 mol/L. In this particular simazine concentration range, the magnetic adsorbents (Fe, H)A800C-0 min and (Fe, H)A600C-90 min showed the best performances. Their adsorption isotherms are well above those of the SiO2(II)400 silica sample and the H-Y zeolite.

Organophosphorus pesticides (OPPs) are a class of chemicals that generally act as cholinesterase inhibitors and have been used as insecticides (Khoshmanesh et al., 2020). Within this field of study, ceramic/carbon nanoparticle nanocomposites (C/C NCs) have been used as a new type of solid-phase extraction (SPE) adsorbent (Heidari & Razmi, 2014), and SEM was applied to characterize the ceramic/carbon nanoparticles. The applicability of this SPE adsorbent was evaluated by extracting and determining, by HPLC-UV, four OPPs in water samples. The results of this study showed that the proposed method had a wide dynamic linear range (0.1–400 ng mL^{-1}) with good linearity ($R^2 > 0.9972$) and low detection limits (9.2–75.0 pg mL^{-1}), obtaining extraction recoveries ranging from 70–99.5%.

Supported heterogeneous nanocatalysts consist of an active nanomaterial dispersed on a less active material called catalytic support. They are widely used in heterogeneous photocatalysis due to their high surface area, considerable photocatalytic activity, and chemical/thermal stability (Bruckmann et al., 2022). The application of supported nanocatalysts (nanocomposites) helps to overcome drawbacks usually encountered when applying isolated nanocatalysts, such as agglomeration of nanoparticles due to their high surface energy and poor dispersion in aqueous solution (Abboud et al., 2021). Moreover, these catalytic supports increase the contact between organic pollutants and the catalytic surface (Bruckmann et al., 2022). Among the supported nanocatalysts used for the degradation of organic pollutants, TiO_2, CuO, Fe_2O_3 and ZnO, supported on zeolites or silica, stand out due to their significant photocatalytic activity, high surface area and thermochemical stability (Chen et al., 2019).

Studies have verified the good dispersibility of the nanocatalysts on the catalytic support, resulting in a nanocomposite with high chemical stability and preserved photocatalytic activity after four to six cycles of heterogeneous photocatalysis (Bruckmann et al., 2022; Kumar et al., 2022). It has also been shown that the range of pesticide degradation percentages is between 80 and 90% after 30 to 180 min, with initial pesticide and nanocatalyst concentrations of 15 to 50 mol L^{-1} and 0.1 to 0.5 g L^{-1}, respectively (Bruckmann et al., 2022).

1.5.3 Metal Matrix Nanocomposites

Titanium oxides, zinc oxide, copper oxides, noble metals (Ag, Au, Pt) and iron-based nanomaterials are widely used for the treatment of various pollutants. The modified forms have better properties due to the synergy of surface properties and the insertion of additional energy levels that help prevent charge recombination (Rani et al., 2022).

A study focused on evaluating two new types of chitosan-metal oxide nanoparticles (Ch-MO NPs), particularly chitosan-copper oxide nanoparticles (Ch-CuO NPs) and chitosan-zinc oxide nanoparticles (Ch-ZnO NPs), analyzed the extraction and effectiveness for removing different pesticides from water (Badawy et al., 2018). The design was implemented with a metal oxide core shell with chitosan as a hard shell after a cross-linking mechanism by glutaraldehyde and epichlorohydrin. The prepared nanoparticles were characterized, and FT-IR confirmed the interaction between chitosan, metal oxide and the cross-linking mechanism. SEM and TEM analysis explained that the nanoparticles have a spherical morphology and a nanometric size of 93.74 and 97.95 nm for Ch-CuO NPs and Ch-ZnO NPs, respectively. The results indicated that the optimal conditions were a pH of 7, a temperature of 25 °C, and a stirring time of 25 min. SPE cartridges were filled with Ch-MO NPs and seven pesticides were extracted from water samples: Abamectin, diazinon, fenamiphos, imidacloprid, lambda-cyhalothrin, methomyl and thiophanate-methyl, which were determined by HPLC. The extraction efficiency of Ch-ZnO NPs was higher than that of Ch-CuO NPs, but both removed a higher amount of most of the pesticides tested than the standard ODS cartridge (C18). The results showed that this method achieves fast and easy extraction in small amounts of adsorbents (Ch-MO NPs) and solvents.

On the other hand, magnetic nanocomposites based on metal-organic frameworks (MOFs) have received great attention in terms of the adsorption of pollutants. (Zhao et al. (2022) assembled hollow MOF-derived NiO/Co@C magnetic NCs by a facile solvothermal reaction, in which cobalt (Co) ions acted as inducers to drive the formation of hollow MOFs by regulating the chemical transformation process.

Benefiting from their coordination effect, the carbonized bimetallic Ni-Co-MOFs were transformed into magnetic-carbon composites. The hollow structure accelerates the mass transfer efficiency, and the magnetism promotes fast and convenient separation. The composite was used as an adsorbent for magnetic SPE and adsorption removal of nitrogenous organic pesticides. NiO/Co@C contained a hollow porous structure, which resulted in a high adsorption performance of pesticides, as high as 62.2 mg g^{-1}. With NiO/Co@C, solid-liquid separation was easily realized, and good reusability and stability were also obtained. After nine cycles, the extraction recovery was still above 90%. Therefore, NiO/Co@C was found to be an effective adsorbent for the pretreatment of aqueous samples and the removal of hazardous contaminants. Table 1.5 shows reported work on nanocomposites on a matrix basis.

1.6 WORLD REGULATORY FRAME FOR NCS USED IN AGRICULTURE: GAPS AND PROSPECTS

Although NCs are available to ameliorate the havoc caused by the excessive use of pesticides, fertilizers and water in agriculture, there is a great disadvantage worldwide regarding the regulatory measures available for the use of these products for human consumption. In this regard, the global regulation in force since 2005 includes ISO Technical Committee (TC) 229, which comprises task groups to explore key concepts related to nanotechnology i.e., ISO/TC 229/TG 02, ISO/TC 229/TG 03 and ISO/TC 229 SG (Trotta & Mele, 2019). ASTM E 56 was formed to develop standards and guidance for nanotechnology and nanomaterials, as standardization related to nanotechnology needs as well. E56 has six technical subcommittees that maintain jurisdiction over these standards, i.e., E56.01, E56.02, E56.03, E56.06, E56.07 and E56.08, regarding informatics and technology, physical and chemical characterization, environment, health, safety, nano-enabled consumer products, education and workforce development and nano-enabled medical products, respectively (Aublant et al., 2020).

Since more than a decade of sustained research and development in nanoscience and nanotechnologies, countries are slowly starting to regulate products that contain nanocomposites. Nevertheless, in the past years, important concerns related to human health, safety and environmental risks associated with nanotechnology have accelerated nanotechnology regulation. The standard regulation to work properly in the agri-industry that public agencies should define are (1) nanomaterials, (2) an international regulatory body, (3) detection, characterization and quantification methods, and (4) exposure and risk assessment of the new products (Javier et al., 2012).

However, the main problem with establishing a regulatory frame for nanocomposite use in the nano agri-food sector is the lack of agreement between the most important agencies and international bodies in the standardization of nanomaterial properties, manipulation, characterization, synthesis, environment and human hazard. Nanotechnology can be used in different fields; therefore, it is so important to standardize protocols (Johnston et al., 2020).

TABLE 1.5
Summary of agrochemical application of nanocomposites based on their matrix

Nanocomposite				
Class	**Matrix material**	**Application**	**Study**	**Reference**
Ceramic/ matrix	Two magnetic adsorbents composed of Fe/Fe_3O_4 nanoparticles embedded in an amorphous silica-alumina ceramic phase	Adsorbent of simazine from water	The pH of the maximum adsorption of simazine was 6.5 and 3.0 for the two adsorbents. At these pH values, simazine adsorption occurred rapidly and massively even from very dilute simazine solutions (0.25 µmol/L)	Pansini et al., 2018
	Zeolitic imidazolate framework based on magnetic multi-walled carbon nanotubes	Extraction of organochlorine pesticide from irrigation water	A zeolitic imidazolate framework based on magnetic multi-walled carbon nanotubes (M-M-ZIF-67) was prepared. The obtained M-M-ZIF-67 material possessed porous surfaces and superparamagnetism due to the utilization of magnetic multi-walled carbon nanotubes as magnetic core and support. The recoveries of all analytes for the method at levels of ten and 100 µg L-1 were 74.9-116.3% and 75.1-112.7%, respectively.	Huang et al., 2018

(Continued)

TABLE 1.5 (CONTINUED)
Summary of agrochemical application of nanocomposites based on their matrix

Nanocomposite				
Class	Matrix material	Application	Study	Reference
Metal/matrix	Magnetic nanoporous carbons (MNPCs) based on Zn/Co-MOFs were synthesized by a one-step carbonization method	Extraction of organophosphorus pesticides from fruit sample	The prepared MNPCs showed high specific surface areas, good chemical stability and high extraction efficiency for OPPs of interest. The reproducibility of the method was in the range of 3.5–9.7%. Recovery experiments were carried out by adding target OPPs at three concentration levels to validate the accuracy of the proposed method, and the recovery of target OPPs was between 84 and 116%	D. Li et al., 2019b
	TiO_2 P25-rGO nanocomposite	Photocatalytic degradation of Atrazine, isoproturon, diuron, alachlor, and	All these pesticides were mostly photodegraded by the hydroxyl radicals (HO%) produced from the photoinduced holes (h(+))	Luna-Sanguino et al., 2019
	Ag-ZnO nanocomposite	Chlorpyrifos	The Ag-ZnO nanocomposites showed excellent photocatalytic activity in the reduction/degradation of coloured (rhodamine B [Rh-B] dye) and colourless (chlorpyrifos pesticide) pollutants when irradiated with sunlight. Among the prepared photocatalysts, the 3 wt.% Ag-ZnO hybrid compound showed the best photocatalytic activity by efficiently degrading the hazardous organic pollutants in a very short time	Choudhary et al., 2019

(Continued)

TABLE 1.5 (CONTINUED)
Summary of agrochemical application of nanocomposites based on their matrix

Nanocomposite				
Class	**Matrix material**	**Application**	**Study**	**Reference**
	Copper nanoparticles	Agricultural pest management, pesticides, phytotoxic activity	Copper is involved in plant growth, metabolism and defense, and has been used in agriculture as a key agent in fungicides to combat plant diseases	Gomes et al., 2022
Polymers/metal oxide	Chitosan-copper oxide nanoparticles (Ch-CuO NPs) and chitosan-zinc oxide nanoparticles (Ch-ZnO NPs	Extraction of abamectin, diazinon, fenamiphos, imidacloprid, lambda-cyhalothrin, methomyl and thiophanate-methyl	The nanoparticles showed a spherical morphology and a nanometric size of 93.74 and 97.95 nm for Ch-CuO and Ch-ZnO respectively. The extraction efficiency of Ch-ZnO NPs was higher than that of Ch-CuO NPs, but both removed a higher amount of most of the pesticides tested. This method achieves fast and simple extraction in small amounts of adsorbents (Ch-MONPs) and solvents. In addition, the method is very sensitive to pesticides and exhibits a high recovery rate	Badawy et al., 2018

1.7 GAPS AND PROSPECTS

Several studies have focused their interest on the beneficial role of nanomaterials, including NCs to agriculture; it has been proved that due to their physical and chemical properties, these novel materials can contribute to the solution of global problems by controlling or reducing the negative effects of biotic and abiotic factors produced for the anthropogenic activities. Contradictorily, many of these properties associated with the nanostructures also are potentially toxic or harmful to human health and can negatively affect the environment. There is still a lack of full understanding of the toxicity of nanomaterials that are present in food. Therefore, more research in this area is required.

Some of the most urgent concerns regarding nanomaterials and their use in agriculture are listed as follows:

- Methods for the detection and characterization of nanomaterials in agri-food are not available.
- Toxicological research on nanocomposite safety is poor.
- The research on the hazard to human health of the ingested nanoparticles is not well developed.
- The development of smart tools for validation of risk assessments of nanocomposites used in the agri-food field.
- The long-term health consequences (if any) of ingestion of insoluble and bio-persistent nanoparticles via food are unknown.
- It is necessary to certify the laboratory practices, for validating and certificating nanocomposites for agricultural applications.
- Overlapping between the different international regulatory entities or agencies is another difficulty related to nanotechnology implementation control.
- It is needed to develop and standardize new methods, preferably based on green synthesis, to widen the offer and applicability of NCs.

An option to solve these requirements is the establishment of international research networks to develop, evaluate and validate novel NCs for applications in the agricultural sector. An appropriate regulatory system at the global level that ensures pre-market evaluation of nano-food products sets liabilities and sets clear limits for any nanomaterial used in food production, related applications and the development of clear and consistent guidelines for risk assessment of nanoproducts would be helpful.

1.8 CONCLUSION

The global demand for sustainable alternatives to properly provide nutrients to crops, efficiently protect the plants against pests and maximize the use of natural resources such as water and soil to guarantee the world's food security has led to intense research related to the design and manufacture of new materials with improved properties compared to the conventional ones. In this context, NCs have been demonstrated to be suitable materials to meet emerging demands in the scientific and technological agricultural segment. The know-how related to the processing methods for different types of NCs is available. However, there are always important opportunities to be improved on in terms of their performance, analyzing the use of less contaminant raw material during their synthesis and making them economically viable to facilitate their access from less-privileged countries around the world.

Although several applications already exist, there are many potential improvement areas, opening the possibility for innovative actions. Similarly, as new nanomaterials are created and released to the market, it is urgent to incentivize the discussion regarding the world normative and legal frames, since in several potential markets this aspect remains unclear and with important gaps regarding the biosafety and environmental and ecological impacts of these nanomaterials.

REFERENCES

Abboud, M., Alnefaie, R., & Alhanash, A. (2021). Unsupported and silica-supported nickel nanoparticles: Synthesis and application in catalysis. *Journal of Nanoparticle Research, 24*, 1–52.

Abd El-Azeim, M., Sherif, M., Hussien, M., Tantawy, I., & Bashandy, S. (2020). Impacts of nano-and non-nanofertilizers on potato quality and productivity. *Acta Ecologica Sinica, 40*, 388–397.

Abd-Elsalam, K. A., Vasil'kov, A. Y., Said-Galiev, E. E., Rubina, M. S., Khokhlov, A. R., Naumkin, A. V., Shtykova, E. V., & Alghuthaymi, M. A. (2018). Bimetallic blends and chitosan nanocomposites: Novel antifungal agents against cotton seedling damping-off. *European Journal of Plant Pathology, 151*, 57–72.

Abdulhameed, M. F., Taha, A. A., & Ismail, R. A. (2021). Improvement of cabbage growth and yield by nanofertilizers and nanoparticles. *Environmental Nanotechnology, Monitoring & Management, 15*, 100437.

Ahmed, T., Noman, M., Jiang, H., Shahid, M., Ma, C., Wu, Z., Nazir, M. M., Ali, M. A., White, J. C., & Chen, J. (2022). Bioengineered chitosan-iron nanocomposite controls bacterial leaf blight disease by modulating plant defense response and nutritional status of rice (*Oryza sativa* L.). *Nano Today, 45*, 101547.

Ahmed, T., Noman, M., Luo, J., Muhammad, S., Shahid, M., Ali, M. A., Zhang, M., & Li, B. (2021). Bioengineered chitosan-magnesium nanocomposite: A novel agricultural antimicrobial agent against *Acidovorax oryzae* and *Rhizoctonia solani* for sustainable rice production. *International Journal of Biological Macromolecules, 168*, 834–845.

Al-Dhabaan, F. A., Shoala, T., Ali, A. A., Alaa, M., & Abd-Elsalam, K. (2017). Chemically-produced copper, zinc nanoparticles and chitosan-bimetallic nanocomposites and their antifungal activity against three phytopathogenic fungi. *International Journal of Agriculture Technology, 13*, 753–769.

Alghuthaymi, M. A., Kalia, A., Bhardwaj, K., Bhardwaj, P., Abd-Elsalam, K. A., & Valis, M., Kuca, K. (2021). Nanohybrid antifungals for control of plant diseases: Current status and future perspectives. *Journal of Fungi, 7*, 48.

Ali, M. A., Ahmed, T., Wu, W., Hossain, A., Hafeez, R., Islam Masum, M. M., Wang, Y., An, Q., Sun, G., & Li, B. (2020). Advancements in plant and microbe-based synthesis of metallic nanoparticles and their antimicrobial activity against plant pathogens. *Nanomaterials, 10*, 1146.

Ardekani, S. R., Rouhaghdam, A. S., & Nazari, M. (2018). N-doped ZnO-CuO nanocomposite prepared by one-step ultrasonic spray pyrolysis and its photocatalytic activity. *Chemical Physics Letters, 705*, 19–22.

Aublant, J.-M., Clifford, C. A., Frejafon, E., Fujimoto, T., Hackley, V. A., Herrmann, J., Hight Walker, A. R., Kaiser, D. L., Koltsov, D. K., & Michael, D. J. (2020). Response to ACS nano editorial "Standardizing nanomaterials". *ACS Nano, 14*, 14255–14257.

Badawy, M. E., El-Nouby, M. A., & Marei, A. E.-S. M. (2018). Development of a solid-phase extraction (SPE) cartridge based on chitosan-metal oxide nanoparticles (Ch-MO NPs) for extraction of pesticides from water and determination by HPLC. *International Journal of Analytical Chemistry, 218*, 1–17.

Baigorria, E., & Fraceto, L. F. (2022). Low-cost biosorbent hybrid hydrogels for paraquat remediation of water. *Journal of Water Process Engineering, 49*, 103088.

Banivaheb, S., Dan, S., Hashemipour, H., & Kalantari, M. (2021). Synthesis of modified chitosan TiO_2 and SiO_2 hydrogel nanocomposites for cadmium removal. *Journal of Saudi Chemical Society, 25*, 101283.

Barbaros, I., Yang, Y., Safaei, B., Yang, Z., Qin, Z., & Asmael, M. (2022). State-of-the-art review of fabrication, application, and mechanical properties of functionally graded porous nanocomposite materials. *Nanotechnology Reviews, 11*, 321–371.

Bharti, S., Thakur, L., Anand, S., & Dutta, V. (2020). Nano-based composites and their synthesis. In I. Bhushan, V. K. Singh, & D. K. Tripathi (Eds.), *Nanomaterials and environmental biotechnology* (pp. 141–161). Springer.

Bhatia, A. K., Dewangan, S., & Singh, A. K. (2023). Nanocomposites: Types and various methods of synthesis. In N. B. Singh (Ed.), *Nanocomposites* (pp. 7–28). Jenny Stanford Publishing.

Bokov, D., Turki Jalil, A., Chupradit, S., Suksatan, W., Javed Ansari, M., Shewael, I. H., Valiev, G. H., & Kianfar, E. (2021). *Nanomaterial by sol-gel method: Synthesis and application* (PhD diss.). Kazan Federal University Digital Repository. Kazan. Russia

Bruckmann, F. S., Schnorr, C., Oviedo, L. R., Knani, S., Silva, L. F., Silva, W. L., Dotto, G. L., & Bohn Rhoden, C. R. (2022). Adsorption and photocatalytic degradation of pesticides into nanocomposites: A review. *Molecules, 27*, 6261.

Chen, Z., Liu, Y., Zhang, C., Pan, Y., Han, R., Chen, Y., Shang, X., Dong, G., & Zhang, J. (2019). Titanium dioxide nanoparticles induced an enhanced and intimately coupled photoelectrochemical-microbial reductive dissolution of As (V) and Fe (III) in flooded arsenic-enriched soils. *ACS Sustainable Chemistry & Engineering, 7*, 13236–13246.

Choudhary, M. K., Kataria, J., Bhardwaj, V. K., & Sharma, S. (2019). Green biomimetic preparation of efficient Ag–ZnO heterojunctions with excellent photocatalytic performance under solar light irradiation: A novel biogenic-deposition-precipitation approach. *Nanoscale Advances, 1*, 1035–1044.

da Silva, T. C. P., Fortes, A. G. D. S., de Abreu, I. R., de Carvalho, L. H., de Almeida, Y. M. B., Alves, T. S., & Barbosa, R. (2022). Development of biodegradable PLA/PBAT-based filaments for fertilizer release for agricultural applications. *Materials, 15*, 6764.

Dehaghi, S. M., Rahmanifar, B., Moradi, A. M., & Azar, P. A. (2014). Removal of permethrin pesticide from water by chitosan–zinc oxide nanoparticles composite as an adsorbent. *Journal of Saudi Chemical Society, 18*, 348–355.

Dhiman, A., Sharma, A. K., & Agrawal, G. (2022). Polymer based engineered materials for sustainable agriculture. *ACS Agricultural Science & Technology, 2*, 693–711.

do Nascimento, D. M., Nunes, Y. L., Feitosa, J. P. A., Dufresne, A., & Rosa, M. D. F. (2022). Cellulose nanocrystals-reinforced core-shell hydrogels for sustained release of fertilizer and water retention. *International Journal of Biological Macromolecules, 216*, 24–31.

Durpekova, S., Bergerova, E. D., Hanusova, D., Dusankova, M., & Sedlarik, V. (2022). Eco-friendly whey/polysaccharide-based hydrogel with poly(lactic acid) for improvement of agricultural soil quality and plant growth. *International Journal of Biological Macromolecules, 212*, 85–96.

Eddarai, E. M., El Mouzahim, M., Boussen, R., Bellaouchou, A., Guenbour, A., & Zarrouk, A. (2022). Chitosan-kaolinite clay composite as durable coating material for slow release NPK fertilizer. *International Journal of Biological Macromolecules, 195*, 424–432.

El Idrissi, A., El Gharrak, A., Achagri, G., Essamlali, Y., Amadine, O., Akil, A., & Zahouily, M. (2022). Synthesis of urea-containing sodium alginate-g-poly(acrylic acid-co-acrylamide) superabsorbent-fertilizer hydrogel reinforced with carboxylated cellulose nanocrystals for efficient water and nitrogen utilization. *Journal of Environmental Chemical Engineering*, *10*, 108282.

Elemike, E. E., Ekennia, A. C., Onwudiwe, D. C., & Ezeani, R. O. (2022). Chapter 8 - Agro-waste materials: Sustainable substrates in nanotechnology. In K. A. Abd-Elsalam, R. Periakaruppan, & S. Rajeshkumar (Eds.), *Agri-waste and microbes for production of sustainable nanomaterials, nanobiotechnology for plant protection* (pp. 187–214). Springer.

El-Ramady, H., Abdalla, N., Alshaal, T., El-Henawy, A., Elmahrouk, M., Bayoumi, Y., Shalaby, T., Amer, M., Shehata, S., & Fári, M. (2018). Plant nano-nutrition: Perspectives and challenges. In K. Gothandam, S. Ranjan, N. Dasgupta, C. Ramalingam, & E. Lichtfouse (Eds.), *Nanotechnology, food security and water treatment* (pp. 129–161). Springer.

Esgario, J. G., Krohling, R. A., & Ventura, J. A. (2020). Deep learning for classification and severity estimation of coffee leaf biotic stress. *Computers and Electronics in Agriculture*, *169*, 105162.

Farhangi-Abriz, S., & Ghassemi-Golezani, K. (2021). Changes in soil properties and salt tolerance of safflower in response to biochar-based metal oxide nanocomposites of magnesium and manganese. *Ecotoxicology and Environmental Safety*, *211*, 111904.

Farooq, M., Rehman, A., & Pisante, M. (2019). Sustainable agriculture and food security. In M. Farooq & M. Pisante (Eds.), *Innovations in sustainable agriculture* (pp. 3–24). Springer.

Fathy, A., Wagih, A., & Abu-Oqail, A. (2019). Effect of ZrO_2 content on properties of Cu-ZrO_2 nanocomposites synthesized by optimized high energy ball milling. *Ceramics International*, *45*, 2319–2329.

Feng, Y., Kreslavski, V. D., Shmarev, A. N., Ivanov, A. A., Zharmukhamedov, S. K., Kosobryukhov, A., Yu, M., Allakhverdiev, S. I., & Shabala, S. (2022). Effects of iron oxide nanoparticles (Fe_3O_4) on growth, photosynthesis, antioxidant activity and distribution of mineral elements in wheat (*Triticum aestivum*) Plants. *Plants*, *11*, 1894.

Fertahi, S., Ilsouk, M., Zeroual, Y., Oukarroum, A., & Barakat, A. (2021). Recent trends in organic coating based on biopolymers and biomass for controlled and slow release fertilizers. *Journal of Controlled Release*, *330*, 341–361.

França, D., Siqueira, G., Nyström, G., Clemens, F., Souza, C. F., & Faez, R. (2022). Charged-cellulose nanofibrils as a nutrient carrier in biodegradable polymers for enhanced efficiency fertilizers. *Carbohydrate Polymers*, *296*, 119934.

Gharbi, B., Taabouche, A., Brella, M., Gheriani, R., Bouachiba, Y., Bouabellou, A., Hanini, F., Barouk, S., Serrar, H., & Rahal, B. (2021). Spray pyrolysis synthesized and ZnO–NiO nanostructured thin films analysis with their nanocomposites for waveguiding applications. *Semiconductors*, *55*, 37–43.

Ghassemi-Golezani, K., Farhangi-Abriz, S., & Abdoli, S. (2021). How can biochar-based metal oxide nanocomposites counter salt toxicity in plants? *Environmental Geochemistry and Health*, *43*, 2007–2023.

Gomes, D. G., Pieretti, J. C., Lourenço, I. M., Oliveira, H. C., & Seabra, A. B. (2022). Copper-based nanoparticles for pesticide effects. In L. Fernandes Fraceto, H. W. Pereira de Carvalho, R. de Lima, S. Ghoshal, & C. Santaella (Eds.), *Inorganic nanopesticides and nanofertilizers* (pp. 187–212). Springer.

González-García, Y., González-Moscoso, M., Hernández-Hernández, H., Méndez-López, A., & Juárez-Maldonado, A. (2021). Induction of stress tolerance in crops by applying nanomaterials. In A. P. Ingle (Ed.), *Nanotechnology in plant growth promotion and protection: Recent advances and impacts* (pp. 129–169). John Wiley & Sons Ltd.

Grusak, M. A. (2001). Plant macro- and micronutrient minerals. *Encyclopedia of Life Sciences, 5*. Retrieved December 20, 2022, from https://web.archive.org/web/20170808165055id_/http://rubisco.ugr.es/fisiofar/pagwebinmalcb/contenidos/Tema11/macro%20y%20micronutrient.pdf

Guha, T., Gopal, G., Kundu, R., & Mukherjee, A. (2020). Nanocomposites for delivering agrochemicals: A comprehensive review. *Journal of Agricultural and Food Chemistry, 68*, 3691–3702.

Hassanpouraghdam, M. B., Mohammadi, L., Gohari, G., & Vojodi Mehrabani, L. (2022). The effects of foliar application of chitosan-salicylic acid nanocomposite on *Mentha spicata* L. under salinity stress in hydroponic conditions. *Journal of Vegetables Sciences, 5*, 35–51.

Heidari, H., & Razmi, H. (2014). Ceramic/carbon nanoparticle nanocomposite as a solid-phase extraction adsorbent for the determination of the organophosphorus pesticides in aquatic samples. Retrieved December 20, 2022, from https: //sid.ir/paper/912689/en

Huang, X., Liu, G., Xu, D., Xu, X., Li, L., Zheng, S., Lin, H., & Gao, H. (2018). Novel zeolitic imidazolate frameworks based on magnetic multiwalled carbon nanotubes for magnetic solid-phase extraction of organochlorine pesticides from agricultural irrigation water samples. *Applied Sciences, 8*, 959.

Idumah, C., Zurina, M., Ogbu, J., Ndem, J., & Igba, E. (2019). A review on innovations in polymeric nanocomposite packaging materials and electrical sensors for food and agriculture. *Composite Interfaces, 27*, 1–72.

Jakhar, A. M., Aziz, I., Kaleri, A. R., Hasnain, M., Haider, G., Ma, J., & Abideen, Z. (2022). Nano-fertilizers: A sustainable technology for improving crop nutrition and food security. *NanoImpact, 27*, 100411.

Javier, F., Luisa, M., Gaton, P., & Rojo, R. (2012). Nanotechnology and food industry. In B. Valdez, M. Achorr, & R. Zlatev (Eds.), *Scientific, health and social aspects of the food industry* (pp. 95–128). InTech.

Jiang, M., Dai, S., Wang, B., Xie, Z., Li, J., Wang, L., Li, S., Tan, Y., Tian, B., & Shu, Q. (2021). Gold nanoparticles synthesized using melatonin suppress cadmium uptake and alleviate its toxicity in rice. *Environmental Science: Nano, 8*, 1042–1056.

Johnston, L. J., Gonzalez-Rojano, N., Wilkinson, K. J., & Xing, B. (2020). Key challenges for evaluation of the safety of engineered nanomaterials. *NanoImpact, 18*, 100219.

Kalia, A., Sharma, S. P., Kaur, H., & Kaur, H. (2020). Novel nanocomposite-based controlled-release fertilizer and pesticide formulations: Prospects and challenges. In K. A. Abd-Elsalam (Ed.), *Multifunctional hybrid nanomaterials for sustainable Agri-food and ecosystems* (pp. 99–134). Elsevier.

Kamrani, S., Penther, D., Ghasemi, A., Riedel, R., & Fleck, C. (2018). Microstructural characterization of Mg-SiC nanocomposite synthesized by high energy ball milling. *Advanced Powder Technology, 29*, 1742–1748.

Kaur, J., Kaur, K., Mehta, S., & Matharu, A. (2020). A novel molybdenum oxide-Starbon catalyst for wastewater remediation. *Journal of Materials Chemistry A, 8*, 14519–14527.

Kaur, P., Duhan, J. S., & Thakur, R. (2018). Comparative pot studies of chitosan and chitosan-metal nanocomposites as nano-agrochemicals against fusarium wilt of chickpea (*Cicer arietinum* L.). *Biocatalysis and Agricultural Biotechnology, 14*, 466–471.

Kenawy, E.-R., Rashad, M., Hosny, A., Shendy, S., Gad, D., & Saad-Allah, K. M. (2022). Enhancement of growth and physiological traits under drought stress in Faba bean (*Vicia faba* L.) using nanocomposite. *Journal of Plant Interactions, 17*, 404–418.

Khoshmanesh, S. M., Hamishehkar, H., & Razmi, H. (2020). Trace analysis of organophosphorus pesticide residues in fruit juices and vegetables by an electrochemically fabricated solid-phase microextraction fiber coated with a layer-by-layer graphenized graphite/graphene oxide/polyaniline nanocomposite. *Analytical Methods, 12*, 3268–3276.

Kim, S. M., Rhie, Y. H., Kong, S. M., Kim, Y. S., & Na, Y. H. (2022). Synthesis of nanocomposite hydrogels for improved water retention in horticultural soil. *ACS Agricultural Science and Technology, 2*, 1206–1217.
Kumar, R., George, L., Jun, Z., & Mukherji, S. (2022). Photocatalytic activity of graphene oxide-TiO_2 nanocomposite on dichlorvos and malathion and assessment of toxicity changes due to photodegradation. *Chemosphere, 308*, 136402.
Kumaraswamy, R., Saharan, V., Kumari, S., Choudhary, R. C., Pal, A., Sharma, S. S., Rakshit, S., Raliya, R., & Biswas, P. (2021). Chitosan-silicon nanofertilizer to enhance plant growth and yield in maize (*Zea mays* L.). *Plant Physiology and Biochemistry, 159*, 53–66.
Lawrencia, D., Wong, S. K., Low, D. Y. S., Goh, B. H., Goh, J. K., Ruktanonchai, U. R., Soottitantawat, A., Lee, L. H., & Tang, S. Y. (2021). Controlled release fertilizers: A review on coating materials and mechanism of release. *Plants, 10*, 238.
Li, D., He, M., Chen, B., & Hu, B. (2019b). Metal organic frameworks-derived magnetic nanoporous carbon for preconcentration of organophosphorus pesticides from fruit samples followed by gas chromatography-flame photometric detection. *Journal of Chromatography A, 1583*, 19–27.
Li, T., Gao, B., Tong, Z., Yang, Y., & Li, Y. (2019a). Chitosan and graphene oxide nanocomposites as coatings for controlled-release fertilizer. *Water Air and Soil Pollution, 230*, 146.
Ludwiczak, A., Osiak, M., Cárdenas-Pérez, S., Lubińska-Mielińska, S., & Piernik, A. (2021). Osmotic stress or ionic composition: Which affects the early growth of crop species more? *Agronomy, 11*, 435.
Luna-Sanguino, G., Tolosana-Moranchel, A., Duran-Valle, C., & Faraldos, M., Bahamonde, A. (2019). Optimizing P25-rGO composites for pesticides degradation: Elucidation of photo-mechanism. *Catalysis Today, 328*, 172–177.
Maksoud, M. A., Bekhit, M., El-Sherif, D. M., Sofy, A. R., & Sofy, M. R. (2022). Gamma radiation-induced synthesis of a novel chitosan/silver/Mn-Mg ferrite nanocomposite and its impact on cadmium accumulation and translocation in brassica plant growth. *International Journal of Biological Macromolecules, 194*, 306–316.
Messa, L. L., Hsieh, Y.-L., & Faez, R. (2022a). Sugarcane bagasse derived phosphorylated cellulose as substrates for potassium release induced by phosphates surface and drying methods. *Industrial Crops and Products, 187*, 115350.
Mezan, S. O., Jabbar, A. H., Hamzah, M. Q., Tuama, A. N., Hasan, N. N., Roslan, M., & Agam, M. A. (2019). Synthesis, characterization, and properties of polystyrene/SiO_2 nanocomposite via sol-gel process. *AIP Conference Proceedings, 2151*, 20034.
Mohamed, R. R., Fahim, M. E., & Soliman, S. M. A. (2022). Development of hydrogel based on carboxymethyl cellulose/poly(4-vinylpyridine) for controlled releasing of fertilizers. *BMC Chemistry, 16*, 52.
Mohanty, P., Mahapatra, R., Padhi, P., Ramana, C. V., & Mishra, D. K. (2020). Ultrasonic cavitation: An approach to synthesize uniformly dispersed metal matrix nanocomposites—A review. *Nano-Structures and Nano-Objects, 23*, 100475.
Nassaj-Bokharaei, S., Motesharezedeh, B., Etesami, H., & Motamedi, E. (2021). Effect of hydrogel composite reinforced with natural char nanoparticles on improvement of soil biological properties and the growth of water deficit-stressed tomato plant. *Ecotoxicology and Environmental Safety, 223*, 112576.
Okey-Onyesolu, C. F., Hassanisaadi, M., Bilal, M., Barani, M., Rahdar, A., Iqbal, J., & Kyzas, G. Z. (2021). Nanomaterials as nanofertilizers and nanopesticides: An overview. *ChemistrySelect, 6*, 8645–8663.
Pansini, M., Sannino, F., Marocco, A., Allia, P., Tiberto, P., Barrera, G., Polisi, M., Battista, E., Netti, P. A., & Esposito, S. (2018). Novel process to prepare magnetic metal-ceramic nanocomposites from zeolite precursor and their use as adsorbent of agrochemicals from water. *Journal of Environmental Chemical Engineering, 6*, 527–538.

Parvej, M. S., Khan, M. I., & Hossain, M. K. (2022). Preparation of nanoparticle-based polymer composites. In S. M. Rangappa, J. Parameswaranpillai, T. G. Yashas Gowda, S. Siengchin, & M. O. Seydibeyoglu (Eds.), *Nanoparticle-based polymer composites* (pp. 55–94). Elsevier.

Pawlik, K., Pawlik, P., Wysłocki, J. J., & Kaszuwara, W. (2020). Structural and magnetic studies of bulk nanocomposite magnets derived from rapidly solidified Pr-(Fe, Co)-(Zr, Nb)-B alloy. *Materials, 13*, 1515.

Rane, A. V., Kanny, K., Abitha, V., & Thomas, S. (2018). Methods for synthesis of nanoparticles and fabrication of nanocomposites. In S. M. Bhagyaraj, O. S. Oluwafemi, & S. Thomas (Eds.), *Synthesis of inorganic nanomaterials* (pp. 121–139). Elsevier.

Rani, M., Choudhary, S., Yadav, J., & Shanker, U. (2022). Metal oxide-based nanocomposites for elimination of hazardous pesticides. In U. Shanker, C. M. Hussain, & M. Rani (Eds.), *Handbook of green and sustainable nanotechnology: Fundamentals, developments and applications* (pp. 1–26). Springer.

Ranjan, R., & Bajpai, V. (2021). Graphene-based metal matrix nanocomposites: Recent development and challenges. *Journal of Composite Materials, 55*, 2369–2413.

Rasmussen, J. J., Wiberg-Larsen, P., Baattrup-Pedersen, A., Cedergreen, N., McKnight, U. S., Kreuger, J., Jacobsen, D., Kristensen, E. A., & Friberg, N. (2015). The legacy of pesticide pollution: An overlooked factor in current risk assessments of freshwater systems. *Water Research, 84*, 25–32.

Rizwan, M., Ali, S., Ali, B., Adrees, M., Arshad, M., Hussain, A., ur Rehman, M. Z., & Waris, A. A. (2019). Zinc and iron oxide nanoparticles improved the plant growth and reduced the oxidative stress and cadmium concentration in wheat. *Chemosphere, 214*, 269–277.

Rubel, R. I., & Wei, L. (2022). Biochar-based controlled release nitrogen fertilizer coated with polylactic acid. *Journal of Polymers and the Environment, 30*, 4406–4417.

Saifuddin, N., Nian, C., Zhan, L., & Ning, K. (2011). Chitosan-silver nanoparticles composite as point-of-use drinking water filtration system for household to remove pesticides in water. *Asian Journal of Biochemistry, 6*, 142–159.

Sarkar, M. M., Pradhan, N., Subba, R., Saha, P., & Roy, S. (2022). Sugar-terminated carbon-nanodots stimulate osmolyte accumulation and ROS detoxification for the alleviation of salinity stress in Vigna radiata. *Scientific Reports, 12*, 1–17.

Seddon, N., Chausson, A., Berry, P., Girardin, C. A., Smith, A., & Turner, B. (2020). Understanding the value and limits of nature-based solutions to climate change and other global challenges. *Philosophical Transactions of the Royal Society B, 375*, 20190120.

Shahryari, F., Rabiei, Z., & Sadighian, S. (2020). Antibacterial activity of synthesized silver nanoparticles by sumac aqueous extract and silver-chitosan nanocomposite against *Pseudomonas syringae* pv. syringae. *Journal of Plant Pathology, 102*, 469–475.

Sharma, N., Singh, A., & Dutta, R. K. (2020). Synthesis of novel nanocomposite based on carboxymethylcellulose (CMC), kaolin and urea fertilizer for controlled release. *Journal of Polymer Materials, 37*, 1–15.

Sheikhalipour, M., Esmaielpour, B., Gohari, G., Haghighi, M., Jafari, H., Farhadi, H., Kulak, M., & Kalisz, A. (2021). Salt stress mitigation via the foliar application of chitosan-functionalized selenium and anatase titanium dioxide nanoparticles in stevia (*Stevia rebaudiana* Bertoni). *Molecules, 26*, 4090.

Sheng, Y., Yang, J., Wang, F., Liu, L., Liu, H., Yan, C., & Guo, Z. (2019). Sol-gel synthesized hexagonal boron nitride/titania nanocomposites with enhanced photocatalytic activity. *Applied Surface Science, 465*, 154–163.

Sheoran, P., Grewal, S., Kumari, S., & Goel, S. (2021). Enhancement of growth and yield, leaching reduction in Triticum aestivum using biogenic synthesized zinc oxide nanofertilizer. *Biocatalysis and Agricultural Biotechnology, 32*, 101938.

Shi, L.-Y., Subramanian, A., Weng, L., Lee, S., Kisslinger, K., Nam, C.-Y., & Ross, C. A. (2022). Selective sequential infiltration synthesis of ZnO in the liquid crystalline phase of silicon-containing rod-coil block copolymers. *Nanoscale*, *14*, 1807–1813.

Sidhu, A., Bala, A., Singh, H., Ahuja, R., & Kumar, A. (2020). Development of MgO-sepoilite nanocomposites against phytopathogenic fungi of rice (Oryzae sativa): A green approach. *ACS Omega*, *5*(23), 13557–13565. https://doi.org/10.1021/acsomega.0c00008

Srivastav, A. L., Dhyani, R., Ranjan, M., Madhav, S., & Sillanpää, M. (2021). Climate-resilient strategies for sustainable management of water resources and agriculture. *Environmental Science and Pollution Research*, *28*, 41576–41595.

Tariq, A., Bhawani, S. A., Asaruddin, M. R., & Alotaibi, K. M. (2021). Introduction to nanocomposites. In S. A. Bhawani, Z. Karim, & M. Jawaid (Eds.), *Polysaccharide-based nanocomposites for gene delivery and tissue engineering* (pp. 15–37). Elsevier.

Trotta, F., & Mele, A. (2019). Nanomaterials: Classification and properties. In F. Trotta & A. Mele (Eds.), *Nanosponges: Synthesis and applications* (pp. 1–26). Wiley-VCH.

Vejan, P., Khadiran, T., Abdullah, R., & Ahmad, N. (2021). Controlled release fertilizer: A review on developments, applications and potential in agriculture. *Journal of Controlled Release*, *339*, 321–334.

Wahyuni, S., Suyanta, S., Sedayu, B. B., Nafisyah, W., & Kartini, I. (2022). Synthesis bioplastic fertilizer of carboxymethyl cellulose/N-P-K/zeolite using twin-screw extruder as a model slow-release. *Key Engineering Materials*, *928*, 89–94.

Wang, X., Cai, A., Wen, X., Jing, D., Qi, H., & Yuan, H. (2017). Graphene oxide-Fe_3O_4 nanocomposites as high-performance antifungal agents against *Plasmopara viticola*. *Science China Materials*, *60*, 258–268.

Wei, H. G., Wang, H., Li, A., Cui, D. P., Zhao, Z. N., Chu, L. Q., Wei, X., Wang, L., Pan, D., Fan, J. C., Li, Y. C., Zhang, J. X., Liu, C. T., Wei, S. Y., & Guo, Z. H. (2020). Multifunctions of polymer nanocomposites: Environmental remediation, electromagnetic interference shielding, and sensing applications. *ChemNanoMat*, *6*, 174–184.

Younes, N., Hassan, H. S., Elkady, M. F., Hamed, A., & Dawood, M. F. (2020). Impact of synthesized metal oxide nanomaterials on seedlings production of three *Solanaceae* crops. *Heliyon*, *6*, e03188.

Younis, S. A., Kim, K.-H., Shaheen, S. M., Antoniadis, V., Tsang, Y. F., Rinklebe, J., Deep, A., & Brown, R. J. (2021). Advancements of nanotechnologies in crop promotion and soil fertility: Benefits, life cycle assessment, and legislation policies. *Renewable and Sustainable Energy Reviews*, *152*, 111686.

Youssef, K., & Hashim, A. F. (2020). Inhibitory effect of clay/chitosan nanocomposite against *Penicillium digitatum* on citrus and its possible mode of action. *Jordan Journal of Biological Sciences*, *13*, 349–355.

Zhang, Y., & Goss, G. G. (2022). Nanotechnology in agriculture: Comparison of the toxicity between conventional and nano-based agrochemicals on non-target aquatic species. *Journal of Hazardous Materials*, *439*, 129559.

Zhao, H., Wu, L., Zhou, Z., Zhang, L., & Chen, H. (2013). Improving the antifouling property of polysulfone ultrafiltration membrane by incorporation of isocyanate-treated graphene oxide. *Physical Chemistry Chemical Physics*, *15*, 9084–9092.

Zhao, J., Huang, P., Wang, X., Yang, J., Zhou, Z., Du, X., & Lu, X. (2022). Efficient adsorption removal of organic nitrogen pesticides: Insight into a new hollow NiO/Co@ C magnetic nanocomposites derived from metal-organic framework. *Separation and Purification Technology*, *287*, 120608.

Zhao, X., Ma, J., Wang, Z., Wen, G., Jiang, J., Shi, F., & Sheng, L. (2012). Hyperbranched-polymer functionalized multi-walled carbon nanotubes for poly (vinylidene fluoride) membranes: From dispersion to blended fouling-control membrane. *Desalination*, *303*, 29–38.

2 Synthesis of Nanofertilizers and Nanopesticides

Current Understanding and Future Prospects

*Aneesa Batool, Saika Bashir, Jebi Sudan, Momina Nazir, Gulam Nabi Yatoo, Anuj Ranjan, Vishnu D. Rajput, Parvaze A. Sofi, Zahoor Ahmad Bhat, Khurshed Hussain, Firdose A. Malik and Sajad Majeed Zargar**

2.1 INTRODUCTION

The potential of agriculture to satisfy the anticipated food demand for the increasing population is widely acknowledged, but there is no agreement on how this can be done sustainably (Alexandratos & Bruinsma, 2012) in terms of land conversion, water consumption and agrochemical pollution of ecosystems; sustainability entails both high yields and agricultural techniques with acceptable environmental effects (Tilman et al., 2002). Using excessive amounts of fertilizers, pesticides and herbicides to increase production per unit area can cause several negative side effects including soil degradation, toxic effects on non-target organisms and environmental pollution (soil, water and air pollution). It can also lead to low input use reliability, decreased food quality, the development of agrochemical resistance among various weeds, pathogens and insects and decreased income from production and soil degradation. Similarly, the efficiency of agrochemicals plays a significant role in selecting the right dose and amount. For example, the world's largest user of nitrogen fertilizer is China, where nearly 50% of nitrogen-based fertilizers are used. 5% of the total application is lost by evaporation, and 10% is lost through washing (FAO, 2002). Improving the efficiency and efficacy would ensure are reduction in the amount of agrochemicals being used for application and also their availability would provide optimal effectiveness.

With the broader goal of solving these problems, research on the application of nanotechnology (NT) in agriculture and, especially in crop production, has recently

* Corresponding author.

DOI: 10.1201/9781003376446-2

attracted a lot of interest. Reducing the agrochemicals required to ensure crop safety is one of the main motivations for the application of NT. One of the techniques to assist, maintain or even boost yields with considerably lower application rates, and concurrently limit harmful effects of agriculture on ecosystems and human health, is targeted delivery of the active ingredient (AIs) to the pest and/or increased fertilizer efficacy (Rajput et al., 2020). This can be done in several ways, including enhanced apparent dissolution rate, sustained release, aimed delivery, enhanced bioavailability, greater leaf adhesion and better sustainability of the AIs in the surroundings (Darzi-Naftchali & Okhtassi-Bidgoli, 2022). The improvement methods have been reported to enhance efficiency by an average of 20% in comparison to conventional equivalents, although nanoformulations can increase the efficiency by up to tenfold, especially in the case of pesticides to their target insect (Kah et al., 2018). Environmental pollution might be greatly reduced by maintaining similar levels of crop protection and nutrition while lowering the usage of agrochemicals by 20–30% (Darzi-Naftchali & Mokhtassi-Bidgoli, 2022). The characteristics of nanomaterials (NMs) or nanoparticles (NPs) are strongly linked to their activity and applications, and characteristics are governed by the types of NMs, size, shape and preparations. To increase the characteristics and reduce the expense of manufacturing, several synthesis techniques are either being invented or refined.

NPs are processed using specific methods to enhance their optic, physical, physical and molecular characteristics (Cho et al., 2013). A significant advancement in equipment has enhanced the characterization of NPs and their further use. Two methods, top-down and bottom-up, are used in the production of NPs, and utilizing NT for a range of applications such as nanopesticides, nanofertilizers, nanoinsecticide and nanoemulsions (Cao et al., 2018; Song et al., 2012; Wibowo et al., 2014). Most commonly used NPs are carbon nanotubes (CNTs) (Sarlak et al., 2014.), graphene oxides hydroxyapatite (Feizi et al., 2013.), double-layered hydroxides intercalated with phosphate ions (Liu & Lal, 2014), nanochitosan-NPK fertilizers (Abdel-Aziz et al., 2016), nanosilicon (Bhat et al., 2022), silica-oxide (SiO_2), zinc oxide (ZnO) nanoparticles (Adhikari et al., 2015), nutrient-loaded zeolite-based (Liu & Lal, 2015) etc.

2.2 NANOFERTILIZERS

Nanofertilizers are manufactured from conventional fertilizers, granular fertilizer materials or materials obtained from various vegetative or reproductive parts of the plant using various chemical, physical, mechanical or biological methods with the aid of NT. They are used to increase soil fertility, productivity and the quality of agricultural products (Singh et al., 2017). The vast surface area of nanofertilizers and particle sizes that are less substantial than pores in the plant's leaves and roots allow for more permeation into the plant from the applied surface and increased absorption and nutrient utilization efficiency (Singh et al., 2017). Reducing the size of the particles increases a fertilizer's effective surface area and the number of particles per unit area, which increases the chance that nanofertilizers will come into touch with one another and boost nutrient penetration and absorption (Liscano et al., 2000).

NPs as nanofertilizers are also reported to enhance the soil microbial dynamics that lead to the effective cycling of nutrients, biocontrol of pathogens and secretion of phytohormones and other useful substances beneficial for the plant's health (Ranjan et al., 2022)

The use of nanofertilizers is highly inventive. In nanofertilizers, micro- or macronutrients can be given as emulsions or NMs, enclosed in nanoshells, or covered with a thin protective coating (DeRosa et al., 2010). By using nanofertilizers, nutrients use efficiency (NUE) can be greatly improved, and they are released in the soil gradually and under regulated conditions, preventing eutrophication and water resource pollution (Naderi & Abedi, 2012., Moaveni & Kheiri, 2011). With increased NUE, crop yield and quality are significantly enhanced, thereby helping in lowering production costs and supporting agricultural sustainability (Usman et al., 2020). The market entry strategy for the key elements N, P, K, Ca, Mg and S, which are used in greater quantities and are now less efficient and more environmentally harmful, should be focused on the development of high-efficiency products for these elements (Bratovcic et al., 2021).The cost of fertilization per unit of fruit output would be greatly decreased if nanofertilizers could be applied at lower kg per hectare and in fewer treatments per season (Ojeda-Barrios et al., 2019).

Because of pH-dependent events, redox potential and the availability of ligands in the soil, nanofertilizers release nutrients at a slower rate than conventional fertilizers while also increasing their stability through aggregation or adsorption without altering chemical speciation (Raliya et al., 2018). As compared to soybean (*Glycine max* L.) plants fed with traditional fertilizer, phosphatic nanofertilizer application has also been associated with an improved growth rate (by 32%) and seed output (by 20%) (Liu & Lal, 2014). The soybean plant's usage of water and fertilizer was improved by the addition of SiO_2 and TiO_2 nanoparticles, which also raised the activity of nitrate reductase and accelerated plant absorption (Changmei et al., 2002). TiO_2 nanoparticles stimulate tobacco seed germination, but they have a detrimental effect on the roots' ability to grow (Frazier et al., 2014).

Shebl et al. (2019) synthesized zinc, manganese, and iron nanoparticles using microwave-assisted hydrothermal process and subsequently applied to squash plants through foliar spray for determining their efficacy. The collected results demonstrated that spraying the plants with MnO_2 NPs produced the best fruit and vegetative development characteristics, yield and photosynthetic pigment content. On the other hand, squash fruits that had been treated with iron oxide NPs had the highest value in terms of their organic matter, protein, lipids and energy content (Bratovcic et al., 2021). The three forms of nanofertilizers include micro-nutrient, macro-nutrient and NMs-enhanced nutrients, which, when combined with nutrients, increase the efficiency of the nutrients but do not themselves contain or give any nutrients (Liu & Lal, 2014). In order to improve the solubility, stability, release in response to an external trigger, effective delivery, economic viability and environmental friendliness of nanomaterials, these properties are being developed and engineered (Rajput et al., 2021). Zeolite-based nanofertilizers, nano-composites, polymer-coated fertilizers, ZnO, silica (SiO_2), iron (Fe), TiO_2, aluminum oxide (Al_2O_3), cerium oxide (CeO_2), quantum dots, gold (Au) nano-rods, ZnCdSe/ZnS core-shell, Mn/ZnSe quantum dots and an indium phosphide-zinc sulfide (InP/ZnS) core-shell are possible NMs suitable for developing as nanofertilizers (Prasad et al., 2017).

Nutrients that the crops need are released into the soil because of the interaction of nano suspension containing active ingredient (AI) with water, commonly referred to as moisture-based trigger release and it is most suitable for nanofertilizers. The polymer coating in nanofertilizers, or the thin film encapsulation of NPs, further delays contact with soil and water to avoid unfavourable nutrient losses (Shang et al., 2019). According to Kottegoda et al. (2017), urea-loaded hydroxyapatite nano-hybrids are potential nano-encapsulated fertilizers for the release of nitrogen over time. Carbon-based NPs such as fullerenes, fullerols, graphene and carbon nanotubes (CNTs) have been reported to increase the germination rate, chlorophyll and protein content, as well as shoot and root lengths and they have become important plant growth regulators. For instance, polyhydroxy fullerene quickens the lengthening of *Hordeum vulgare* roots (Panova et al., 2016.) and fullerol encourages the production of chlorophyll in *Triticum aestivum* (Wang et al., 2016). Compared to conventional bulk fertilizers, urea-hydroxyapatite nano-hybrid fertilizers gently and uniformly release N to enhance plant growth and development (Kottegoda et al., 2011). The necessary nutrients are combined with the zeolites, which aid in the prolonged release of nutrients. For instance, nutrients were made available for up to 60 days using nano zeolite that included potassium sulphate, urea, calcium and hydroxyapatite (Kalia et al., 2020; Lawrencia et al., 2021). After 42 days of *Zea mays*plant growth in NH_4 zeolites-enriched soils, NUE was found to be between 72–95%, compared to 30–76 % in $(NH_4)_2SO_4$-enriched soils (Perrin et al., 1998).

2.3 NANOPESTICIDES

Pesticides are frequently employed in agriculture to increase crop productivity. The use of nanopesticides provides a means of both controlling pesticide distribution and enhancing effects while using fewer chemicals. The first type of nanopesticides can be predicted as having AIs that are either manufactured NMs like metal NPs (e.g., Ag and Cu) (Bratovcic, 2020; Guilger-Casagrande et al., 2019) or metal oxide NPs (e.g., ZnO, CuO, MnO_2, SiO_2, and TiO_2) (Elmer et al., 2018; Shukla et al., 2020). The manufacturing and use of nanopesticides seek to increase their effectiveness and availability while also minimizing or completely eradicating any potential risks and reducing the number of active components used in the process (Bratovcic et al., 2021). Reformulations of approved AIs with insecticidal, fungicidal or herbicidal qualities make up the various types of nanopesticides currently being developed, as well as others that are already available on the market (Walker et al., 2017). In order to increase the stability of the pesticide, improve its dispersion in aqueous media, and enable a controlled release of the active ingredient, nanoencapsulation is also used for pesticides, especially hydrophobic ones (their insolubility in water is a limiting factor) which improves its dispersion in aqueous media, and allows a controlled release of the active ingredient (de Oliveira et al., 2015; Grillo et al., 2016). Alginate, carrageenan, xanthan gum, chitosan and modified starch are just a few examples of the polymers used in nanoencapsulation for nanopesticides, while other examples include proteins and polysaccharides like chicken egg albumin, zein, casein, lactalbumin, lactoglobulin, collagen and gelatin (Pestovsky & Martínez-Antonio, 2017). The regulated release and delayed breakdown of the AIs in nanocapsules make them more effective at preventing and treating plant diseases and pests. Chitosan, polyethylene glycol (PEG), starch, cellulose and polyester material are some examples of the various types of polysaccharide materials that have been used in synthetic processes (Kashyap et al., 2015; Loha et al., 2012). Temifos and imidacloprid-containing PEG-encapsulated nanopesticides were created by Bhan et al. (2014), using the melt-dispersion process and were found to be more effective against *Culex quinquefasciatus* larvae. Poly (epsilon-caprolactone) (PCL) nanocapsules have been created for the controlled release of atrazine in an effort to reduce the adverse effects of herbicides and have proven to have increased herbicidal activity compared to atrazine's commercial formulation (de Oliveira et al., 2015).

Nanoemulsions (NEs) are also known to increase solubility, spreadability, wettability and higher mechanical stability over conventional emulsions (Chhipa, 2017). To increase the effectiveness against various pests, a nanoemulsion comprising Tebuconazole (TBZ), organosilicon, alkyl glucosides, and various fatty acid methyl esters were created (Díaz-Blancas et al., 2016; Jiang et al., 2012; Lim et al., 2012). Insecticide-coated liposomes (Adak et al., 2012), neem oil (Sarkar et al., 2012.), garlic essential oil (Pankaj et al., 2012), *Artemisia arborescens* essential oil (Kaushik et al., 2013), carbofuran (Grillo et al., 2016), mancozeb (Amenta et al., 2015), atrazine and simazine (Ocsoy et al., 2013), carbofuran (Yin et al., 2012), thiram (Fraceto et al., 2016) and beta-cyfluthrin (Loha et al., 2012.) are a few examples of nanoformulations that were developed to increase efficiency, stability and reduce the concentration of pesticides. SiO_2 NPs coated with (3-mercaptopropyl) triethoxysilane were reported to be significantly effective against insects (Chhipa, 2017). In the case of regulated releases, validamycin has been demonstrated to be more active for extended periods when used with calcium carbonate ($CaCO_3$) and porous SiO_2 particles (Qian et al., 2011). At a concentration 10,000 times lower than that advised, copper (Cu) in nanoform showed four orders better action against bacterial blight on pomegranates (Mondal & Mani, 2012).

Thus, the above studies exhibit the feasibility of the application of NPs in the formulation of nanopesticides, and their convincing application also demonstrates the reduced dependency of higher doses, targeted application and controlled release. Hence, better availability, and most important thing is its lesser effect the environment and non-target organisms.

2.4 SYNTHESIS OF NANOAGROCHEMICALS

For the synthesis of NPs, top-down or bottom-up techniques have both been functionalizing. The bottom-up strategy is recognized as the constructive technique. The bottom-up approach contrasts sharply with the top-down approach (Abid et al., 2022). Figure 2.1 briefly describes the formation of NPs through these approaches. Various types of bottom-up approaches have been described as under:

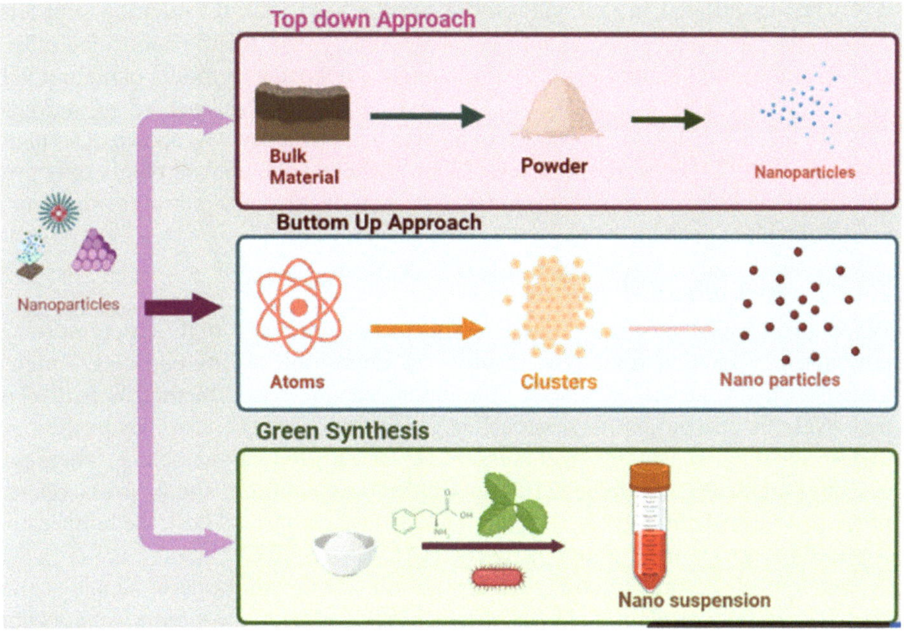

FIGURE 2.1 Different methods used for the synthesis of nanoparticles

2.4.1 Bottom-up Strategies

2.4.1.1 Chemical Vapour Deposition (CVD)

A thin film of gaseous reactants is deposited onto a substrate using chemical vapour deposition. By mixing gas molecules, the deposition is carried out in a reaction chamber at room temperature. When a heated substrate makes touch with the mixed gas, a chemical reaction happens. On the surface of the substrate, this reaction creates a thin coating of product that is collected and utilized. The temperature of the substrate is a determining element in CVD (Abid et al., 2022).

CVD has got many advantages in the synthesis of NPs. By controlling process parameters, it is possible to control orientations of nanostructure, surface morphology and crystal structure, as well as the substrate's homogeneity and material adhesion, which enable the creation of intricately structured coatings. It also enables large-scale manufacturing of 2D NMs with purity, good crystal quality and few substrate faults and has the highest degree of control (Pottathara et al., 2019; Sayago et al., 2020). Compared to other methods utilized for the same purposes, the CVD approach has a number of benefits; however, there are significant drawbacks. Precursors used in the process are poisonous, flammable, explosive and extremely rare. The cost of synthesis is higher; it also leads to the higher manufacturing cost of some CVD variants. The high deposition temperatures of certain CVD variants, along with their production of highly poisonous gaseous byproducts, limit the kinds of substrates that can be used (Pottathara et al., 2019; Sayago et al., 2020).

CVD was utilized in the synthesis of novel PVA-starch formulation-supported Cu-Zn NPs carrying carbon nanofibers as a nanofertilizer for the controlled release of micro-nutrients (Kumar et al., 2018). Using a chickpea as a model plant and various doses, including 0.25, 0.50, 1.0, 2.0 and 4.0 g of PBMC per kg of soil (garden), for up to 30 days, the produced nanofertilizer's efficacy was shown. The plants grown in PBMC (1-1) were the tallest (33 cm), whereas the control plants only grew at a length of 18 cm, demonstrating the efficiency of the prepared micro-nutrient in promoting plant growth.

2.4.1.2 Hydrothermal Process

When a solid substance reacts with an aqueous solution at a high temperature and pressure in a reaction vessel, the result is the deposition of tiny particles, which is known as hydrothermal synthesis. A solution reaction-based method is hydrothermal. Water is utilized in this approach as a solvent, hence the term "hydrothermal" to describe the process. In a steel pressure vessel, mostly in an autoclave, where processing parameters are managed by modifying temperatures and/or pressures, the process is carried out. Beyond the point at which water starts to boil, the temperature is raised until vapour saturation is reached (Abid et al., 2022).

Due to uniform precipitation, low cost, a welcoming atmosphere, simple scaling up and pure final product, the hydrothermal process has made significant contributions to modern science and technology. The hydrothermal method can also be split into hydrothermal synthesis, treatment, crystal growth, organic waste treatment, and the preparation of functional ceramic powder (Abid et al., 2022).

The hydrothermal technique may have some advantages over other parallel processes, such as precise control of the NPs size, the ability to produce materials with high vapour pressure, low melting points, and pyrolysis propensities, and the ability to produce high-crystallinity, nanocrystalline materials (Asim et al., 2014; Jamkhande et al., 2019). The synthesis of NMs using this method has some limitations as compared to other processes, such as the inability to directly observe crystal growth, the expense of the autoclave, and the difficulty of controlling the process (Asim et al., 2014; Jamkhande et al., 2019).

Template-free microwave-assisted hydrothermal was utilized in the synthesis of MnZn ferrite as a nanofertilizer for squash plants (*Cucurbita pepo* L.). The produced NMs were used for foliar application on *C. pepo* L. planting at various concentrations (0, 10, 20 and 30 ppm). For two consecutive successful planting seasons, the effects of applying the produced ferrite NPs on the *C. pepo* L. plant yielded mineral content, proximate analysis and vegetative growth. The nanoferrite produced at 160 °C and applied to the growing plants at a concentration of 10 ppm resulted in the highest yield increase (49.3 and 52.9%) compared to the untreated *C. pepo* L. for the two consecutive seasons, whereas plants treated with 30 ppm ferrite NPs produced at 180°C produced *C. pepo* L leaves with the highest organic matter content (73.0 and 72.5%) and total energy (260 and 258.3 kcal g^{-1}). On the other hand, plants treated with 20 ppm ferrite NPs produced at 160°C were able to produce squash fruits with the highest levels of organic matter (76.6 and 76.3%) and total energy (253.6 and 250.3 kcal g^{-1}) (Shebl et al., 2020).

In another study, *Paulownia tomentosa* leaves were used to green synthesize a new nanofertilizer utilizing the hydrothermal approach, and its efficacy in reducing drought stress in *Ocimum basilicum* was subsequently examined. Three irrigation regimes, comprising 100% of field capacity (IRF100), 75% of field capacity (IRF75) and 50% of field capacity (IRF50), were used to test five concentrations of nanofertilizer (0, 50, 70, 90 and 110 ppm). The study reported improved levels of soluble proteins and photosynthetic pigments increased after the treatment of paulownia-based nanofertilizers. Under extreme stress conditions, NFP-90 application decreased the formation of proline and malondialdehyde, respectively, by 51.8% and 30.8% compared to the control, which suggests that it ameliorates the negative effects of drought stress (Sohrabi et al., 2022).

2.4.1.3 Co-precipitation Method

According to (Abid et al., 2022), co-precipitation is one of the earliest wet chemical techniques used to create NPs. The most fundamental and often applied method for creating a variety of NPs is this one. Although impurities precipitate with the product in this process, it is simple to separate those using techniques like filtration and washing. The co-precipitation strategy entails using a precipitation reaction to produce a homogenous solution with a constant composition of two or more cations. This method has the advantage of immediately creating homogenous NPs with small diameters and a narrow size dispersion through multiple chemical reactions in the solution. For the purpose of maintaining a required pH, the solution is mixed directly or drop-by-drop with another solution that contains dissolved precipitation agents, such as sodium hydroxide, ammonia and many others (Abid et al., 2022).

The co-precipitation technique has some benefits over other processes, including low temperature, energy efficiency, ease of fabrication, a wide range of possibilities for modifying particle surface state and overall uniformity (Vaseghi & Nematollahzadeh, 2020).

There are a lot of drawbacks to consider, including the need for using a lot of chemicals and the possibility of producing potentially dangerous waste. This method also won't work if the reactants have precipitation rates that are noticeably different from one another. Additionally, stabilizers and a range of post-processing techniques are needed for the majority of chemically precipitated NPs in order to maximize the purity of the finished products (Vaseghi & Nematollahzadeh, 2020).

The co-precipitation method was used to create iron oxide NPs as nanofertilizers on some quality characteristics of ginger (*Zingiber officinale*). In contrast to iron-chelated supplementation, ginger plants were fed with iron oxide NPs, and their performance was recorded. The plants demonstrated a satisfactory response, indicating good cellular health of the plants. The current method of employing nanofertilizers as plant supplements is a smart farming strategy due to its easiness and adaptability. As it improves Fe storage in edible plants, increasing the Fe level in food may also aid in biofortification. This will assist to address the issue of Fe deficiency anaemia, one of the most common micro-nutrient deficits in humans worldwide (Siva et al., 2016).

Additionally, magnetic Fe_3O_4 NPs were created using the chemical co-precipitation method, and they were then used as fertilizer on coriander (*Coriandrum sativum*) plants for efficient mitigation of cadmium (Cd) and lead (Pb) toxicity. Using 10 mgL^{-1} Fe_3O_4 NPs, 100 mgL^{-1} NPK, and no fertilizer, the mitigation percentage in *C. sativum*was determined to be 97.37, 81.12, 77.79 (Cd) and 98.45, 80.77, 75.89 (Pb), respectively. On the other hand, employing comparable doses of Fe_3O_4 NPs, NPK, and without fertilizer, growth impacts were seen in length, measuring 53.3, 33.02, 25.4 cm (Cd) and 48.26, 43.18 and 38.1 cm (Pb) correspondingly. Additionally, effective Fe_3O_4 NPs were found to have outstanding antibacterial action against gram-positive and gram-negative micro-organisms (Fahad et al., 2022).

2.4.1.4 Sol-gel Method

A colloidal suspension of particles in a liquid phase is known as a sol. A solid macromolecule dissolved in a liquid is the gel. Due to its simplicity and the ease with which most NPs may be synthesized, sol-gel is the most used bottom-up approach. A chemical solution that serves as a precursor for an integrated system of discrete particles is used in this wet chemical process. Precursors utilized most frequently in the sol-gel method are metal oxides and chlorides. The precursor is subsequently mixed with the host liquid by sonication, shaking or stirring, resulting in a system with a liquid and solid phase. Different techniques, including sedimentation, filtering and centrifugation, are used to recover the nanoparticles after phase separation, and drying is then used to further remove the moisture (Ealia & Saravanakumar, 2017).

One of the many advantages of the sol-gel technique is its cost-effectiveness. Other advantages include the homogeneous nature of the manufactured material, the low process temperature, the method's simplicity in producing complex nanostructures and composite materials, the higher purity of the end products and the ability to add small amounts of dopants to the sol that are evenly distributed in the final product. However, there are some drawbacks to utilizing this strategy, including the fact that reaction times are a little longer, organic chemicals are harmful to one's health, and post-treatment is required after sample purification.

The sol-gel process has been used to produce engineered urea-doped hydroxyapatite NMs as nitrogen and phosphorus fertilizers for *O. sativa* L. A recommended dose of commercial bulk urea fertilizer (BUF) and a control treatment was compared with the synthesized urea-doped hydroxyapatite NPs (UHNP) at recommended (UHNP RD) and half-recommended (UHNP HRD) doses for nutrient availability in soil and yield of *O. sativa* L. When compared to the control treatment, the grain yield increased by 61, 60 and 65% in the BUF, UHNP HRD and UHNP RD treatments, respectively. This suggests that the UHNP can produce a yield that is comparable to that of commercial fertilizers. In comparison to BUF, the concentrations of NH_4-N, NO_3-N and PO_4-P in the leachate decreased under UHN HRD by 69.11, 16.3 and 87%, respectively, and under UHNP RD by 60, 10.6 and 66.3%. The UHNP HRD had the maximum NUE for the nutrients nitrogen (76.5%) and phosphorus (14.27%) (Pohshna & Mailapalli, 2022).

The zinc hydroxide nitrate (HNO_4Zn) complex was also synthesized by a sol-gel method for use as foliar nanofertilizers. The product's masses might be enhanced roughly 60 times to around 70g in each batch in comparison to the theoretical value computed for the mentioned process. To confirm the possible use of products as foliar nanofertilizers, further scaling up of the precursors' quantities and assessment of the product's surface charge is needed (Le et al., 2016). The sol-gel method was also used to create high throughput SiO_2-NPs for soil application of nano-Silica on *Z. mays* to improve yield and to test its insecticidal activity against some stored grain insects after the post-harvest (El-Naggar et al., 2020).

2.4.1.5 Green Synthesis

Nanoparticle preparation using physical and chemical methods is sometimes time-consuming, requires sophisticated instruments and may not be environmentally friendly. To overcome such issues, there is a new advancement in NT called "green synthesis", which uses biological systems like plants and microbes for the preparation of NPs. Additionally, the green synthesis of NPs are not hazardous and expensive. In this approach, plant extracts are utilized to synthesize metal NPs because they are thought to be commercially viable and cost-effective for large-scale manufacture. Researchers have hypothesized that a variety of biomolecules such as vitamins, amino acids, proteins, phenolic acids, alkaloids, etc. are found in plants and micro-organisms and contribute to the biological reduction of NPs (Kandasamy & Sorna Prema, 2015).

Due to their rapid growth, simplicity of cultivation and propensity to thrive under ambient settings of temperature, pH and pressure, micro-organisms have taken a distinctive place among the numerous biological sources for the green synthesis of metal NPs. Micro-organisms can produce metal NPs by a similar reduction mechanism, which involves enzymatically capturing target metal ions from the environment and transforming them into the elemental form (Dashora & Sharma, 2018).

The metal NPs produced through green synthesis have a wide range of potential uses in agriculture to boost crop yield. Products including nanopesticides, nanofungicides, nanobiosensors and nanofertilizers are frequently produced using metal NPs. This technology can improve agricultural goods in terms of quality and productivity and lessen chemical contamination, or even shield crops from environmental stressors (Usman et al., 2020).

The numerous benefits of green synthesis include energy efficiency, low-cost, hazard-free production, secure products and economically lesser waste. Green NPs have several applications apart from agriculture. They have found their greater application in the public health sector, especially in the biomedical industry and pharmaceutical applications.

For example, AuNPs and AgNPs were synthesized using the *Averrhoa bilimbi* fruit extract, using the salts chloroauric acid and silver nitrate ($AgNO_3$), respectively. Au and Ag NPs production was indicated by a colour change to yellow and violet. Lower salt concentrations resulted in a sluggish rate of NPs production, which lowers UV spectrometer absorption. Additionally, it was hypothesized that the *A. bilimbi* fruit's phenols, alcohols, and amines help reduce and cap NPs (Issac et al., 2013). Additionally, Au NPs were optimized and stabilized by using *Hibiscus rosa-sinensis*a medicinal plant extract with microwave heating. Thus, green Au NPs made using a process mediated by *H.rosa-sinensis* remarkably stable when exposed to various pollutants and upon being sterilized. For biological and sensing applications, Au NPs could be used because of their stability characteristics (Yasmin et al., 2014).

2.4.2 Top-Down Approach

The top-down or disruptive process involves breaking down a large substance into little pieces. Some of the most popular techniques for creating nanoparticles include mechanical milling, laser ablation, sputtering and thermal breakdown (Ealia & Saravanakumar, 2017).

2.4.2.1 Laser Ablation

A popular technique for creating NPs from different solvents is called Laser Ablation Synthesis in Solution (LASiS). A laser beam condenses a plasma plume that creates NPs when it strikes a metal immersed in a liquid solution (Amendola & Meneghetti, 2009). To create NPs, a solid target (usually metal) is exposed to a pulsed laser beam while submerged in a liquid (acetone, ethanol, water, chloroform, isopropanol, etc.). The laser settings and liquid medium may be changed to vary the NPs' shape, composition and other attributes (Patil & Burungale, 2020). It is a reputable top-down approach that offers an alternate solution to the usual chemical reduction of metals to synthesize metal-based NPs. It is a "green" technique since LASiS offers stable nanoparticle production in organic solvents and water without the need for chemical stabilizers or other chemicals (Ealia & Saravanakumar, 2017).

2.4.2.2 Thermal Decomposition

Heat that ruptures the molecular bonds in the substance causes thermal decomposition, an endothermic chemical breakdown (Salavati-Niasari et al., 2008). The decomposition temperature is the precise temperature at which an element begins to chemically break down. The metal is broken down into NPs at certain temperatures through a chemical process that also yields secondary compounds (Ealia & Saravanakumar, 2017).

2.4.2.3 Mechanical Milling

Mechanical milling is the top-down technique for creating different NPs since it is the most widely employed. Different components are crushed in an inert environment using mechanical milling for milling and post-annealing NPs during synthesis (Yadav et al., 2012). In mechanical milling, plastic deformation determines particle shape, fracture determines particle size reduction and cold welding determines particle size expansion (Ealia & Saravanakumar, 2017). Based on the movement of the milling balls and vials, some examples include the planetary mill, attrition mill, horizontal mill, vibratory mill and rotatory mill. Preparation of seaweed nanopowder particles using planetary ball milling and its effect was observed on date palm (*Phoenix dactylifera* L.) seedlings when NPs in the form of a solution were applied to the leaves, the number of branches, dry weight, total soluble sugars and other secondary metabolites increased. Additionally, it was shown that nanopowders had an impact on the stimulation of chlorophylls, antioxidant activity (DPPH and chelating activity) and several antioxidant enzyme activities (POD, APX, and CAT). In another study, ball milled nano-sized zeolite loaded with zinc sulphate ($ZnSO_4$), indicates that the nanozeolite-based fertilizers provide Zn for up to 40 days. This study makes the case for the use of nanozeolite-based Zn fertilizers as a method to accomplish the delayed release of Zn, which would enhance crops' ability to use Zn (Subramanian & Rahale, 2012). A few examples of nanoagrochemicals that have been synthesized by various approaches have been tabulated in Table 2.1.

TABLE 2.1
A few examples of nanoagrochemicals that have synthesized by various approaches

Techniques	Examples
Chemical vapour deposition (CVD)	Synthesis of novel PVA-starch formulation supported Cu-Zn nanoparticle carrying carbon nanofibers as a nanofertilizers for the controlled release of micro-nutrients (Kumar et al., 2018). Biosynthesis of zinc oxide nanomaterials from plant extracts (Huang et al., 2021).
Hydrothermal process	Template-free microwave-assisted hydrothermal synthesis of manganese zinc ferrite as a nanofertilizer for squash plant (*Cucurbita pepo* L) (Abid et al., 2022). Plant-based nanofertilizer prepared from *Paulownia tomentosa*: fabrication, characterization and application on *Ocimum basilicum;* Biogenic synthesis of iron oxide nanoparticles using *Moringa oleifera* and chitosan and its evaluation on cornGermination (Shebl et al., 2020).
Co-precipitation	Synthesis, characterization of iron oxide; nanoparticles and their applications as nanofertilizers on some quality characters of ginger (*ZingiberofficinaleRosc.*) (Fahad et al., 2022). Zinc aluminosilicate which enabled high yield of *oryza sativa* (Naseem et al., 2020).
Sol-gel	Soil application of nano-silica on maize yield and its insecticidal activity against some stored insects after the post-harvest (El-Naggar et al., 2020).
Green synthesis	Synthesis of gold and silver nanoparticles using the *Averrhoa bilimbi* fruit extract; synthesis of ZnO nanoparticles by *Calotropis gigantea* (Vidya et al., 2013). Optimization and stabilization of gold nanoparticles by using herbal plant extract with microwave heating (Issac et al., 2013).
Ball milling	Preparation of seaweed nano powder
thermal decomposition	Particles; its effect was observed on date palm seedlings.
laser ablation	Metal, oxide and polymer-based (Ealia et al., 2017).
	Carbon and metal oxide-based (Ealia et al., 2017).
	Carbon based and metal oxide-based (Ealia et al., 2017).

2.5 CRITICAL ANALYSIS AND THE FUTURE DEVELOPMENTAL POSSIBILITIES OF NANOAGROCHEMICALS

2.5.1 Nanopesticides

Nanopesticides demonstrated higher solubility and stabilities of the AIs for efficient pest control (Venugopal & Sainadh, 2016). However, there are some factors that identify the need for improvement in order to make a substantial contribution to agricultural operations. Some of the important points are being highlighted in this context (Ocsoy et al., 2013):

a. To maximize the effectiveness of NPs, green chemistry and environmental sustainability principles should be preferably used.
b. The creation of a process for scaling up NPs for commercial use.
c. Field-level comparison of nanoformulation activities with already available commercial products to assess the usefulness of the technology.
d. Nanopesticide toxicity level assessment based on environmental impact.
e. The regulations governing the use of NPs in agriculture need to be improved.
f. In terms of solubility of the active ingredient, stability, controlled release and targeted delivery of the active ingredient, the development of smart nanopesticides will offer many solutions to the agrochemical industry. However, much more research is needed tou nderstand the fate of nanopesticides in the environment.
g. To understand the methods by which nanoformulations may alter the behaviour of the AIs, comparative studies of such formulations with pure AI are required.
h. It has been convincing that a number of nanopesticides are more effective than their commercial counterparts, and in some circumstances, they are also less hazardous to non-target species. However, the underlying processes are still not completely understood, thus further study is needed before any generalizations can be formed (Kah & Hofmann, 2014).

Making environmentally friendly nanopesticides that are less toxic than traditional formulations are currently being encouraged more. This was accomplished by taking into account AIs with natural sources (such as pheromones and essential oils) as well as possibly friendlier adjuvants (such as the use of biodegradable polymers with natural origins and the avoidance of surfactants). The preparation of the majority of those potential goods is still in its infancy. Future studies must determine if they can perform as well as or better than present formulations while being competitive in terms of price (Kah & Hofmann, 2014).

Studies on the environmental destiny of nanopesticides are still rare and have often solely concentrated on the specific processes that the formulation was designed to prevent (such as a breakdown in a formulation designed to prevent the AI from premature decay). In general, unanticipated modifications have not been taken into account. The advantages and risks associated with nanopesticides cannot yet be reliably assessed given the existing level of understanding. Extensive research is needed over the ensuing years, and it will be necessary to:

- Develop experimental protocols to produce reliable fate properties.
- Look into the bioavailability and longevity of nanopesticides (e.g., release profiles and agglomeration behaviour under realistic conditions).
- Evaluate and, where necessary, improve current environmental hazard evaluation methodologies.

However, findings could not be replicated across a variety of hosts or pest species. In certain situations, the toxicity of the nanopesticide was shown to be almost an order of magnitude greater than that of the traditional formulation (Balaji et al., 2015; Pankaj et al., 2012; Saini et al., 2014; Sasson et al., 2007). The need for critical comparisons with pertinent existing commercial equivalents is highlighted in one instance where the nanopesticide, when compared to one commercial product, was found to have substantially higher toxicity to the target organism, but it had a similar (or even lower) toxicity in comparison to another commercial product (Sasson et al., 2007).

2.5.2 Nanofertilizers

In comparison to other minerals, plant macro-nutrients, particularly N and P, are crucial for agricultural output, food security and environmental protection. Therefore, it is essential and a top research priority to develop and apply nanofertilizers that are highly effective (low leaching and/or gasification rates, low soil immobilization rates and high plant uptake rates) and pose minimal environmental risks (low eutrophication potential and low N leaching rates to groundwater) (Singh et al., 2017).

Micro-nutrient research should concentrate on increasing the usability (plant absorption rate) of these fertilizers to solve the field leaching (e.g., Mo) or soil fixation concerns (e.g., Fe, Zn or Cu) associated with the traditional micro-nutrient fertilizers (Liu & Lal, 2015). More significantly, research is required to evaluate the advantageous effects of these micro-nutrient nanofertilizers with their commercially available micro-nutrient equivalents (for example, FeNPs vs. $FeCl_3$ or Fe (EDTA) as Fe sources) under field conditions (Liu & Lal, 2015).

Another issue with employing NMs as fertilizers is nanotoxicity. To allay public concerns regarding the related nanotoxicity of a newly created nanofertilizer, more study on its toxicity must be done. There are several different kinds of naturally occurring NPs in ecosystems, which is one of two key issues regarding nanotoxicity studies (Liu & Lal, 2015):

(a) Crops require only minute quantities of micro-nutrients to sustain normal physiological functions, whereas excessively high concentrations of any micro-nutrient would cause phytotoxicity.
(b) Plants should have unique methods for controlling these zeolite NPs to be a potential means for effectively enhancing N usage and lowering N leaching. Research is still needed to evaluate the increased expenses associated with zeolite procurement and the N-loading procedure with the yield and

environmental advantages of zeolite nanofertilizers. Other nutrient-rich nanoparticle forms, such as SiO_2-NPs, Fe_2O_3/ Fe_3O_4NPs and CNTs should also be studied and developed.

According to a recent study, agronomic yields can be greatly raised by utilizing TiO_2-NPs or CNTs as a plant growth stimulant. For example, when used to farm production scale, CNT treatment may be more than double tomato (*Solanum lycopersicum*) output compared to controls, which is a significantly larger harvest improvement (Khodakovskaya et al., 2013). Data on the application of nanofertilizers application and their outcomes are very sparse and mostly based on laboratory or greenhouse experiments. For its application in field conditions, more evidence is required to support along with their risk assessment in terms of environmental health. Furthermore, since some of these NMs may fail to provide the plants with any nutrients, specific growth-promoting pathways must be investigated and defined. Any advancement in crop production stimulators other than nutrient delivery would be a turning point not just for NT research but also for the start of a brand-new "Green Revolution".

2.6 CONCLUSION

In an attempt to ensure food security, an excessive amount of synthetic agrochemicals are used, which is detrimental to both environment, human health and economy. A significant amount of these goes unused, which is again a worrisome problem. Thus, nanotechnology is offering a way out in the form of nanoagrochemicals which behave differently than the commercial agrochemicals. The nanoagrochemicals have bioavailability, targeted and controlled release, higher solubility and are required in lesser quantities. There are several methods to synthesize nanoagrochemicals, but there is still need for improvement to make a substantial contribution to agricultural operations. To maximize the effectiveness of NPs, green chemistry and environmental sustainability principles should be preferably used. The fate of nanoagrochemicals and their comparison with the commercial forms to evaluate their advantageous effects with their commercially available counterpart's micro-nutrient needs an in depth study.

New technologies should go from the laboratory stage to controlled field testing in order to overcome the issues and difficulties related to nanoagrochemicals. Such methods might significantly enhance environmental quality, food security, sanitation and nutritional health.

REFERENCES

Abdel-Aziz, H. M. M., Hasaneen, M. N. A., & Omer, A. M. (2016). Nano chitosan-NPK fertilizer enhances the growth and productivity of wheat plants grown in sandy soil. *Spanish Journal of Agricultural Research*, *14*, e0902. https://doi.org/10.5424/sjar/2016141-8205

Abid, N., Khan, A. M., Shujait, S., Chaudhary, K., Ikram, M., Imran, M., Haider, J., Khan, M., Khan, Q., & Maqbool, M. (2022). Synthesis of nanomaterials using various top-down and bottom-up approaches, influencing factors, advantages, and disadvantages: A review. *Advances in Colloid and Interface Science*, *300*, 102597. https://doi.org/10.1016/j.cis.2021.102597

Adak, T., Kumar, J., Shakil, N. A., & Walia, S. (2012). Development of controlled release formulations of imidacloprid employing novel nano-ranged amphiphilic polymers. *Journal of Environmental Science and Health, Part B*, *47*, 217–225. https://doi.org/10.1080/03601234.2012.634365

Adhikari, T., Kundu, S., Biswas, A. K., Tarafdar, J. C., & Subba Rao, A. (2015). Characterization of zinc oxide nano particles and their effect on growth of maize (*Zea mays* L.) plant. *Journal of Plant Nutrition*, *38*, 1505–1515. https://doi.org/10.1080/01904167.2014.992536

Alexandratos, N., & Bruinsma, J. (2012). World agriculture towards 2030/2050. *Land Use Policy*, *20*.1-147

Amendola, V., & Meneghetti, M. (2009). Laser ablation synthesis in solution and size manipulation of noble metal nanoparticles. *Physical Chemistry Chemical Physics*, *11*, 3805–3821. https://doi.org/10.1039/b900654k

Amenta, V., Aschberger, K., Arena, M., Bouwmeester, H., Botelho Moniz, F., Brandhoff, P., Gottardo, S., Marvin, H. J., Mech, A., Pesudo, L. Q., & Rauscher, H. (2015). Regulatory aspects of nanotechnology in the agri/feed/food sector in EU and non-EU countries. *Regulatory Toxicology and Pharmacology*, *73*, 463–476. https://doi.org/10.1016/j.yrtph.2015.06.016

Asim, N., Ahmadi, S., Alghoul, M. A., Hammadi, F. Y., Saeedfar, K., & Sopian, K. (2014). Research and development aspects on chemical preparation techniques of photoanodes for dye sensitized solar cells. *International Journal of Photoenergy*. https://doi.org/10.1155/2014/518156

Balaji, A. P. B., Mishra, P., Suresh Kumar, R. S., Ashu, A., Margulis, K., Magdassi, S., Mukherjee, A., & Chandrasekaran, N. (2015). The environmentally benign form of pesticide in hydrodispersive nanometric form with improved efficacy against adult mosquitoes at low exposure concentrations. *Bulletin of Environmental Contamination and Toxicology*, *95*, 734–739. https://doi.org/10.1007/s00128-015-1661-y

Bhan, S., Mohan, L., & Srivastava, C. N. (2014). Relative larvicidal potentiality of nano-encapsulated Temephos and Imidacloprid against Culex quinquefasciatus. *Journal of Asia-Pacific Entomology*, *17*, 787–791. https://doi.org/10.1016/j.aspen.2014.07.006

Bhat, K. A., Batool, A., Mansoor, M., Manzoor, M., Bashir, Z., Nazir, M., & Zargar, S. M. (2022). Silicon- and nanosilicon-mediated disease resistance in crop plants. In *Silicon and nano-silicon in environmental stress management and crop quality improvement: Progress and prospects*. https://doi.org/10.1016/B978-0-323-91225-9.00028-5

Bratovcic, A. (2020). Biosynthesis of green silver nanoparticles and its UV-vis characterization. *International Journal of Innovative Science, Engineering & Technology*, *7*, 170–176.

Bratovcic, A., Hikal, W. M., Said-Al Ahl, H. A. H., Tkachenko, K. G., Baeshen, R. S., Sabra, A. S., & Sany, H. (2021). Nanopesticides and nanofertilizers and agricultural development: Scopes, advances and applications. *Open Journal of Ecology*, *11*, 301–316. https://doi.org/10.4236/oje.2021.114022

Cao, L., Zhou, Z., Niu, S., Cao, C., Li, X., Shan, Y., & Huang, Q. (2018). Positive-charge functionalized mesoporous silica nanoparticles as nanocarriers for controlled 2,4-dichlorophenoxy acetic acid sodium salt release. *Journal of Agricultural and Food Chemistry*, *66*, 6594–6603. https://doi.org/10.1021/acs.jafc.7b01957

Changmei, L., Chaoying, Z., Junqiang, W., Guorong, W., & Mingxuan, T. (2002). Research of the effect of nanometer materials on germination and growth enhancement of Glycine max and its mechanism. *Soybean Science*, *21*, 168–171.

Chhipa, H. (2017). Nanopesticide: Current status and future possibilities. *Agriculture Research Technology Open Access Journal*, *5*, 1–4. https://doi.org/10.19080/artoaj.2017.05.555651

Cho, E. J., Holback, H., Liu, K. C., Abouelmagd, S. A., Park, J., & Yeo, Y. (2013). Nanoparticle characterization: State of the art, challenges, and emerging technologies. *Molecular Pharmaceutics*, *10*, 2093–2110. https://doi.org/10.1021/mp300697h

Darzi-Naftchali, A., & Mokhtassi-Bidgoli, A. (2022). Saving environment through improving nutrient use efficiency under intensive use of agrochemicals in paddy fields. *Science of The Total Environment*, *822*, 153487. https://doi.org/10.1016/j.scitotenv.2022.153487

Dashora, A., & Sharma, K. (2018). Green synthesis of nanoparticles and their applications. *Advanced Science, Engineering and Medicine*, *10*, 523–541. https://doi.org/10.1166/asem.2018.2190

De Oliveira, J. L., Campos, E. V. R., Gonçalves Da Silva, C. M., Pasquoto, T., Lima, R., & Fraceto, L. F. (2015). Solid lipid nanoparticles co-loaded with simazine and atrazine: Preparation, characterization, and evaluation of herbicidal activity. *Journal of Agricultural and Food Chemistry*, *63*, 422–432. https://doi.org/10.1021/jf5059045

Delgadillo-Vargas, O., García-Ruiz, R., & Forero-Álvarez, J. (2016). Fertilising techniques and nutrient balances in the agriculture industrialization transition: The case of sugarcane in the Cauca river valley (Colombia), 1943–2010. *Agriculture, Ecosystems & Environment*, *218*, 150–162. https://doi.org/10.1016/j.agee.2015.11.003

DeRosa, M. C., Monreal, C., Schnitzer, M., Walsh, R., & Sultan, Y. (2010). Nanotechnology in fertilizers. *Nature Nanotechnology*, *5*, 91. https://doi.org/10.1038/nnano.2010.2

Díaz-Blancas, V., Medina, D. I., Padilla-Ortega, E., Bortolini-Zavala, R., Olvera-Romero, M., & Luna-Bárcenas, G. (2016). Nanoemulsion formulations of fungicide tebuconazole for agricultural applications. *Molecules*, *21*, 1271. https://doi.org/10.3390/molecules21101271

Ealias, A. M., & Saravanakumar, M. P. (2017). A review on the classification, characterisation, synthesis of nanoparticles and their application. In *IOP conference series: Materials science and engineering*. https://doi.org/10.1088/1757-899X/263/3/032019

El-Naggar, M. E., Abdelsalam, N. R., Fouda, M. M. G., Mackled, M. I., Al-Jaddadi, M. A. M., Ali, H. M., Siddiqui, M. H., & Kandil, E. E. (2020). Soil application of nano silica on maize yield and its insecticidal activity against some stored insects after the post-harvest. *Nanomaterials*, *10*, 739. https://doi.org/10.3390/nano10040739

Elmer, W., De Latorre-Roche, R., Pagano, L., Majumdar, S., Zuverza-Mena, N., Dimkpa, C., et al. (2018). Effect of metalloid and metal oxide nanoparticles on fusarium wilt of watermelon. *Plant Disease*, *102*, 1394–1401. https://doi.org/10.1094/PDIS-10-17-1621-RE

Fahad, B. A., Agheem, M. H., Memon, S. A., Baloch, A. R., Tunio, A., et al. (2022). Efficient mitigation of cadmium and lead toxicity in coriander plant utilizing magnetite (Fe_3O_4) nanofertilizer as growth regulator and antimicrobial agent. *International Journal of Environmental Analytical Chemistry*, *102*, 3868–3879. https://doi.org/10.1080/03067319.2020.1776861

Feizi, H., Kamali, M., Jafari, L., & Rezvani Moghaddam, P. (2013). Phytotoxicity and stimulatory impacts of nanosized and bulk titanium dioxide on fennel (*Foeniculum vulgare* Mill). *Chemosphere*, *91*, 506–511. https://doi.org/10.1016/j.chemosphere.2012.12.012

Food and Agricultural Organization the United Nations. (2002). *World agriculture: Towards 2015/*2030. Food and Agricultural Organization the United Nations.

Fraceto, L. F., Grillo, R., de Medeiros, G. A., Scognamiglio, V., Rea, G., & Bartolucci, C. (2016). Nanotechnology in agriculture: Which innovation potential does it have? *Frontiers in Environmental Science*, *4*, 20. https://doi.org/10.3389/fenvs.2016.00020

Frazier, T. P., Burklew, C. E., & Zhang, B. (2014). Titanium dioxide nanoparticles affect the growth and microRNA expression of tobacco (*Nicotiana tabacum*). *Functional & Integrative Genomics*, *14*, 75–83. https://doi.org/10.1007/s10142-013-0341-4

Grillo, R., Abhilash, P. C., & Fraceto, L. F. (2016). Nanotechnology applied to bio-encapsulation of pesticides. *Journal of Nanoscience and Nanotechnology, 16*, 1231–1234. https://doi.org/10.1166/jnn.2016.12332

Guilger-Casagrande, M., Germano-Costa, T., Pasquoto-Stigliani, T., Fraceto, L. F., & de Lima, R. (2019). Biosynthesis of silver nanoparticles employing *Trichoderma harzianum* with enzymatic stimulation for the control of *Sclerotinia sclerotiorum. Science Report, 9*, 14351. https://doi.org/10.1038/s41598-019-50871-0

Huang, Y., Haw, C. Y., Zheng, Z., Kang, J., Zheng, J. C., & Wang, H. Q. (2021). Biosynthesis of zinc oxide nanomaterials from plant extracts and future green prospects: A topical review. *Advanced Sustainable Systems, 5*, 2000266. https://doi.org/10.1002/adsu.202000266

Isaac, R. S., Sakthivel, G., & Murthy, C. H. (2013). Green synthesis of gold and silver nanoparticles using *Averrhoa bilimbi* fruit extract. *Journal of Nanotechnology, 2013*. https://doi.org/10.1088/1757-899X/263/3/032019

Jamkhande, P. G., Ghule, N. W., Bamer, A. H., & Kalaskar, M. G. (2019). Metal nanoparticles synthesis: An overview on methods of preparation, advantages and disadvantages, and applications. *Journal of Drug Delivery Science and Technology, 53*, 101174. https://doi.org/10.1016/j.jddst.2019.101174

Jiang, L. C., Basri, M., Omar, D., Rahman, M. B. A., Salleh, A. B., Rahman, R. N. Z. R. A., & Selamat, A. (2012). Green nano-emulsion intervention for water-soluble glyphosate isopropylamine (IPA) formulations in controlling *Eleusine indica* (*E. indica*). *Pesticide Biochemistry and Physiology, 102*, 19–29. https://doi.org/10.1016/j.pestbp.2011.10.004

Kah, M., & Hofmann, T. (2014). Nanopesticide research: Current trends and future priorities. *Environmental International, 63*, 224–235. https://doi.org/10.1016/j.envint.2013.11.015

Kah, M., Kookana, R. S., Gogos, A., & Bucheli, T. D. (2018). A critical evaluation of nanopesticides and nanofertilizers against their conventional analogues. *Nature Nanotechnology, 13*, 677–684. https://doi.org/10.1038/s41565-018-0131-1

Kalia, A., Sharma, S. P., Kaur, H., & Kaur, H. (2020). Novel nanocomposite-based controlled-release fertilizer and pesticide formulations: Prospects and challenges. In *Multifunctional hybrid nanomaterials for Sustainable agri-food and ecosystems*. https://doi.org/10.1016/B978-0-12-821354-4.00005-4

Kandasamy, S., & Sorna Prema, R. (2015). Methods of synthesis of nano particles and its applications. *Journal of Chemical and Pharmaceutical Research, 7*, 278–285.

Kashyap, P. L., Xiang, X., & Heiden, P. (2015). Chitosan nanoparticle based delivery systems for sustainable agriculture. *International Journal of Biological Macromolecules, 77*, 36–51. https://doi.org/10.1016/j.ijbiomac.2015.02.039

Kaushik, P., Shakil, N. A., Kumar, J., Singh, M. K., & Yadav, S. K. (2013). Development of controlled release formulations of thiram employing amphiphilic polymers and their bioefficacy evaluation in seed quality enhancement studies. *Journal of Environmental Science and Health, Part B, 48*, 677–685. https://doi.org/10.1080/03601234.2013.778614

Khodakovskaya, M. V., Kim, B. S., Kim, J. N., Alimohammadi, M., Dervishi, E., Mustafa, T., & Cernigla, C. E. (2013). Carbon nanotubes as plant growth regulators: Effects on tomato growth, reproductive system, and soil microbial community. *Small, 9*, 115–123. https://doi.org/10.1002/smll.201201225

Kottegoda, N., Munaweera, I., Madusanka, N., & Karunaratne, V. (2011). A green slow-release fertilizer composition based on urea-modified hydroxyapatite nanoparticles encapsulated wood. *Current Science, 101*, 73–78.

Kottegoda, N., Sandaruwan, C., Priyadarshana, G., Siriwardhana, A., Rathnayake, U. A., Berugoda Arachchige, D. M., Kumarasinghe, A. R., Dahanayake, D., Karunaratne, V., & Amaratunga, G. A. (2017). Urea-hydroxyapatite nanohybrids for slow release of nitrogen. *ACS Nano, 11*, 1214–1221. https://doi.org/10.1021/acsnano.6b07781

Kumar, R., Ashfaq, M., & Verma, N. (2018). Synthesis of novel PVA–starch formulation-supported Cu–Zn nanoparticle carrying carbon nanofibers as a nanofertilizer: Controlled release of micronutrients. *Journal of Material Science, 53*, 7150–7164. https://doi.org/10.1007/s10853-018-2107-9

Lawrencia, D., Wong, S. K., Low, D. Y. S., Goh, B. H., Goh, J. K., Ruktanonchai, U. R., Soottitantawat, A., Lee, L. H., & Tang, S. Y. (2021). Controlled release fertilizers: A review on coating materials and mechanism of release. *Plants, 10*, 238. https://doi.org/10.3390/plants10020238

Le, C. T., & , T. M. (2016). Synthesis of zinc hydroxide nitrate complex by a sol-gel method for use as foliar fertilizers – A scale up approach. *VNUHCM Journal of Science and Technology Development, 19*, 138–145. https://doi.org/10.32508/stdj.v19i3.587

Lim, C. J., Basri, M., Omar, D., Abdul Rahman, M. B., Salleh, A. B., & Raja Abdul Rahman, R. N. Z. (2012). Physicochemical characterization and formation of glyphosate-laden nano-emulsion for herbicide formulation. *Industrial Crops and Products, 36*, 607–613. https://doi.org/10.1016/j.indcrop.2011.11.005

Liscano, J. F., Wilson, C. E., Norman, R. J., Jr., & Slaton, N. A. (2000). *Zinc availability to rice from seven granular fertilizers* (Vol. *963*). Arkansas Agricultural Experiment Station.

Liu, R., & Lal, R. (2014). Synthetic apatite nanoparticles as a phosphorus fertilizer for soybean (*Glycine max*). *Scientific Reports, 4*, 5686. https://doi.org/10.1038/5686

Liu, R., & Lal, R. (2015). Potentials of engineered nanoparticles as fertilizers for increasing agronomic productions. *Science Total Environment, 514*, 131–139. https://doi.org/10.1016/j.scitotenv.2015.01.104

Loha, C., Chattopadhyay, H., & Chatterjee, P. K. (2012). Assessment of drag models in simulating bubbling fluidized bed hydrodynamics. *Chemical Engineering Science, 75*, 400–407. https://doi.org/10.1016/j.ces.2012.03.044

Moaveni, P., & Kheiri, T. (2011). TiO_2 Nano Particles Affected on Maize (Zea mays L). 2nd International Conference on Agricultural and Animal Science, Maldives, 25-27 November 2011, Vol. 22, IACSIT Press, Singapore, 160–163.

Mohamed, A. A., Sameeh, M. Y., & El-Beltagi, H. S. (2022). Preparation of seaweed nanopowder particles using planetary ball milling and their effects on some secondary metabolites in date palm (*Phoenix dactylifera* L.) seedlings. *Life, 13*, 39. https://doi.org/10.3390/life13010039

Mondal, K. K., & Mani, C. (2012). Investigation of the antibacterial properties of nanocopper against *Xanthomonas axonopodis* pv. punicae, the incitant of pomegranate bacterial blight. *Annals of Microbiology, 62*, 889–893. https://doi.org/10.1007/s13213-011-0382-7

Naderi, M. R., & Abedi, A. (2012). Application of nanotechnology in agriculture and refinement of environmental pollutants. *Journal of Nanotechnology, 11*, 18–26.

Naseem, F., Zhi, Y., Farrukh, M. A., Hussain, F., & Yin, Z. (2020). Mesoporous $ZnAl_2Si_{10}O_{24}$ nanofertilizers enable high yield of *Oryza sativa* L. *Scientific Reports, 10*, 10841. https://doi.org/10.1038/s41598-020-67611-4

Ocsoy, I., Paret, M. L., Ocsoy, M. A., Kunwar, S., Chen, T., You, M., & Tan, W. (2013). Nanotechnology in plant disease management: DNA-directed silver nanoparticles on graphene oxide as an antibacterial against *Xanthomonas perforans*. *ACS Nano, 7*, 8972–8980. https://doi.org/10.1021/nn4034794

Ojeda-Barrios, D. L., Morales, I., Juárez-Maldonado, A., Sandoval-Rangel, A., Fuentes-Lara, L. O., & Benavides-Mendoza, A. (2019). Importance of nanofertilizers in fruit nutrition. In *Fruit crops: Diagnosis and management of nutrient constraints*. https://doi.org/10.1016/B978-0-12-818732-6.00035-6

Pankaj, S. N. A., Kumar, J., Singh, M. K., & Singh, K. (2012). Bioefficacy evaluation of controlled release formulations based on amphiphilic nano-polymer of carbofuran against Meloidogyne incognita infecting tomato. *Journal of Environmental Science and Health, Part B*, *47*, 520–528. https://doi.org/10.1080/03601234.2012.665667

Panova, G. G., Ktitorova, I. N., Skobeleva, O. V., Sinjavina, N. G., Charykov, N. A., & Semenov, K. N. (2016). Impact of polyhydroxy fullerene (fullerol or fullerenol) on growth and biophysical characteristics of barley seedlings in favourable and stressful conditions. *Plant Growth Regulation*, *79*, 309–317. https://doi.org/10.1007/s10725-015-0135-x

Patil, S. P., & Burungale, V. V. (2020). Physical and chemical properties of nanomaterials. In *Nanomedicines for breast cancer theranostics*. https://doi.org/10.1016/b978-0-12-820016-2.00002-17070.

Perrin, T. S., Drost, D. T., Boettinger, J. L., & Norton, J. M. (1998). Ammonium-loaded clinoptilolite: A slow-release nitrogen fertilizer for sweet corn. *Journal of Plant Nutrition*, *21*, 515–530. https://doi.org/10.1080/01904169809365421

Pestovsky, Y. S., & Martínez-Antonio, A. (2017). The use of nanoparticles and nanoformulations in agriculture. *Journal of Nanoscience Nanotechnology*, *17*, 8699–8730. https://doi.org/10.1166/jnn.2017.15041

Pimentel, D. (2005). Environmental and economic costs of the application of pesticides primarily in the United States. *Environment, Development and Sustainability*, *7*, 229–252. https://doi.org/10.1007/s10668-005-7314-2

Pohshna, C., & Mailapalli, D. R. (2022). Engineered urea-doped hydroxyapatite nanomaterials as nitrogen and phosphorus fertilizers for rice. *ACS Agricultural Science & Technology*, *2*, 100–112. https://doi.org/10.1021/acsagscitech.1c00191

Pottathara, Y. B., Grohens, Y., Kokol, V., Kalarikkal, N., & Thomas, S. (2019). Synthesis and processing of emerging two-dimensional nanomaterials. In *Nanomaterials synthesis* (pp. 1–25). Elsevier. https://doi.org/10.1016/B978-0-12-815751-0.00001-8

Prasad, R., Bhattacharyya, A., & Nguyen, Q. D. (2017). Nanotechnology in sustainable agriculture: Recent developments, challenges, and perspectives. *Frontiers in Microbiology*, *8*, 1014. https://doi.org/10.3389/fmicb.2017.01014

Qian, K., Shi, T., Tang, T., Zhang, S., Liu, X., & Cao, Y. (2011). Preparation and characterization of nano-sized calcium carbonate as controlled release pesticide carrier for validamycin against *Rhizoctonia solani*. *Microchimica Acta*, *173*, 51–57. https://doi.org/10.1007/s00604-010-0523-x

Rajput, V., Minkina, T., Mazarji, M., Shende, S., Sushkova, S., Mandzhieva, S., Burachevskaya, M., Chaplygin, V., Singh, A., & Jatav, H. (2020). Accumulation of nanoparticles in the soil-plant systems and their effects on human health. *Annals of Agricultural Sciences*, *65*, 137–143. https://doi.org/10.1016/j.aoas.2020.08.001

Raliya, R., Saharan, V., Dimkpa, C., & Biswas, P. (2018). Nanofertilizer for precision and sustainable agriculture: Current state and future perspectives. *Journal of Agricultural and Food Chemistry*, *66*, 6487–6503. https://doi.org/10.1021/acs.jafc.7b02178

Ranjan, P., Abubakar Sadique, M., Yadav, S., & Khan, R. (2022). An electrochemical immunosensor based on gold-graphene oxide nanocomposites with ionic liquid for detecting the breast cancer CD44 biomarker. *ACS Applied Materials & Interfaces*, *14*, 20802–20812. https://doi.org/10.1021/acsami.2c03905

Saini, P., Gopal, M., Kumar, R., & Srivastava, C. (2014). Development of pyridalyl nanocapsule suspension for efficient management of tomato fruit and shoot borer (*Helicoverpa armigera*). *Journal of Environmental Science and Health, Part B*, *49*, 344–351. https://doi.org/10.1080/03601234.2014.882168

Salavati-Niasari, M., Davar, F., & Mir, N. (2008). Synthesis and characterization of metallic copper nanoparticles via thermal decomposition. *Polyhedron*, *27*, 3514–3518. https://doi.org/10.1016/j.poly.2008.08.020

Sarkar, D. J., Kumar, J., Shakil, N. A., & Walia, S. (2012). Release kinetics of controlled release formulations of thiamethoxam employing nano-ranged amphiphilic PEG and diacid based block polymers in soil. *Journal of Environmental Science and Health, Part A*, *47*, 1701–1712. https://doi.org/10.1080/10934529.2012.687294

Sarlak, N., Taherifar, A., & Salehi, F. (2014). Synthesis of nanopesticides by encapsulating pesticide nanoparticles using functionalized carbon nanotubes and application of new nanocomposite for plant disease treatment. *Journal of Agricultural and Food Chemistry*, *62*, 4833–4838. https://doi.org/10.1021/jf404720d

Sasson, Y., Levy-Ruso, G., Toledano, O., & Ishaaya, I. (2007). Nanosuspensions: Emerging novel agrochemical formulations. In *Insecticides design using advanced technologies*. https://doi.org/10.1007/978-3-540-46907-0_187.

Sayago, I., Honta˜n´on, E., & Aleixandre, M. (Eds.). (2020). Preparation of tin oxide nanostructures by chemical vapor deposition. In: Orlandi MO (ed) *Tin oxide materials* (pp.247–280). Elsevier.

Shang, Y., Kamrul Hasan, M., Ahammed, G. J., Li, M., Yin, H., & Zhou, J. (2019). Applications of nanotechnology in plant growth and crop protection: A review. *Molecules*, *24*, 2558. https://doi.org/10.3390/molecules24142558

Shebl, A., Hassan, A. A., Salama, D. M., Abd El-Aziz, M. E., & Abd Elwahed, M. S. A. (2019). Green synthesis of nanofertilizers and their application as a foliar for *Cucurbita pepo* L. *Journal of Nanomaterials*, *2019*, 1–11. https://doi.org/10.1155/2019/3476347

Shebl, A., Hassan, A. A., Salama, D. M., Abd El-Aziz, M. E., & Abd Elwahed, M. S. A. (2020). Template-free microwave-assisted hydrothermal synthesis of manganese zinc ferrite as a nanofertilizer for squash plant (*Cucurbita pepo* L). *Heliyon*, *6*, e03596. https://doi.org/10.1016/j.heliyon.2020.e03596

Shukla, G., Gaurav, S. S., & Singh, A. (2020). Synthesis of mycogenic zinc oxide nanoparticles and preliminary determination of its efficacy as a larvicide against white grubs (*Holotrichia* sp.). *International Nano Letters*, *10*, 131–139. https://doi.org/10.1007/s40089-020-00302-0

Singh, M. D., Gautam, C., Prakash Patidar, O., & Meena, H. M. (2017). Nano-fertilizers is a new way to increase nutrients use efficiency in crop production. *International Journal of Agriculture Sciences*, *9*, 975–3710.

Siva, V., Gremyr, I., Bergquist, B., Garvare, R., Zobel, T., & Isaksson, R. (2016). The support of quality management to sustainable development: A literature review. *Journal of Cleaner Production*, *138*, 148–157. https://doi.org/10.1016/j.jclepro.2016.01.020

Sohrabi, Y., Sharifi Kalyani, F., Heydari, M., Yazdani, M., Omer, K. M., & Yousefi, A. R. (2022). Plant-based nano-fertilizer prepared from Paulownia Tomentosa: Fabrication, characterization, and application on *Ocimum basilicum*. *Chemical and Biological Technologies in Agriculture*, *9*, 1–13. https://doi.org/10.1186/s40538-022-00352-w

Song, M. R., Cui, S. M., Gao, F., Liu, Y. R., Fan, C. L., Lei, T. Q., & Liu, D. C. (2012). Dispersible silica nanoparticles as carrier for enhanced bioactivity of chlorfenapyr. *Journal of Pesticide Science*, *37*, 258–260. https://doi.org/10.1584/jpestics.D12-027

Subramanian, K. S., & Rahale, C. S. (2012). Ball milled nanosized zeolite loaded with zinc sulfate: A putative slow release Zn fertilizer. *International Journal of Innovative Horticulture*, *1*, 33–40.

Tilman, D., Cassman, K. G., Matson, P. A., Naylor, R., & Polasky, S. (2002). Agricultural sustainability and intensive production practices. *Nature*, *418*, 671–677. https://doi.org /10.1038/nature01014

Usman, M., Farooq, M., Wakeel, A., Nawaz, A., Cheema, S. A., Rehman, H. U., Ashraf, I., & Sanaullah, M. (2020). Nanotechnology in agriculture: Current status, challenges and future opportunities. *Science of the Total Environment, 721*, 137778. https://doi.org/10.1016/j.scitotenv.2020.137778

Vaseghi, Z., & Nematollahzadeh, A. (2020). Nanomaterials: Types, synthesis, and characterization. *Green Synthesis of Nanomaterials for Bioenergy Applications*, 23–82. https://doi.org/10.1002/9781119576785.ch2

Venugopal, N. V. S., & Sainadh, N. V. S. (2016). Novel polymeric nanoformulation of mancozeb - An eco-friendly nanomaterial. *International Journal of Nanoscience, 15*, 1650016. https://doi.org/10.1142/S0219581X16500162

Vidya, C., Hiremath, S., Chandraprabha, M. N., Antonyraj, M. A. L., Gopal, I. V., Jain, A., & Bansal, K. (2013). International journal of current engineering and technology green synthesis of ZnO nanoparticles by Calotropis Gigantea. *International Journal of Current Engineering and Technology, 1*, 118–120.

Walker, G. W., Kookana, R. S., Smith, N. E., Kah, M., Doolette, C. L., Reeves, P. T., Lovell, W., Anderson, D. J., Turney, T. W., & Navarro, D. A. (2017). Ecological risk assessment of nano-enabled pesticides: A perspective on problem formulation. *Journal of Agricultural and Food Chemistry, 66*, 6480–6486. https://doi.org/10.1021/acs.jafc.7b02373

Wang, C., Chang, X. L., Shi, Q., & Zhang, X. (2018). Uptake and transfer of 13 C-fullerenols from scenedesmus obliquus to daphnia magna in an aquatic environment. *Environmental Science Technology, 52*, 12133–12141. https://doi.org/10.1021/acs.est.8b03121

Wang, C., Zhang, H., Ruan, L., Chen, L., Li, H., Chang, X. L., Zhang, X., & Yang, S. T. (2016). Bioaccumulation of 13 C-fullerenol nanomaterials in wheat. *Environmental Science: Nano, 3*(4), 799–805.

Wibowo, D., Zhao, C. X., Peters, B. C., & Middelberg, A. P. J. (2014). Sustained release of fipronil insecticide in vitro and in vivo from biocompatible silica nanocapsules. *Journal of Agricultural and Food Chemistry, 62*, 12504–12511. https://doi.org/10.1021/jf504455x

Yadav, T. P., Yadav, R. M., & Singh, D. P. (2012). Mechanical milling: A top down approach for the synthesis of nanomaterials and nanocomposites. *Nanoscience and Nanotechnology, 2*, 22–48. https://doi.org/ 10.5923/j.nn.20120203.01

Yasmin, A., Ramesh, K., & Rajeshkumar, S. (2014). Optimization and stabilization of gold nanoparticles by using herbal plant extract with microwave heating. *Nano Convergence, 1*, 12. https://doi.org/10.1186/s40580-014-0012-8

Yin, Y. H., Guo, Q. M., Yun, H. A. N., Wang, L. J., & Wan, S. Q. (2012). Preparation, characterization and nematicidal activity of lansiumamide B nano-capsules. *Journal of Integrative Agriculture, 11*, 1151–1158. https://doi.org/10.1016/S2095-3119(12)60109-9

3 Applications of Nanotechnology in Abiotic Stress Management and Biofortification

Debojyoti Moulick, Fiza Rashida Khan, Nafisa Siddiqui, Jyotirmayee Pradhan, Pragnya Khamari and Subrata Mahanta**

3.1 INTRODUCTION

With the ever-increasing population and progressing climate change, new challenges have emerged. These include depleting fossil reserves, deteriorating air quality and most crucially, shortage of food. It has been estimated that the world population will reach 9.1 billion by 2050 and for feeding such a large population, food production must expand by up to 70% (Ehrlich & Harte, 2015). This can be achieved if we increase the yield and cropping intensities of major staple crops, but this is not the only problem to address to solve the food crisis. Plants are sessile organisms that cannot relocate to protect themselves from the ill effects of bad weather, nutrient-deficient soil or bacterial infections (Choudhury & Moulick, 2022; Choudhury et al., 2022; Saxena et al., 2016). So, in order to keep the agricultural sector high in supplies, management of these environmental factors is very crucial. With escalating levels of soil and air pollution, elevated UV radiation and increasing CO_2 and O_3 levels, the nutrient content and crop yield have suffered a major blow, and hence mitigating the effect of these stresses is an ongoing concern (Pattnaik & Sahu, 2022; Saha et al., 2022; Das et al., 2022; Majumdar, 2022; Mukherjee & Hazra, 2022; Moulick et al., 2019, 2021, 2023b). Stress in the case of plants can be defined as any factor(s) beyond the permissible limit that restricts growth and hampers metabolism (Umar et al., 2021). Plants largely encounter two types of stress, named abiotic and biotic stress, which ultimately decrease their yield (Wani & Sah, 2014). Plants encounter a number of abiotic stresses due to conditions like drought, high amounts of dissolved salts, high temperature etc. (Sagar et al., 2022; Dash et al., 2022; Panda et al., 2022; Wani et al., 2016). It has been found that abiotic stress results in the production of reactive oxygen

* Corresponding author.

 DOI: 10.1201/9781003376446-3

TABLE 3.1
Applications of nanotechnology in agriculture

Products	Applications	Examples	References
Plant protection products	Nanocapsules, nanoparticles, nanoemulsions and viral capsids as smart delivery systems of active ingredients for disease and pest control in plants	Neem oil (*Azadirachta indica*)	Anjali et al. (2012)
Fertilizers	Nanocapsules, nanoparticles and viral capsids for the enhancement of nutrients absorption by plants and the delivery of nutrients to specific site	Macro-nutrient fertilizers coated with zinc oxide nanoparticles	Milani et al. (2015)
Water/liquid retention	Nanomaterials, e.g. zeolites and nanoclays, for water or liquid agrochemicals retention in the soil for their slow release to the plant	Soil-enhancer product, based on a nanoclay component, for water retention and release	McMurray et al. (2006)
Nanosensors and diagnostic devices	Nanomaterials and nanostructures (e.g. electrochemically active carbon nanotubes, nanofibers and fullerenes) that are highly sensitive biochemical sensors to closely monitor environmental conditions, plant health and growth	Pesticide detection with a liposome-based nanobiosensor	Vamvakaki and Chaniotakis (2007)

species which cause oxidative damage and alter plant metabolic pathways (Choudhury et al., 2022a, 2022b, 2022c; He et al., 2018). In today's time, sustainable crop production and increasing crop yield are not possible without the use of chemicals like fertilizers, herbicides, pesticides etc. but these traditional crop practices might not be sufficient to fulfil the exponential demand for food and to manage stress in plants. Moreover, their continued use over a long time causes much harm to the environment by contaminating water sources and bioaccumulating and then transmitting through the food chain. For this reason, different types of technologies have been employed to tackle all these problems so that they are effective as well as non-detrimental to the environment.

In this regard, the use of nanotechnology to mitigate these stresses and increase crop yield is unmatched. Nanotechnology has varying implementations in the agricultural sector (Sarkar et al., 2022; Moulick et al., 2023). It is an innovative and efficient method to generate nanosized particles of size from 1 to 100 nm. In view of the fact that they are minute, their surface area to volume ratio is increased (Feregrino-Perz et al., 2018), which enhances their biochemical reactivity (Ali et al., 2014). It also imparts beneficial properties which can be exploited to develop metabolic responses inside the cells. Nanotechnology is being employed for food production, increasing the fortification of nutrients and pathogen invasion (Chand Mali et al., 2020). A number of nanoparticles developed are nanopesticides, nanoinsecticides, nanoemulsions and nanofertilizers (Parisi et al., 2015). Some nanoparticle products are given in Table 3.1.

Another issue that is being addressed by the use of nanoparticles in agriculture is micro-nutrient deficiencies in plants. Even by the use of excessive fertilizers, plants are not able to assimilate them properly, which results in the crop being nutrient deficient, leading to decreased bioavailability to humans. The use of nanoparticles in biofortification has been emerging as a promising field. Nutrients like boron (Hussain et al., 2020a), copper (El-Ramady et al., 2021), selenium, iron, calcium and zinc are biofortified. Some vitamins can also be fortified like vitamin B1, riboflavin (vitamin B2) and cobalamin (vitamin 12). Biofortification can be achieved in many ways like seed priming and foliar applications (Torre-Roche et al., 2020). Using biopolymers such as cellulose-based nanocarriers improves the nutrient conversion ratio. Also, soil flora degrades them slowly resulting in their controlled release (Hess, 2013). Biofortification can also be accomplished by conventional breeding practices by selecting genotypes with the highest micro-nutrient content (Zhang et al., 2019). These are some of the ways by which nanotechnology has resulted in increased nutrient content in plants.

3.2 ROLE OF NANOBIOTECHNOLOGY IN ABIOTIC STRESS

Molecules having dimensions of 1–100 nm are called nanoparticles (Tariq et al., 2021). Nanomolecules are obtained by regulation and manipulation using a chemical reaction in the presence of a catalyst. These have the potential to change possessions of any material at the micro-level. In short, these can act as micro-bullets (Gouda et al., 2018). The application of nanotechnology to eradicate abiotic stress is very common nowadays (Saharan et al., 2022), Table 3.2.

There are various approaches to managing abiotic stress. Accumulation of polysugars, proline, glycine-betaine and the ascorbic acid act as stress metabolites. Other than this, superoxide dismutase (SOD), ascorbate peroxidase (APX), glutathione reductase and catalase (CAT) are some common enzymes which act as antioxidants.

PGPR plays a major role in managing water stress. This application can be studied using strains of *Pseudomonas putida* and *Pseudomonas fluorescens* (Gouda et al., 2018). Salinity stress tolerance in okra is reduced by an enzyme called ROS scavenger. Crops like soybean, wheat and chickpea are saved by the combined application of PGPR and drought resistance (Gouda et al., 2018). TU-induced technique i.e., Thiourea (TU) having major constituents as 36% nitrogen and 42% sulphur, helps to improve abiotic stress. E3-ubiquitin ligases have positive as well as negative impacts on abiotic stress. To overcome these things plants undergo several molecular mechanisms involving detection, an indication making of RNA copy, translation, transcript management and fluctuations after translational (Saharan et al., 2022).

3.2.1 Mechanism of Abiotic Stress Regulation Using Nanoparticles (NPs)

The defence mechanism of plants helps them to sustain in such adverse conditions. It has been reported that the concentration-dependent factor on plant growth and development contributes to resistance against abiotic stress (Tariq et al., 2021). NPs are associated with plants based on chemical composition, size, reactivity and surface

TABLE 3.2
Effects of nanoparticles on plants in abiotic stress

Nanoparticles	Types of abiotic stresses	Source of nanoparticles	Effects in plants	References
Zinc oxide (17 mg/L)	250 mM salt stress	*Carica papaya*	Less production of cytotoxic metabolities such as malondialdehyeand other harmful ROS. Increment in enzymatic activity of antioxidants and maintains the level of important osmolytes like proline. It increases the plant yield.	Yasmin et al. (2021)
Gold (67.20 nm)	100 mM salt stress	*Triticum aestivum*	Increase in the level of non-enzymatic antioxidants and enhanced activity of enzymatic antioxidants. More accumulation of nitrogen and maintainance of ions like K+ and Na+ homeostasis	Wahid et al. (2022)
Titanium oxide (around 50-60 nm)	Arsenic toxicity (10 μM)	*Musa paradisiaca*	More production of overall protein content, increased enzymatic activity of antioxidants that leads to decreased ROS mediated injuries	Katiyar et al. (2020)
Iron (10–18 nm)	10% polyethylene glycol mediated drought stress	*Chaetomorpha antennina*	More activities of enzymatic antioxidants and non-enzymatic antioxidants. Advancement in plant growth and photosynthetic parameters	Sreelakshmi et al. (2020)
Silver	25, 100 mM salt stress	*Triticum aestivum, Capparis spinosa*	Increases photosynthetic rate and overall plant growth. Maintains auxin and cytotokinin content and fall in level of abscisic acid	Wahid et al. (2020)

covering which results in various morphological and physiological transformations. Some benefits of using NPs in plants are, for example, zinc oxide having a huge impact on seed germination but the excess concentration of it leads to impaired germination. Development of shoot, root and leaf areas is managed by silver. Other than this, it also reduces the phenolic concentration. Carbon nanotubes can act as carriers for DNA and dye molecules in plant cellular systems. Toxicity in crops is regulated by gold. It also helps in the sprouting of the seed. Titanium dioxide has a major contribution towards the propagation of radicles and plumules. It is also involved in nitrogen metabolism through enzymatic activities. It also does oxidation and reduction. The photocatalytic activity of titanium dioxide is known. The growth of seedlings, the chlorophyll of leaves and the gathering of proline are done by silicon dioxide. Nanotechnology has brought a revolution in medicine, industries, the environment, agriculture and many other fields. It is an emerging tool in the current scenario to get rid of hunger and poverty (Tariq et al., 2021).

Stress can be defined as a stimulus which upsets the normal functioning of any system. Plants are consistently exposed to various kinds of biotic and abiotic stress. Water scarcity in India is quite common. Like humans, plants have been suffering from this problem for years. Vascular plants are primarily drought-stricken. Consequently, they adapt themselves to such stressful conditions. Drought is one of the major hurdles in agriculture. Drought is not limited only to decreased rainfall. It also signifies low moisture availability in soil, high evaporative weight, high temperature, harsh solar radiation, hardness of the soil, nutrient deficiency and salt build up in the soil surface (Tariq et al., 2021). Drought being multi-dimensional, affects 45% of agricultural land in the world. Lack of water in plants leads to decreased turgor pressure. As a result, the solute concentration increases in the cytosol of a cell and even in the extracellular matrix. It has several consequences, such as cellular growth being inhibited, accretion of proline and abscisic acid taking place, and plants failing to reproduce. The formation of scavengers such as glutathione and ascorbate, which destroys radicals and also increased production of ROS i.e. reactive oxygen species, is also seen. Decreased transpiration, stomatal closure and disturbance in gaseous exchange and photosynthesis is due to drought stress. Enzymatic activities are interrupted, which includes RuBisCO. Calvin cycle and ATP formation are also disturbed (Maswada et al., 2020).

To overcome this stress, plants go through a variety of mechanisms. Since drought stress mainly involves osmotic imbalance, plants produce many compounds like inorganic ions like K^+, Na^+ and certain molecules known as osmolytes (Abid et al., 2018). These include amino acids like proline, and glutamate along with sugars like mannitol, sorbitol, fructans, trehalose and sucrose (Hoekstra et al., 2001). Accumulation of these solutes helps to maintain the membrane's structural integrity and cell-hydrated state. These small molecules maintain structural integrity by substituting the hydroxyl group of sugar OH group with the OH group of water. These osmolytes also protect cellular structures by stabilizing interactions with proteins by making hydrogen bonds with macro-molecules and inhibiting intra-molecular hydrogen bond formation which disrupts the protein tertiary structure and hence their function (Szabados & Savouré, 2010). They are also responsible for scavenging

ROS produced due to drought stress. For instance, proline acts as a free radical scavenger, pH buffer and a cell redox balancer during osmotic stress (Trovato et al., 2008). It also reduces atomic oxygen production and hence also decreases the incidental photo-induced damage in thylakoid membranes. When proline concentration increases to a large extent, photosynthesis is reduced significantly (Lawlor & Cornic, 2002). Three prominent plant breeding techniques are being used to enhance drought tolerance in crops. These include conventional breeding, marker-assisted breeding and genetic engineering. (Ashraf, 2010)

3.2.2 Different Nanomaterials and Their Use in Drought Stress Tolerance in Plants

The most widely used nanomaterial in the agriculture field is carbon-based nanomaterials (CNMs). These are stable molecules having a uniform dispersion medium throughout. They are quite popular due to their small surface area and high chemical reactivity. Chlorophyll index, membrane stability index, antioxidant capacity and total phenolics is managed by the usage of MWCNTs (multi-walled carbon nanotubes) at 50mg/l to a plant named *Salvia mirzayanii* when drought stress is bearable. MWCNTs at 30 mg/L are used in caucasian alder for seedling growth. But there are some drawbacks to this nanoparticle. Its high concentration can badly affect seed incubation and seedling growth in cucumbers. The optimum amount of fullerol NPs help to get rid of drought stress. A foliar spray of it at 10–100 mg/L encourages seed germination. Metallic oxides like CeO_2, Fe oxides, TiO_2, ZnO and metalloids such as Si, SiO_2 have so much contribution towards crop. Some non-metallic NPs are also known. K increases the concentration of K^+ ion cells which plays a major role in balancing osmotic potential in cotton plants affected by drought pressure (Maswada et al., 2020). In the case of sorghum plants, the treatment of flag leaf, improvement of grain nutrient and translocation under stress conditions is done by using ZnO-NPs at 5 mg/kg. To fight against abiotic stresses, the application of Si is very well-known. Phosphorus helps in the enhancement of biomass and chlorophyll content and increases the rate of transport and adsorption in plants. Jute seeds are first treated along with hydroxyapatite NPs to tackle the adverse effects of drought stress (Maswada et al., 2020). Silver nanoparticles are also widely used because of their repressive effect on some organisms as well as bacteria. The use of AgNPs on lentils in drought stress shows a positive impact (Ahmad et al., 2020).

The unique physio-chemical properties are the USP (unique selling proposition) of nanoparticles. Sustainable development in the agricultural field is achieved by these biosynthesized nanoparticles (Ahmad et al., 2020). They assure the delivery of nutrients at the target site only. Hence it is not wrong to say that the impact of nanoparticles is different on different plants. It depends upon their concentration, size and application manner. Horticultural studies require emphasis on anatomical, phenological, ecological, cytological as well as molecular mechanisms. Solving the problem of removing drought stress and improving crop performance requires great effort and it is an expensive procedure. So, the application of NPs should be done with proper care and it should be used in optimum concentration (Maswada et al., 2020).

Heat stress, which is one of the abiotic stresses, adversely influences the morphological growth of plants by getting in the way of the photosynthesis process. This photosynthesis process may be put down to increased levels of oxidative stress (Iqbal et al., 2019). Thermal stress, which is another name for the negative impact of heat stress on plants, is considerable, detrimental and often accounts for decreased crop yield and productivity. A great challenge for plants to sustain and survive is mainly due to adverse thermal effects. This includes a climatic change that is a whole increase in the global temperature of the earth along with changed patterns of precipitation that appreciate enhancing thermal stress on different parts of plants and also in other organisms (Tariq et al., 2021). Due to climate change and global warming, economic losses are seen for agriculture. Heat stress decreases the productivity of crops because of yield reduces which include biochemical, morphological, anatomical and physiological processes (Shalaby et al., 2022). It is well-known that the application of nanotechnology in the agriculture sector will be a promising achievement in the near future.

3.2.3 Different Nanomaterials and Their Use in Heat Stress Tolerance in Plants

Under heat stress, AgNPs (silver nanoparticles) protect and regulate the growth of wheat. Different levels of AgNPs work for the improvement of the morphological influencer like root shoot length, root number, plant fresh and dry weight, leaf area and the number of wheat plants. AgNPs improve morphology in wheat plants under heat stress. *Moringa oleifera* plant extract is used for AgNPs synthesis ensured by characterization through UV-visible spectroscopy, XRD (x-ray diffraction), SEM (scanning electron microscope) and AFM (atomic force microscopy) (Iqbal et al., 2019).

For the reduction of heat stress intensity and good yield available Si and Si nanoparticles are helpful. Reduction of protein level also improves yield quality. Particularly in glutenin, additionally, it lowers the activity of alpha-amylase. It is clear that nanoparticles improve the steps of grain quality in wheat grains. As well, Si and nano Si importantly increase the oil absorption capacity (OAC) of the flour of stressed wheat grains. Additionally, it provokes an enhancement of the cellulose content (glucose/fructose), pectin content (gelatin), starch and soluble protein contents, total phenol flavonoid and oxalic acid. Moreover, Ca and K levels in heat-stressed wheat plant straw are connected with a decline in the lignin and phytic acid pigments (Helal et al., 2022).

Magnetite nanoparticles, like Fe_2O_4 which is biocompatible, are used as to lessen heat stress in wheat plants. This magnetite nanoparticle is synthesized by using the hydrothermal method, then that is characterized by advanced techniques before field application (Borai et al., 2017). The enhancement of yield parameter and quality are due to an increase of determined biological markers (GSH and GST). Also, one of the positive effects is that it increases concentration on cell organelles which was observed by transmission electron microscopy to counteract heat stress (Kumar & Sharma, 2020). Pretreatment of crop plants by zinc oxide and TiO_2 nanoparticles mainly improve resistance against heat stress along with increase in degree of SOD (superoxide dismutase) and GPX (glutathione peroxidase) activity, which also leads to lowering H_2O_2 levels.

Salinity is a major type of abiotic stress factor observed in drought-affected areas and regions where drainage systems of irrigated farms are poor. Salt stress deteriorates crop production and demeans the quality of food crops. Salinity stress in plants interferes with osmotic homeostasis, causing changes in the primary metabolism (Parida & Das, 2005). Salt stress is a major limitation due to which sustainable crop production cannot be attained. The Food and Agriculture Organization (FAO, 2016) reported that salinity stress influences more than 6% of total global land and 19.5% of total irrigated land.

Disturbance to plant growth occurring due to salinity stress is caused by excess sodium and chloride ion accumulation, which are found in high quantities in coastal and arid region soils. This leads to an alteration of cellular metabolism and an imbalance of ionic equilibrium (Munns & Tester, 2008). An excess amount of sodium chloride causes major problems, that include:

(i) Increase in osmotic pressure of external solution.
(ii) Interference in the transportation of ions like potassium and calcium.
(iii) Disturbance in the enzyme systems caused due to toxic effects exerted by sodium and chloride ions on the plant membrane.

Apart from the abovementioned problems, salinity stress upsets photosynthesis in plants, interferes with synthesis of protein and disturbs lipid metabolism. Oxidative stress is also affected by salinity stress (Parida & Das, 2005).

Different species have different tolerance levels for salinity stress. Cereal crops like barley can resist up to 250 mM. Crops like bread wheat, maize and sorghum are moderately tolerant towards salinity stress. Meanwhile, wheat is intolerant towards salinity stress (Maas & Hoffman, 2008). Nanoparticles having plant-based origins are found to be quite effective in the control of salinity stress.

3.2.4 Different Nanomaterials and Their Use in Salinity Stress Tolerance in Plants

Selenium nanoparticles obtained from leaves of *Hordeum vulgare* are found to be efficient in managing soil salinity by increasing the plant growth and lowering plant stress markers that mainly involve hydrogen peroxide and malondialdehyde (Habibi & Aleyasin, 2020).

Different levels of Zn nanoparticles have been found to be associated with increasing the radical scavenging activity of DPPH (2,2-diphenyl-1-picryl-hydrazyl-hydrate) in the root and shoot of *Camelina sativa L* grown in saline soil. Zinc nanoparticles improved the yield of *Solanum melongena* by 16% in saline soil in the presence of limited water supply. Zinc nanoparticles derived from *Sorghum bicolor* leaf extract activated the antioxidative defence mechanisms and enhanced overall plant growth and photosynthesis during salinity stress in *Abelmoschus esculentus*.

The application of zinc oxide nanoparticles obtained from *Phoenix dactylifera* leaf extracts on *Vigna unguiculata* and *Abelmoschus esculentus* resulted in administration of Fe_3O_4 nanoparticles generated from *Hevea Brasiliense* bark extracts, which sparked the biomass and photosynthetic pigment levels and also caused reduction in the oxidative damage (Sebastian et al., 2018). *Mangifera indica L* when treated with Zn and SiNPs under salinity conditions showed an increase in fruit production, both in terms of quantity and quality. This was found to be associated with role of Zn in the development of floral parts and the production of phytohormones in plants. Hussein and Abou Baker (2018) studied the foliar application of nanozinc to ease the harsh effects of salinity and confirmed that diluted sea water can be used for irrigation of cotton plants. They showed that an additional amount of P-fertilizer with nanozinc could be used to avoid P/Zn ratio.

Titanium nanoparticles obtained using leaves of *Chenopodium quinoa* and *Trianthema portulacastrum* enhanced the chlorophyll level and yield in wheat plants (Irshad et al., 2021). Khan studied the effect of nano TiO_2 in plant development processes including defence mechanisms against environmental stresses (Khan, 2016). They proved that nano TiO_2 at the rate of 20mg/L was the best at improving the quality of tomatoes produced.

Sulphur nanoparticles obtained from *Ocimum basilicum* increased resistivity in *Helianthus annuus* against 100 mM manganese toxicity (Saad-Allah & Ragab, 2020). AgNPs also enhanced root and shoot length, and improved biomass in *Satureja hortensis L* facing salinity stress.

Avestan et al. reported that salinity stress treatment in strawberry plants were dependent on morphological and physiological parameters. They proposed that $nSiO_2$ treatments subdued the negative impact of salinity by increasing the epicuticular wax layer (EWL) (Avestan et al., 2019). $nSiO_2$ helped salinity stressed plants to maintain chlorophyll content, relative water protection (RWP) and relative water control (RWC) levels. They investigated

i. Influence of silicon treatments on EWL accumulation by testing EWL on various plant stages.
ii. Presence of genetic variation for EWL accumulation in strawberries.
iii. Benefit of more EWL deposition in saline conditions relative to the benefit of the other signalling and psychological changes that are linked to increase silicon uptake.

Yassen and associates (Yaseen et al., 2017) investigated the effect of silicon dioxide nanoparticles on cucumber and proved that silicon dioxide nanoparticles caused an enhancement of nitrogen and phosphorous content and subdued the sodium content. They reported that silicon dioxide nanofertilizers improved the yield of cucumber.

Soil health is directly associated with the yield of crops. However, there has been a decay in the health of soil in the past few decades. This deterioration of soil health is linked with increased industrialization and extensive use of chemical fertilizers (Lal, 2015). The major soil pollutants include greenhouse gases, pesticides and toxic heavy metal (HM). The problem with these contaminants is that they are persistent and toxic in nature. Among these, heavy metals are the contaminants which negatively affect human health. Elements with density higher than 5g/cm^3 and having atomic mass greater than 20 are known as heavy metals. They are produced from erosion of parental materials of soil called "*Paedogenesis*", and have very low concentration in the soil which is increasing gradually. The increase in paedogenesis is a result of various human activities such as dumping waste produced from textile mills, metal mining industries and releasing industrial waste in the water bodies which are deposited in the soil indirectly. HM such as lead, chromium, arsenic, zinc, mercury and cadmium are highly toxic in nature. These HMs in the soil influence plant growth and cause the formation of ROS (reactive oxygen species), which damages vital macro-molecules inside the organisms, thereby influencing the nutrition quality (Moulick et al., 2023c). Nanoremediation protects plants from such heavy metals by enhancing the microbial activity of plants. This is called "nano- bioremediation" (Amin et al., 2021). NPs enter the cells by penetrating the plasma membrane. These NPs trigger the plant defence system and induce the genes of stress signalling. NPs reduce the mobility of heavy metals and thereby absorb the HMs present in the soil (Sebastian et al., 2019). Nanoparticles make complexes in the cell walls of plants after binding with the heavy metals and make them unavailable. The complex gets adsorbed on the plant cell surface, which prevents their migration and reduces their biological activity (Wang et al., 2021).

Heavy metal stress can be eliminated by activating plant oxidation defence mechanisms. Biochemical reaction in plants like photosynthesis and respiration produce ROS as a by- product. However, in stress conditions, ROS levels increase, which damages the plant cell membrane. Antioxidant enzymes such as peroxidase (POD), superoxide dismutase (SOD), ascorbate peroxidase (APX), catalase (CAT) and glutathione peroxidase (GPX) can maintain or reduce the ROS levels in the plant cell and the level of this enzyme can be increased by using NPs (Mathur et al., 2022).

3.3 BIOFORTIFICATION

Crop production in traditional ways or by using technologies could increase the amount of nutrients in food which lead to increased agricultural productivity. Application of nanotechnology in agriculture includes nanobiotechnology to uplift the value of production, boost farmer's economy and secure and protect crops for a long time. The primary source of food plants and the quality of food plants decide the health condition of the majority of people. In poor communities, the lack of essential micro-nutrients in food is becoming a global issue with major adverse effects (Elemike et al., 2019). Around two billion people globally suffer from macronutrient deficiencies, also known as a "*hidden hunger*" (Szerement et al., 2022). According to Clemens, iron and zinc deficiencies cause a high level of child death and global disease. Several concerns such as the fight against nutrient insufficiencies such as dietary heterogeneity and the application of drug and technical fortification. Industrial fortification of food has not been fully successful; however, it is unaffordable to poor people. Biofortification is the process of enhancement in supplements at the time of cultivation (Elemike et al., 2019). Another important part of biofortification is that it provides a comparatively sustainable, cost-effective and long term delivery of nutrients to people in remote rural areas to overcome malnutrition and increase the nutritional quality of their daily diets (Lal et al., 2020). It is used for combating macro-nutrient deficiencies and is affordable and available for everyone.

3.3.1 Biofortification in Crops

Biofortification is a technique to improve food calories and the development of a new variety of sweet potato plants that are rich in vitamin A. On the basis of developing varieties, there are various techniques of biofortification: agronomic biofortification, conventional breeding and nutritional genetic modification.

Biofortifiction is more beneficial in developing countries which are rich in minerals and vitamins by cultivation. The success of biofortification programmes depends on the availability of micro-elements which nourish the soil and help plants to consume it, available via fertilizer having minute size. By the correlation of soil and fertilizer, nutrient is transported to the consumer. This is confirmation of improving nutrients and crop production through the usage of micro-nutrient fertilizer. The use of fertilizer generally in salt form or ionic form results readily in the loss of soil fertility.

Reproduction of crops and changes in genetics are made to upgrade the capacity of intake and supplement content. Plant multiplication method reduces the cost, makes crops sustainable and tends to increase the macro-nutrient content of some cereals, leguminous plants and tubes. It also increases the main part of food content which is vitamins. Orange fleshed sweet potato (OFSP) and quality protein maize are quite popular and are interesting examples of conventionally bred biofortified crops.

Nutritional genetic alteration is integrating sufficient micro-nutrient traits in plants that are proved to be highest producing species, which encourages farmers to grow more crops. Ameliorating the amount of Fe, Zn and Se in plants by maintaining the plant genetic composition and use of biotechnological methodologies can maintain nutritional deficiency in human food. But this is a cost-effective approach and takes a lot of time. Genetically modified micro-nutrients that are rich in crops should not be acquired by cultivators. Breeding is a more acceptable process for biofortification than those gene modification processes (Elemike et al., 2019).

3.3.2 Examples of Biofortification

As the body can't keep excessive amounts of amino acids, their consumption must be regular, which can be overcome by biofortification. Appropriate protein applicants for biofortification deal with the storage of protein. Sporamin A from sweet potato in seed albumin AmA1 protein from prince's father (*Amaranthus hypochondriacus*) and ASPI, a synthetic deposited protein enriched in vital amino acids. Orange sweet potato (OSP), rich in vitamin A, is the first-ever biofortified crop. Golden rice 1 contains the phytoene synthesis (PSY) gene from daffodil and the phytoene desaturase (CRTI) gene from the bacterium *Erwinia uredovora*, both found only in the rice seed. By substituting PSY with genes from maize and rice, it increases the level of β-carotene by 23 times in golden rice (Yashveer et al., 2014).

Multi-electron field trials of wheat plants transformed with a rice nicotiana mine synthesis gene (OsNAS2) under the regulatory control of the maize ubiquitin promoter, Ubi1, provide proof of concept of the transgenic approach in wheat which is developed and evaluated in commercially important wheat (Yashveer et al., 2014).

Moreover, the growth and development of humans are influenced by the zinc in their diet. The Zn deficiency problem can be substituted by the fortification of food crops. The first zinc rice variety is similar to that of the other popular, conventional varieties like Bangladesh Rice Research Institute (BRRI) dhan33 and BINA dhan7. BRRI dhan62 can escape terminal drought because of its short duration (Yashveer et al., 2014).

3.4 CONCLUSION

Nanotechnology indeed has led to a revolution in the field of agriculture. In the past two decades, it has shown considerable remediation of the effects faced by agroindustry due to various environmental factors including abiotic stress. Nanotechnology is an emerging field which is showing promising results for solving the food crisis in the world. It not only increases crop productivity but also helps plants to overcome abiotic stress. This chapter attempts to deliver dynamic ongoing advancements about the use of nanotechnology in plant responses to different types of abiotic stress and their use in biofortification. Like all other organisms, plants have also evolved to fight against these abiotic stresses. Their response to these adverse conditions is a result of complex biochemical pathways which leads to their adaptation and help them to withstand harsh stress. These responses can be mechanical, chemical or biochemical and they can also be induced by the use of nanoparticles. This is how nanoparticles mitigate the effect of abiotic stress in plants. Nanotechnology has also enabled the use of nanocarriers for micro-nutrient delivery for the biofortification of plants. They are being employed as nanofertilizers. Although they may pose the side effect of decreasing photosynthetic efficiency, nanoparticles are effective in delivering nutrients like zinc. Their controlled release makes them good fertilizers which avoid toxic nutrient conditions. Thus, these nanofertilizers can be designed according to the need. The use of these NPs has been shown to enhance growth and is also effective in suppressing disease, but some issues are still to be addressed. For example, in case of crop damage by fungal infections, a nano-based nutrient transport system is not developed. The use of nanoparticles in managing abiotic stress and biofortification is a field that needs to be researched thoroughly as it can provide knowledge about the mechanism and molecular chemistry of how plants deal with stress. This will be helpful when we want to develop synthetic ways for crop management. This field may prove to be a way to solve the ongoing food crisis in the near future.

REFERENCES

Abid, M., Ali, S., Qi, L. K., Zahoor, R., Tian, Z., Jiang, D., Snider, J. L., & Dai, T. (2018). Physiological and biochemical changes during drought and recovery periods at tillering and jointing stages in wheat (*Triticum aestivum* L.). *Science Reports*, *8*, 4615. https://doi.org/10.1038/s41598-018-21441-7

Ahmad, J., Qamar, S., Kausar, N., & Qureshi, M. I. (2020). Nanoparticles: The magic bullets in mitigating drought stress in plants. In K. Hakeem & T. Pirzadah (Eds.), *Nanobiotechnology in agriculture. Nanotechnology in the life sciences*. Springer. https://doi.org/10.1007/978-3-030-39978-8_8

Ali, M. A., Rehman, I., Iqbal, A., Din, S., Rao, A. Q., Latif, A., & Husnain, T. (2014). Nanotechnology, a new frontier in Agriculture. *Advances in Life Sciences*, *1*(3), 129–138.

Amin, I., Nazir, R., & Rather, M. A. (2021). Nano-bioremediation: An innovative approach for remedying heavy metals using fungi. *Journal of Bioremediation & Biodegradation*, *12*(487), 2.

Anjali, C. H., Sharma, Y., Mukherjee, A., & Chandrasekaran, N. (2012). Neem oil (*Azadirachta indica*) nanoemulsion—A potent larvicidal agent against Culex quinquefasciatus. *Pest Management Science*, *68*(2), 158–163.

Ashraf, M. (2010). Inducing drought tolerance in plants: Recent advances. *Biotechnology Advances*, *28*(1), 169–183. https://doi.org/10.1016/j.biotechadv.2009.11.005

Avestan, S., Ghasemnezhad, M., Esfahani, M., & Byrt, C. S. (2019). Application of nano-silicon dioxide improves salt stress tolerance in strawberry plants. *Agronomy*, *9*(5), Art. No. 5. https://doi.org/10.3390/agronomy9050246

Borai, I. H., Hendawey, M. H., Salah Eldin, T. A., Hassan, N. S., & Mahdi, A. A. (2017). Magnetite (Fe_3O_4) nanoparticles mitigated heat stress hazards in wheat. *Journal of Biological Chemistry and Environmental Sciences*, *12*(2), 557–574.

Chand Mali, S., Raj, S., & Trivedi, R. (2020). Nanotechnology a novel approach to enhance crop productivity. *Biochemistry Biophysics Report*, *24*, 100821. https://doi.org/10.1016/j.bbrep.2020.100821

Choudhury, S., Mazumder, M. K., Moulick, D., Sharma, P., Tata, S. K., Ghosh, D., & Hossain, A. (2022a). A computational study of the role of secondary metabolites for mitigation of acid soil stress in cereals using dehydroascorbate and mono - Dehydroascorbate reductases. *Antioxidant*, *11*(3), 458. https://doi.org/10.3390/antiox11030458

Choudhury, S., & Moulick, D. (2022). *Response of field crops to abiotic stress: Current status and future prospects* (1st ed.). CRC Press.

Choudhury, S., Moulick, D., Ghosh, D., Soliman, M., Alkhedaide, A., Gaber, A., & Hossain, A. (2022c). Drought-induced oxidative stress in pearl millet (*Cenchrus americanus* L.) at seedling stage: Survival mechanisms through alteration of morphophysiological and antioxidants activity. *Life*, *12*(8), 1171.

Choudhury, S., Moulick, D., Mazumder, M. K., Pattnaik, B. K., Ghosh, D., Vemireddy, L. R., & Hossain, A. (2022b). An in vitro and in silico perspective study of seed priming with zinc on the phytotoxicity and accumulation pattern of arsenic in rice seedlings. *Antioxidants*, *11*(8), 1500. https://doi.org/10.3390/antiox11081500

Das, D., Mondal, B., Dey, A., Mridha, D., & Roychowdhury, T. (2022). Overview on the effects of heavy metals on the biological activities of leafy vegetables. In S. Choudhury & D. Moulick (Eds.), *Response of field crops to abiotic stress* (pp. 67–78). CRC Press.

Dash, G. K., Sahoo, S. K., Jena, J., Barik, M., Parida, S., Baig, M. J., & Swain, P. (2022). Drought and high-temperature stress tolerance in field crops. In S. Choudhury & D. Moulick (Eds.), *Response of field crops to abiotic stress* (pp. 103–109). CRC Press.

Ehrlich, P. R., & Harte J. (2015). Opinion: To feed the world in 2050 will require a global revolution. *Proceedings of the National Academy of Sciences of the United States of America*, *112*(48), 14743–14744. https://doi.org/10.1073/pnas.1519841112

Elemike, E. E., Uzoh, I. M., Onwudiwe, D. C., & Babalola, O. O. (2019). The role of nanotechnology in the fortification of plant nutrients and improvement of crop production. *Applied Science*, *9*(3), 499.

El-Ramady, H., Abdalla, N., Elbasiouny, H., Elbehiry, F. Elsakhawy, T., Omara, A. E., & Zia-Ur-Rehman, M. (2021). Nano-biofortification of different crops to immune against COVID-19: A review 2021. *Ecotoxicology and Environmental Safety*, *222*, 112500. https://doi.org/10.1016/j.ecoenv.2021.112500

Feregrino-Perez, A. A., Magaña-López, E., Guzmán, C., & Esquivel, K. (2018). A general overview of the benefits and possible negative effects of the nanotechnology in horticulture. *Scientia Horticulturae*, *238*, 126–137.

FAO. 2016. Agroindustry policy brief. www.fao.org/3/a-i3950e.pdf

Gouda, S., Kerry, R. G., Das, G., Paramithiotis, S., Shin, H.-S., & Patra, J. K. (2018). Revitalization of plant growth promoting rhizobacteria for sustainable development in agriculture. *Microbiological Research*, *206*, 131–140. https://doi.org/10.1016/j.micres.2017.08.016

Habibi, G., & Aleyasin, Y. (2020). Green synthesis of Se nanoparticles and its effect on salt tolerance of barley plants. *International Journal of Nano Dimensions*, *11*(2), 145–157.

He, M., He, C.-Q., & Ding, N.-Z. (2018). Abiotic stresses: General defenses of land plants and chances for engineering multistress tolerance. *Frontiers in Plant Science*, *9*, 1–18. https://doi.org/10.3389/fpls.2018.01771

Helal, N. M., Khattab, H. I., Emam, M. M., Niedbała, G., Wojciechowski, T., Hammami, I., & Hassan, H. M. (2022). Improving yield components and desirable eating quality of two wheat genotypes using Si and NanoSi particles under heat stress. *Plants*, *11*, 1819. https://doi.org/10.3390/plants11141819

Hess, S. Y. (2013). Zinc: Deficiency disorders and prevention programs. In B. Caballero, L. Allen & A. Prentice (Eds.), *Encyclopedia of human nutrition* (3rd ed., pp. 431–436). Academic Press.

Hoekstra, F. A., Golovina, E. A., & Buitink, J. (2001). Mechanisms of plant desiccation tolerance. *Trends Plant Sciences*, *6*(9), 431–438.

Hussain, M., Mehboob, N., Naveed, M., Shehzadi, K., & Yasir, T. A. (2020a). Optimizing boron seed coating level and boron-tolerant bacteria for improving yield and biofortification of chickpea. *The Journal of Soil Science and Plant Nutrition*, *20*(4), 2471–2478.

Hussein, M. M., & Abou-Baker, N. H. (2018). The contribution of nano-zinc to alleviate salinity stress on cotton plants. *Royal Society Open Science*, *5*(8), 171809.

Iqbal, M., Raja, N. I., Mashwani, Z.-U.-R., Hussain, M., Ejaz, M., & Yasmeen, F. (2019). Effect of silver nanoparticles on growth of wheat under heat stress. *Iranian Journal of Science and Technology, Transaction A, Science*, *43*(2), 387–395.

Irshad, M. A., ur Rehman, M. Z., Anwar-ul-Haq, M., Rizwan, M., Nawaz, R., Shakoor, M. B., & Ali, S. (2021). Effect of green and chemically synthesized titanium dioxide nanoparticles on cadmium accumulation in wheat grains and potential dietary health risk: A field investigation. *Journal of Hazardous Materials*, *415*, 125585.

Katiyar, P., Yadu, B., Korram, J., Satnami, M. L., Kumar, M., & Keshavkant, S. (2020). Titanium nanoparticles attenuates arsenic toxicity by up-regulating expressions of defensive genes in *Vigna radiata* L. *Journal of Environmental Sciences*, *92*, 18–27. https://doi.org/10.1016/j.jes.2020.02.013

Khan, M. N. (2016). Nano-titanium dioxide (nano-TiO_2) mitigates NaCl stress by enhancing antioxidative enzymes and accumulation of compatible solutes in tomato (*Lycopersicon esculentum* Mill.). *Journal of Plant Sciences*, *11*(1/3), 1–11.

Kumar, P., & Sharma, P. K. (2020). Soil salinity and food security in India. *Frontiers in Sustainable Food Systems*, *4*, 533781. https://doi.org/10.3389/fsufs.2020.533781

Lal, M. K., Kumar, A., Kardile, H. B., Raigond, P., Changan, S. S., Thakur, N., & Singh, B. (2020). Biofortification of vegetables. *Advances in Agri-Food Biotechnology*, *22*, 1129–1159.

Lal, R. (2015). Restoring soil quality to mitigate soil degradation. *Sustainability*, *7*(5), 5875–5895.

Lawlor, D. W., & Cornic, G. (2002). Photosynthetic carbon assimilation and associated metabolism in relation to water deficits in higher plants. *Plant, Cell & Environment*, *25*(2), 275–294.

Maas, E. V., & Hoffman, G. J. (1977). Crop salt tolerance—Current assessment. *Journal of the Irrigation and Drainage Division*, *103*(2), 115–134.

Majumdar, J. (2022). Stress in plants: A curse in plant productivity and blessing in food security. In S. Choudhury & D. Moulick (Eds.), *Response of field crops to abiotic stress* (pp. 167–176). CRC Press.

Maswada, H. F., Mazrou, Y. S., Elzaawely, A. A., & Eldein, S. M. A. (2020). Nanomaterials. Effective tools for field and horticultural crops to cope with drought stress: A review. *Spain Journal of Agricultural Research*, *18*(2), 15.

Mathur, S., Singh, D., & Ranjan, R. (2022). Remediation of heavy metal (loid) contaminated soil through green nanotechnology. *Frontiers in Sustainable Food Systems*, *6*, 932424.

McMurray, T. A., Dunlop, P. S. M., & Byrne, J. A. (2006). The photocatalytic degradation of atrazine on nanoparticulate TiO_2 films. *Journal of Photochemistry Photobiology A: Chemistry*, *182*(1), 43–51.

Milani, N., Hettiarachchi, G. M., Kirby, J. K., Beak, D. G., Stacey, S. P., & McLaughlin, M. J. (2015). Fate of zinc oxide nanoparticles coated onto macronutrient fertilizers in an alkaline calcareous soil. *PLoS One*, *10*(5), e0126275. https://doi.org/10.1371/journal.pone.0126275

Moulick, D., Bhutia, K. L., Sarkar, S., Roy, A., Mishra, U. N., Pramanick, B., & Brestic, M. (2023b). The intertwining of Zn-finger motifs and abiotic stress tolerance in plants: Current status and future prospects. *Frontiers in Plant Sciences*, *13*, 1083960. https://doi.org/10.3389/fpls.2022.1083960

Moulick, D., Chowardhara, B., & Panda, S. K. (2019a). Agroecotoxicological aspect of Arsenic (As) and Cadmium (Cd) on field crops and its mitigation: Current status and future prospect. In S. Srivastava, A. Srivastava, & P. Suprasanna (Eds.), *Plant-metal interactions*. Springer. https://doi.org/10.1007/978-3-030-20732-8_11

Moulick, D., Ghosh, D., Mandal, J., Bhowmick, S., Mondal, D., Choudhury, S., & Biswas, J. K. (2023c). A cumulative assessment of plant growth stages and selenium supplementation on arsenic and micronutrients accumulation in rice grains. *Journal of Cleaner Production*, *386*, 135764. https://doi.org/10.1016/j.jclepro.2022.135764

Moulick, D., Hazra, S., Mukherjee, A., Sinha, S., Mahanta, S., Das, A., & Biswas, J. K. (2023). Application of nanotechnology in mitigating arsenic stress and accumulation in crops: Where we are and where we are moving towards. In: Niazi, N.K., Bibi, I., Aftab, T. (eds) Global Arsenic Hazard. Environmental Science and Engineering. Springer, Cham. https://doi.org/10.1007/978-3-031-16360-9_12

Moulick, D., Samanta, S., Sarkar, S., Mukherjee, A., Pattnaik, B. K., Saha, S., & Santra, S. C. (2021). Arsenic contamination, impact and mitigation strategies in rice agro-environment: An inclusive insight. *Science of the Total Environment*, *800*, 149477. https://doi.org/10.1016/j.scitotenv.2021.149477

Moulick, D., Santra, S. C., Ghosh, D., & Panda, S. K. (2019). An assessment of efficiency of zinc priming in rice (cv. MTU-7029) during germination and early seedling growth. In M. Hasanuzzaman & V. Fotopoulos (Eds.), *Priming and pretreatment of seeds and seedlings*. Springer. https://doi.org/10.1007/978-981-13-8625-1_24

Mukherjee, A., & Hazra, S. (2022). Impact of elevated CO_2 and O_3 on field crops and adaptive strategies through agro-technology. In S. Choudhury & D. Moulick (Eds.), *Response of field crops to abiotic stress* (pp. 177–190). CRC Press.

Munns, R., & Tester, M. (2008). Mechanisms of salinity tolerance. *Annual Review of Plant Biology*, *59*, 651.

Panda, D., Barik, J., & Behera, P. K. (2022). Improving submergence tolerance in rice: Recent progress and future perspectives. In S. Choudhury & D. Moulick (Eds.), *Response of field crops to abiotic stress* (pp. 111–122). Taylor & Francis Group.

Parida, A. K., & Das, A. B. (2005). Salt tolerance and salinity effects on plants: A review. *Ecotoxicology Environmental Safety*, *60*(3), 324–349. https://doi.org/10.1016/j.ecoenv.2004.06.010

Parisi, C., Vigani, M., & Rodríguez-Cerezo, E. (2015). Agricultural nanotechnologies: What are the current possibilities? *Nano Today*, *10*(2), 124–127. https://doi.org/10.1016/j.nantod.2014.09.009

Pattnaik, B. K., & Sahu, C. (2022). Chromium dynamics in soil-plant system. In S. Choudhury & D. Moulick (Eds.), *Response of field crops to abiotic stress* (pp. 55–65). CRC Press.

Saad-Allah, K. M., & Ragab, G. A. (2020). Sulfur nanoparticles mediated improvement of salt tolerance in wheat relates to decreasing oxidative stress and regulating metabolic activity. *Physiology and Molecular Biology of Plants*, *26*(11), 2209–2223.

Sagar, L., Praharaj, S., Singh, S., Attri, M., Pramanick, B., Maitra, S., Hossain, A., & Sahoo, U. (2022). Drought and heat stress tolerance in field crops: Consequences and adaptation strategies. In S. Choudhury & D. Moulick (Eds.), *Response of field crops to abiotic stress* (pp. 91–102). CRC Press.

Saha, B., Chowardhara, B., Chowra, U., & Panda, C. K. (2022). Aluminum toxicity and ionic homeostasis in plants. In S. Choudhury & D. Moulick (Eds.), *Response of field crops to abiotic stress* (pp. 79–90). CRC Press.

Saharan, B. S., Brar, B., Duhan, J. S., Kumar, R., Marwaha, S., Rajput, V. D., & Minkina, T. (2022). Molecular and physiological mechanisms to mitigate abiotic stress conditions in plants. *Life*, *12*, 1634. https://doi.org/10.3390/life12101634

Sarkar, M., Paul, A., & Das, D. R. (2022). Availing engineered and biosynthesized metal-based nanoparticles to alleviate abiotic stress. In S. Choudhury & D. Moulick (Eds.), *Response of field crops to abiotic stress* (pp. 15–33).

Saxena, R., Tomar, R. S., & Kumar, M. (2016). Exploring nanobiotechnology to mitigate abiotic stress in crop plants. *Journal of Pharmaceutical Sciences and Research*, *8*(9), 974–980. Taylor & Francis Group.

Sebastian, A., Nangia, A., & Prasad, M. N. V. (2018). A green synthetic route to phenolics fabricated magnetite nanoparticles from coconut husk extract: Implications to treat metal contaminated water and heavy metal stress in *Oryza sativa* L. *Journal of Cleaner Production*, *174*, 355–366.

Sebastian, A., Nangia, A., & Prasad, M. N. V. (2019). Cadmium and sodium adsorption properties of magnetite nanoparticles synthesized from *Hevea brasiliensis* Muell. Arg. bark: Relevance in amelioration of metal stress in rice. *Journal of Hazardous Materials*, *371*, 261–272.

Sreelakshmi, B., Induja, S., Adarsh, P. P., Rahul, H. L., Arya, S. M., Aswana, S., Haripriya, R., Aswathy, B. R., Panicker, M., & Vishnudasan, D. (2021). Drought stress amelioration in plants using green synthesised iron oxide nanoparticles. *Materials Today Proceedings*, *41*, 723–727.

Shalaby, T. A., Taha, N. A., Rakha, M. T., El-Beltagi, H. S., Shehata, W. F., Ramadan, K. M. A., & Bayoumi, Y. A. (2022). Can grafting manage fusarium wilt disease of cucumber and increase productivity under heat stress? *Plants*, *11*, 1147. https://doi.org/10.3390/plants11091147

Szabados, L., & Savouré, A. (2010). Proline: A multifunctional amino acid. *Trends in Plant Science*, *15*(2), 89–97.

Szerement, J., Szatanik-Kloc, A., Mokrzycki, J., & Mierzwa-Hersztek, M. (2022). Agronomic biofortification with Se, Zn, and Fe: An effective strategy to enhance crop nutritional quality and stress defense—A review. *Journal of Soil Science and Plant Nutrition*, *22*, 1129–1159. https://doi.org/10.1007/s42729-021-00719-2

Tariq, M., Choudhary, S., Singh, H., Siddiqui, M. A., Kumar, H., Amir, A., & Kapoor, N. (2021). *Role of nanoparticles in abiotic stress*. Technology in Agriculture.

Torre-Roche, R., Jesus, C., Tamez, C., Zuverza-Mena, N., Hamdi, H., Adisa, I., & White, J. (2020). Seed biofortification by engineered nanomaterials: A pathway to alleviate malnutrition? *Journal of Agricultural and Food Chemistry*, *68*, 12189–12202. https://doi.org/10.1021/acs.jafc.0c04881

Trovato, M., Mattioli, R., & Costantino, P. (2008). Multiple roles of proline in plant stress tolerance and development. *Rendiconti Lincei*, *19*(4), 325–346.

Umar, O. B., Ranti, L. A., Abdulbaki, A. S., Bola, A. L., Abdulhamid, A. K., Biola, M. R., & Victor, K. O. (2021). Stresses in plants: Biotic and abiotic. In: Mahmood-ur-Rahman Ansari (Eds.), *Current trends in wheat research* (pp. 1–8). IntechOpen.

Vamvakaki, V., & Chaniotakis, N. A. (2007). Pesticide detection with a liposome-based nano-biosensor. *Biosensors and Bioelectronics*, *22*(12), 2848–2853.

Wahid, I., Rani, P., Kumari, S., Ahmad, R., Hussain, S. J., Alamri, S., Tripathy, N., & Khan, M. I. R. (2022). Biosynthesized gold nanoparticles maintained nitrogen metabolism, nitric oxide synthesis, ions balance, and stabilizes the defense systems to improve salt stress tolerance in wheat. *Chemosphere*, *287*(Pt 2), 132142. https://doi.org/10.1016/j.chemosphere.2021.132142

Wang, K., Wang, Y., Wan, Y., Mi, Z., Wang, Q., Wang, Q., & Li, H. (2021). The fate of arsenic in rice plants (*Oryza sativa* L.): Influence of different forms of selenium. *Chemosphere*, *264*(1), 128417.

Wani, S. H., Kumar, V., Shriram, V., & Sah, S. K. (2016). Phytohormones and their metabolic engineering for abiotic stress tolerance in crop plants. *The Crop Journal*, *4*(3), 162–176.

Wani, S. H., & Sah, S. K. (2014). Biotechnology and abiotic stress tolerance in rice. *Journal of Rice Research*, *2*, e105. https://doi.org/10.4172/jrr.1000e105

Yashveer, S., Singh, V., Kaswan, V., Kaushik, A., & Tokas, J. (2014). Green biotechnology, nanotechnology and bio-fortification: Perspectives on novel environment-friendly crop improvement strategies. *Biotechnology and Genetic Engineering Reviews*, *30*(2), 113–126.

Yasmin, H., Mazher, J., Azmat, A., Nosheen, A., Naz, R., Hassan, M. N., Noureldeen, A., & Ahmad, P. (2021). Combined application of zinc oxide nanoparticles and biofertilizer to induce salt resistance in safflower by regulating ion homeostasis and antioxidant defence responses. *Ecotoxicology and Environmental Safety*, *218*, 112262. https://doi.org/10.1016/j.ecoenv.2021.112262

Yassen, A., Abdallah, E., Gaballah, M., & Zaghloul, S. (2017). Role of silicon dioxide nano fertilizer in mitigating salt stress on growth, yield and chemical composition of cucumber (*Cucumis sativus* L.). *International Journal of Agricultural Research*, *22*, 130–135.

Zhang, H., Demirer, G. S., Zhang, H., Ye, T., Goh, N. S., Aditham, A. J., & Landry, M. P. (2019). DNA nanostructures coordinate gene silencing in mature plants. *Proceedings of the National Academy of Sciences of the United States of America*, *116*(15), 7543–7548. https://doi.org/10.1073/pnas.1818290116

4 Prospects of Nanotechnology for Abiotic and Biotic Stresses Amelioration in Field Crops

Zahoor Ahmad, Ejaz Ahmad Waraich, Celaleddin Barutçular, Sarfraz Ahmed, Muhammad Aamir Iqbal, Amir Aziz, Muhammad Adnan Bukhari, Amber Raza, Muhammad Ahmad and Shehar Bano*

4.1 INTRODUCTION

Globally, crop plants are facing numerous abiotic and biotic stresses triggered by climate change and environmental degradation. The salient abiotic stresses include salinity, drought, heat, chilling, heavy metals toxicity and nutrient imbalance, while insect-pests and weeds are the most pronounced biotic stresses for crop plants. These stresses impart numerous deleterious impacts on plants and reduce economic yield (Ahmad et al., 2021, 2022; Khalid et al., 2019). For the reduction in the negative effect of climate change, different strategies have been adopted including the use of organic fertilizers, plant growth regulators, tolerant genotypes etc. Recently, nanotechnology which utilizes nanoparticles (NPs) having a size of 1 to 100 nanometers has emerged as a promising stress mitigation approach (Jeevanandam et al., 2018).

The NPs' properties are the result of their small size which might be exploited to impart tolerance against abiotic and biotic stresses. Some NPs that are used to treat the plant are carbon nano-tubes, copper, magnesium oxide, aluminium oxide, molybdenum NPs, zinc oxide, zinc, copper oxide, iron oxide, manganese, selenium MPs, silver, silicon, silicon oxide, magnesium, calcium, caesium, cobalt, titanium dioxide (CTNs, Cu, MgO, AIO4, Mo, Zn, ZnO, CuO, FeO, Mn, Se, Ag, Si, SiO_2, Mg, Ca, Ce, Co, TiO_2). The plants developed efficient molecules that rapidly sense the stresses in the environment. Plants developed a wide range of molecules. The plant

* Corresponding author.

DOI: 10.1201/9781003376446-4

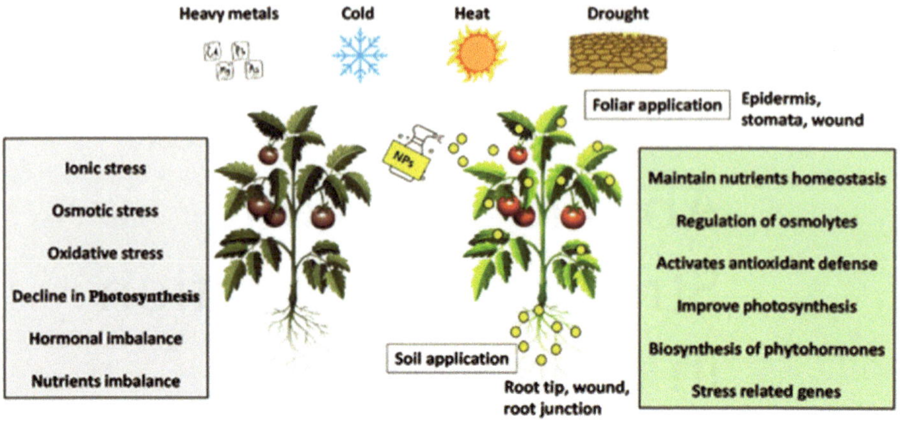

FIGURE 4.1 Modulation of plant responses by nanoparticles

adapted according to the stresses and improved its growth rate (Vanwallendael et al., 2019). The NPs in the plants show an important role. The NPs enhance the response against stress and increase growth. The crop yield might be increased by improving tolerance against stresses under the NPs treatments (Arif et al., 2020). The NPs modulate a variety of biochemical, molecular and physiological processes that tend to improve the tolerance in plants (Figure 4.1).

4.2 MITIGATION OF ABIOTIC STRESSES THROUGH NANOTECHNOLOGY

4.2.1 Physiological Mechanisms

The use of nanotechnology by employing NPs can strengthen various physiological processes which enable crop plants to survive environmental extremes and biotic stresses. Photosynthesis, which is the most vital physiological process in plants, might be enhanced by employing nanotechnology and NPs. The NPs alter the enzyme activity in the C3 cycle and modulate photosynthesis pigment (Prasad et al., 2011). The effectiveness of NPs on seed germination and growth varies depending on the concentration and host plant species. Nanowax foliar spraying on sorghum plants under drought circumstances improved seed output in plants compared to water spraying. The TiO_2 NPs significantly alter the morphological, biochemical and physiological characteristics of plants which enable crop plants to survive abiotic stresses (Haghighi et al., 2014). Exogenous applications of NPs throughout the growth period of cucumber plants increased photosynthetic rate, chlorophyll production and rubisco activase activity, which increased plant dry mass (Hartikainen et al., 2000). As a result of improved photosynthesis, it was also discovered that foliar NPs administration might boost soybean seed production (Hasanuzzaman et al., 2014). The underlying mechanism is NPs target the cellular organelles in plants and release different contents from them (Thomashow, 1999), subsequently controlling the activity of the antioxidant enzymes SOD, CAT and POD (Siddiqi & Husen, 2016). Increased SOD activity in plants exposed to TiO_2 NP demonstrated this effect (Khalid et al., 2022). Some elements have been used for incremental drought stress resistance, together with oxides as NPs (Siddiqui et al., 2020). When various crops were treated with silica and ZnO NPs, an increase in growth, physiology and biochemistry was seen (Savicka & Škute, 2010). The Si NPs also reduced the effects of drought stress on wheat plants (Rizwan et al., 2019). Likewise, ZnO NPs mitigated the detrimental effects of salt and drought stress on plants (Khan et al., 2019). Excessive NPs application was shown to produce oxidative stress, which resulted in protein regulation, cell cycle arrest, programmed cell death and the production of antioxidant enzymes (Hussein & Abou-Baker, 2018). Under drought circumstances, NP-treated plants showed a considerable decrease in MDA levels along with free radicals such as H_2O_2 and O_2^-. However, it was also noted that the application of TiO_2 increased the activities of the antioxidant enzymes POD and CAT, while MDA levels were decreased as a result of the activation of the plant's antioxidant system (Giraldo et al., 2014). Plants treated with CuO NPs continued to produce more anthocyanin under drought stress, and it was also demonstrated that proline production increased under drought stress. The free radical buildup was increased in wheat roots exposed to CuO NPs, which is consistent with plants adapting to the challenge of NP-induced ROS bursts. Higher ROS levels may cause transcriptional alterations that result in tolerance, further indicating that drought stress is a cause of elevated ABA. Fullerenol NPs [C60(OH)24] may have numerous benefits for reducing the oxidative impact of drought stress, albeit this will rely on the concentration of NPs applied and the amplification of several antioxidant enzymes (GR, SOD, GPX, APX, and CAT) in plants (Abbasi et al., 2022; Aurangzaib et al., 2022; Qi et al., 2013; uz Zaman et al., 2022). Depending on the type of plant, salinity level and other conditions, one of the activities most impacted by salinity stress is photosynthesis (Hnilickova et al., 2021). By applying NPs to the leaves, which increases photosynthesis and light absorption, plants may manufacture more complexes for light absorption. Numerous studies show that NPs greatly enhance the amount of chlorophyll in plants (Ali et al., 2021). It has been revealed that NPs increased photosynthesis by increasing the number of photosynthetic pigments in stressed plants (Alabdallah & Alzahrani, 2020; Singh et al., 2021). Manganese (Mn) is included in several cell components such as mitochondria, chloroplasts, enzyme structures etc. It is also in charge of speeding up oxygen evolution and the photosynthetic electron transport chain process. Similarly, the Mn NPs are also capable of sustaining photosynthetic rates in the presence of abiotic stress (Ye et al., 2020). According to a study, Mn supplementation increased the *Vigna radiata* plants' nitrate reductase activity, chlorophyll content and membrane stability index under salt stress (Shahi & Srivastava, 2018). Previous studies have demonstrated that adding Cu to maize plants can mitigate the detrimental effects of salt on water relations and photosynthesis (Iqbal et al., 2018). The impact of NPs on nutrient absorption, transport and final allocation may be crucial for enhancing plant nutrition (Kopittke et al., 2019). One of the most important variables for plant resistance to salt stress is the high potassium-to-sodium ratio, which is disturbed by salinity stress. By adding NPs to the plant and thereby raising the osmotic potential inside the plant, it is possible to enhance plant growth under salt stress (Tahjib-Ui-Arif et al., 2019). According to Farhangi-Abriz and Torabian (Farhangi-Abriz & Torabian, 2018), the nano-SiO_2 increased leaf K^+ content, which improved soybean seedling development in a saline environment. Moreover, Cu NPs applied topically improved tomato plant growth and Na^+/K^+ ratio (Pérez-Labrada et al., 2019). Using Fe2O3 NPs, it was also discovered that *Trachyspermum ammi* plants were less vulnerable to salt stress. Using pepper plants that have been exposed to salt, Ye et al. (2020) studied how Mn NPs affected the development of pepper plants. It is generally known that cold and heat stress harm plant development and physiology. Plant growth and functioning can be maintained by applying NPs during stressful situations. Plant productivity was enhanced by the use of biological selenium NPs at a concentration of 100 g/mL. This was accomplished by enhancing plant growth, photosynthetic rate and gas exchange at high temperatures (El-Saadony et al., 2021). Kareem et al. (2022) reported that the application of nano-ZnO NPs at a high temperature boosted the enzymatic balance, gas exchange parameters and chlorophyll activity, which increased the number, size and total grain production of the pods. Application of ZnO and TiO_2 increased antioxidant defence mechanisms and membrane stability in both roots and shoot parameters in heat stressed wheat seedlings (Thakur et al., 2022). Higher plant production was ensured by nano-ZnO NPs' capacity to control osmatic potential and reduce thylakoid damage by activating antioxidant defence. In rice, ZnO NPs have also been linked to cold tolerance; their application boosted root, shoot and dry bulk while decreased oxidative stress and enhanced photosynthetic activity (Song et al., 2021). Applying NPs (nSiO2, nSe, nZnO, and GNRs) to sugarcane reduced the effects of oxidative activity and gas exchange parameters while improving the photosynthetic rate and reducing the effects of cold stress by raising chlorophyll content (Elsheery et al., 2020). Nanoparticles (NPs) dosage plays a crucial role in preserving ion concentration within plants, thereby enhancing membrane stability and regulating osmotic control.. Plant productivity is increased when it can continue to transmit water and nutrients under pressure. Under stressful environments, enzymatic oxidative stress must be avoided to ensure optimum growth and economic yield. Under conditions of heat and cold stress, the production of enzymatic anti-oxidizers, such as SOD, POD, CAT and APX, is promoted by the application of NPs. However, the concentration of NPs application is crucial because, at larger concentrations, NPs toxicity can be seriously detrimental (Rajput et al., 2021; Wang et al., 2016). In plants exposed to heavy metals, NPs retain membrane stability, increase chloroplast pigments and speed up photosynthesis (Ahmed et al., 2021a). The FeO NPs were used, according to Hussain et al. (Alejandro & Rubiales, 2009), to alleviate the effects of cadmium (Cd) on wheat. The effects of Cd toxicity on growth and yield were reduced, and improvements were made to the wheat's morphological characteristics, photosynthetic pigments and dry biomass. The deleterious effects of Cd toxicity were mitigated, and plant growth and photosynthetic rate both rose. Similar outcomes were found by Sardar (Sardar et al., 2022) in the remediation of Cd by nano-TiO_2 in coriander, including a decrease in Cd content, a reduction in oxidative damage brought on by Cd stress and an improvement in agronomic parameters. The application of nano-TiO_2 improved the photosynthetic rate and growth metrics in soybean (Singh & Lee, 2016). Memari-Tabrizi et al. found that applying Si NPs caused a decrease in Cd stress, which resulted in a rise in summer savoury biomass (Farajzadeh Memari-Tabrizi et al., 2021). To reduce the effects of Cd in rice, graphite carbon nitride was synthesized (Hao et al., 2021), plant biomass was increased and Cd-induced toxicity significantly decreased. It has been demonstrated that various NPs treatments can reduce MDA and H_2O_2 levels in plants by controlling antioxidant enzymes such as SOD, CAT, guaiacol and ascorbate peroxidases (Azeez et al., 2021). Among HMs, lead (Pb) is another hazardous metal and requires complicated phytoextraction; lead (Pb) phytoremediation is extremely important. Fatemi et al. reported on the mediation of Pb in coriander (Munns & Tester, 2008) with various Si NP concentrations. Pb stress reduced plant biomass and vitamin C levels while increasing flavonoid levels. Increased antioxidant enzyme activity lessened the negative consequences of Pb poisoning. Jin found that the application of nanohydroxyapatite significantly increased ryegrass biomass that was significantly reduced by Pb toxicity (Jin et al., 2016). Plant growth is inhibited by chromium (Cr) phytotoxicity due to a decrease in photosynthetic pigments and chlorophyll fluorescence (Tripathi et al., 2015). Thus, NPs foliar application against different HM toxicity have exhibited encouraging results; however further investigations are awaited to explore the underlying mechanisms.

4.2.2 Molecular Mechanisms

The impacts of NPs on various plant species at the molecular level have been extensively studied using transcriptomic and proteomic techniques. The kind, size and form of NPs as well as the dose utilized have been reported to have a significant impact on morphological and physiological outcomes (Avestan et al., 2019; Khan, 2016). Since the P5CS gene encodes for proline biosynthesis, expression of this gene increases plants' tolerance to many environmental stress situations, including biotic and abiotic challenges. In addition to Ca21 and ROS, MAPK2, a member of the MAP kinase gene family, is crucial for controlling phytohormones and modulating antioxidant defence mechanisms under various stress conditions (Khan, 2016). The transcriptional regulators AREB/ABF are essential for the control of the AREB gene, which encodes abscisic acid, and they play a crucial role in promoting resilience to harsh situations like drought and salt stress (Khan, 2016; Yassen et al., 2017). In response to salt and drought stress, the ZFHD gene is down-regulated, which is regulated by the abscisic acid (ABA) production pathway. Conversely, osmotic pressure is decreased and solute aggregation is improved when the TAS14 gene is down-regulated, making plant species more resilient to salt and drought stress (Khan, 2016). Glucosinolate and phenolic-related genes, which are also connected to biotic and abiotic stressors, had their metabolic pathways altered by the application of Ag NPs (5 and 10 mg/L) to rape plants, while carotenoid genes were suppressed (Thiruvengadam et al., 2015). The ABA production pathway regulates ZFHD gene down-regulation, which lessens the detrimental consequences of salt and drought stress. Ag and Ag1 NP treatment of *Arabidopsis* plants led to the down-regulation of ethylene and auxin-related genes and the up-regulation of genes linked to metal response and oxidative stress (Khan, 2016). The molecular processes dictate the vital biological processes of crop plants. It is critical to assess potential pathways and the molecular impacts of NPs on plants, and this has already been done (Ali et al., 2021).

Without interfering with cellular functions and gene expression, NPs cannot be effective. This is because salinity stress alters gene expression in different areas of the cell products, which in turn influences plant growth. MiR164 expression is down-regulated during NP-mediated root development, which is associated with auxin hormone signalling. Furthermore, lateral roots can form and blooming can be sped up as a result of increased miR169 expression and decreased miR167 expression (Tolaymat et al., 2017).

A foliar application of Zn NPs on rapeseed plants (*Brassica napus* L.) under salinity stress reduced the expression of some genes, such as SKRD2, MYC and MPK4. Contrastingly, it increased the expression of ARP genes and MPK associated with physiological and hormonal responses and transcription factors, and MYC and SKRD2 which are involved in abiotic stress tolerance in plant cells (Hezaveh et al., 2019). The Si NPs applied to *Cannabis sativa* boosted the plant's growth and molecular alterations under salinity stress (Saffie-Siebert et al., 2021). The regulation of plant stress response is controlled by gene expression and subsequently affects the production of enzymes and proteins. Tolerance to heat and cold stress is regulated by several genes, transcription factors and proteins. Plant survival is directly impacted by the activation of expression, whether it be up-regulation or down-regulation. ZnO nanoparticles applied topically to rice seedling leaves enhanced the production of genes related to the antioxidative system (OsCu/ZnSOD1, OsCu/ZnSOD2, OsCu/ZnSOD3, OsPRX11, OsPRX65, OsPRX89, OsCATA and OsCATB) and chilling-response transcription factors (OsbZIP52, OsMYB4, OsM) (Song et al., 2021). The EREB, R2R3MYB, HSF-34, WRKY1, MAPK1, HDA3 and CAT genes were transactionally elevated in soybean seedlings, increasing the number of photosynthetic pigments, proline concentration, antioxidant enzyme activity and plant yields (Mirakhorli et al., 2021). According to transcriptome analysis, (50 nm) Cu-based NPs altered the expression of genes involved in oxidative stress response, brassinosteroid production and root development (Ali et al., 2019). Accumulation of secondary metabolites (acetyl glucosamine, phenyl lactate and 4-aminobutyrate) involved in cell signalling and immune responses was investigated in Cu nanoparticles with a size of 40 nm. It was discovered that metabolites involved in production and defence responses destroyed flavonoids, fatty acids, riboflavin and amino acids (Zhao et al., 2016). When silicon was used on wheat seedlings instead of silicon nanoparticles, TaPIP1 and TaNIP2 aquaporin genes were shown to be overexpressed during heat stress, increasing the relative water content of the plants (Younis et al., 2020). ZnO NPs considerably improved heat stress-induced relief of TGS-GUS genes in *Arabidopsis thaliana* seedlings exposed to heat stress (37°C) and growing in a greenhouse (Wu & Wang, 2020). Because each plant species has a unique defence mechanism to deal with stress, the interactions of heavy metals on the plant system mechanically and chemically vary depending on the kind of plant. Si NPs have been shown to have a positive impact on lowering Cd absorption and toxicity in rice, according to Cong (Cong et al., 2019). Si NPs suppress the expression of the genes low-affinity cation transporter (LCT1) and natural resistance-associated macro-phage protein 5, which are involved in the movement and uptake of Cd from root to shoot (NRAMP5). Both the silicon absorption gene low silicon rice 1 (LSI1) and the heavy metal ATPase 3 (HMA3) gene, which transports Cd into the vacuoles, are increased. Si NPs are used to improve silicon uptake from roots and decrease Cd uptake. In the xylem and phloem, Ahmed et al. (2021b) discovered the Cd transporter gene, including OsHMA2 and OsHMA3, which are in charge of heavy metal transport. All three genes expressed much less in rice after the application of hydrogel and FeO nanoparticles. Heavy metals are transported through the natural resistance-associated macro-phage protein (NRAMP) gene family in plant species such as rice, potato, pepper, tomato, *Arabidopsis* and soybean (Tian et al., 2021). By reducing Cd absorption and transport genes, Si NPs treatment enhanced wheat development and reduced the stress caused by heavy metals (Khan et al., 2019).

4.3 MITIGATION OF BIOTIC STRESSES THROUGH NANOTECHNOLOGY

Nanoparticles can be designed with unique chemical, physical and biological properties to distinctively differ from those of their molecular and bulk counterparts (Vlot et al., 2008). Nanoparticles alone have the potential to be directly applied to plant seeds, foliage or roots for protection against pests and pathogens, such as insects, bacteria, fungi and viruses. Metal nanoparticles such as silver, copper, zinc oxide and titanium dioxide have been intensively researched for their antibacterial and antifungal properties and are known for their antiviral properties. This section gives a brief overview and an update on current literature reviews about the individual nanoparticles that already exist (Kim et al., 2018).

Recently, silver nanoparticles have increased in popularity, due to "green synthesis" production in plants, bacteria, fungi or yeast. Silver nanoparticles have shown antifungal inhibition of *Alternaria alternata, Sclerotinia sclerotiorum, Macrophomina phaseolina, Rhizoctonia solani, Botrytis cinerea and Curvularia lunata* by well diffusion assay. When silver nanoparticles were sprayed onto bean leaves, complete suppression of sun-hemp rosette virus was observed. Elbeshehy et al. (2015) further showed that faba bean plants challenged with bean yellow mosaic virus, and sprayed with silver nanoparticles, produced remarkably better results when the nanoparticles were applied 24 hours post-infection, compared to spray application before infection or simultaneously at the time of inoculation. Silver nanoparticles have shown immense potential for plant disease management against fungal and bacterial pathogens, but there are significant hurdles associated with them, such as production, toxicity and soil interaction (Sadeghi et al., 2017).

Other commonly used metal nanoparticles include copper, titanium dioxide and gold. Nanoparticles, such as copper and titanium dioxide, are more commonly utilized for fertilizer with little research into plant disease management. The research covering titanium dioxide, silver, copper nanoparticles for antimicrobial properties and aluminium nanoparticles as insecticidal dust has been highlighted. Furthermore, titanium dioxide nanoparticles in fertilizers have produced protection from bacteria and inactivation of viruses (Sadeghi et al., 2017). Introducing poly-dispersed gold nanoparticles via a mechanical abrasive was seen to melt and dissolve the *Barley yellow mosaic virus* particles conferring resistance to the plant (Alkubaisi et al., 2014).

Chitosan is another popular nanoparticle that has favourable biological properties, such as biodegradability, biocompatibility, non-allergenicity and antimicrobial activity, with low toxicity to animals and humans. Chitosan nanoparticles induce viral resistance in plant tissues by protecting them against infections caused by the mosaic virus of alfalfa, snuff, peanut, potato and cucumber (Kashyap et al., 2015). Chitosan nanoparticles have shown antimicrobial properties, such as controlling *Fusarium crown*, root rot in tomato, *Botrytis* bunch rot in grapes and *Phyricularia grisea* in rice but are less effective against bacteria. Malerba and Cerana summarized potential mechanisms that lead to the antimicrobial effects of chitosan, such as agglutination, disruption of the cell membrane, inhibition of H^+-ATPase activity, inhibition of toxin production and microbial growth, inhibition of the synthesis of messenger RNA and proteins and blockage of nutrient flow. Antiviral effects have been observed in beans against *Bean mild mosaic virus* and, in tobacco, against *Tobacco mosaic virus* and *Tobacco necrosis virus*. Chitosan has also shown effectiveness against oleander aphid (*Aphis nerii*), cotton leafworm (*Spodoptera littoralis*), root-knot nematode (*Meloidogyne javanica*) and nymphs of the spear psylla (*Cacopsylla pyricola*). Chitosan holds promise as both nanoparticles alone and has shown to have enormous potential as a nanocarrier (Malerba & Cerana, 2016).

4.3.1 Physiological Mechanisms

Inorganic nanoparticles are non-toxic, hydrophilic biocompatible and highly stable. Examples of inorganic nanoparticles are calcium phosphate nanoparticles, gold nanoparticles, iron oxide nanoparticles, zinc oxide nanoparticles, silver nanoparticles, etc. (Paul & Sharma, 2010). Inorganic nanoparticles have ions, and these ions are antimicrobial agents. Zinc and copper lead to reactive oxygen species (ROS) generation to kill the engulfed pathogen. Gold and silver are very toxic to bacteria at low concentrations. The application of nanoparticles leads to a disruption in the cell membranes of pathogens. A nanoparticle application also leads to the generation of reactive oxygen species (ROS) and free radicals and metal ions. In addition to this, nanoparticles of very small size also intercalate into the DNA. Metals are known to bind to thiol or the amine moiety of cellular proteins and lead to the deactivation and precipitation of proteins. The high attraction of metal ions by proteins leads to an increase in cellular concentrations and eventually cell death (Mittapally et al., 2018). Polymeric nanoparticles (PNPs) are used for the production of nanocapsules or nanospheres and are composed of various natural or synthetic polymers. Many of these polymers are biodegradable and non-toxic (Barros et al., 2018). Examples of polymers used to prepare PNPs include cellulose, gelatine, chitosan, alginate, etc. PNPs range in size from 10 nm to 1 µm. They are colloidal particles and solid in nature. Nanospheres have a matrix system where the drug is uniformly dispersed. Nanocapsules have the drug embedded in a cavity which, in turn, is surrounded by a polymeric membrane. These polymers that make up the PNPs are hydrophilic, are made up of carbohydrates or proteins and are hydrophobic (Sharma et al., 2019). PNPs provide an excellent vehicle for the delivery of inorganic nanoparticles. PNPs can be easily modified according to the target site and can also be controlled by their size, morphology and surface charge. Also, inorganic nanoparticles inside a PNP show low aggregation and reduce concerns regarding cytotoxicity (Zare et al., 2017). Nanoparticles can be directly applied to the seeds, foliage or roots in order to protect the plants from a pathogen invasion. These nanoparticle actions can be comparable to chemical pesticides. ZnO nanoparticles were fungicidal to *B. cinerea* and *Penicillium expansum*. The nanoparticles were seen to prevent the development of fungal hyphae and also prevented the development of the conidia and conidiophores. Platelet-shaped ZnO particles have been seen to have more antifungal activity when compared to rods and nanoparticles against *Fusarium solani* (Pariona et al., 2020). Chitosan has been studied for the formation of micro-particles and nanoparticles owing to its biodegradable and biocompatible properties, its non-toxicity to animals and humans and its antimicrobial activity (Zhou et al., 2011).

4.3.2 Molecular Mechanisms

To reduce the losses to crops caused by plant pathogens, plant biologists have used many methods to engineer resistant plants. RNA silencing-based resistance has been a powerful tool that has been used to engineer resistant crops during the last two decades. Based on this mechanism, diverse approaches have been developed thus far. RNA silencing has been utilized to produce plants that are resistant to plant viruses, such as RNA and DNA viruses, viroids, insects and to fungal pathogens. RNA interference (RNAi) is a process through which the synthesis of targeted proteins is down-regulated by interfering in the expression of the genetically encoded message. The interference of the genetically encoded message is an RNA-mediated and post-transcriptional gene regulation process (Fire et al., 1998). Usually, viral vectors are used for the delivery of siRNAs or miRNAs into plants. Viral vectors transiently and directly express the siRNA in the plant and do not rely on plant transformation. The most commonly used viral vector for the delivery of siRNAs is the bipartite tobacco rattle virus (Dinesh-Kumar et al., 2003). In addition to the tobacco rattle virus, the cabbage leaf curl virus vector has also been developed for expressing synthetic and endogenous miRNAs in plants (Tang et al., 2010). There are various disadvantages in using viral vectors for post-transcriptional gene silencing (PTGS); one of them being that viruses are limited in their host range. Viral vectors do not infect all the tissues uniformly, although they might spread systemically. In addition to these drawbacks, the phenotype attributed to PTGS is mixed with the onset of virus symptoms like the mosaic pattern of the disease and mild leaf curling. These disadvantages of using viral vectors make it imperative to look for an alternative, more efficient method that can be used for PTGS in plants in order to confer greater pathogen resistance to crops. Nanoparticles offer a solution to overcoming numerous challenges in siRNA delivery. Gold nanoparticles provide particularly attractive scaffolds for the creation of transfection agents (Rosi et al., 2006). Gold colloids are bio-inert, non-toxic and readily synthesized and functionalized. Recently, gold nanoparticles with a core diameter of 2 nm and an overall diameter of 6 nm have been designed. These gold nanoparticles resemble histone octamers, where the nucleosome core proteins are 6 nm in diameter. These gold nanoparticles also resemble histones in shape, size and surface functionality (Mitter et al., 2017). Besides, dsRNAs, siRNAs have also been delivered either to cell cultures through nanosecond pulsed laser-induced stress or to protoplasts as conjugated with polymer nanoparticles have been seen to serve as attractive materials for DNA and RNA delivery (Roca & Haes, 2008).

4.3.3 Agronomic Management

Better crop management practices like tillage, planting time, planting geometry, quality seed, optimum seed rate, method of sowing, mulching and nutrient, weed and water management provide greater opportunities to increase pulse productivity in India by mitigating the abiotic and biotic stresses. Tillage is necessary for obtaining ideal conditions for seed germination, seedling establishment and growth of crops (Das & Roychoudhury, 2014). For dryland pulses, deep ploughing results in better moisture conservation which leads to better root proliferation and higher productivity over shallow cultivation. The conservation tillage system is one such tillage option designed to conserve soil moisture in moisture-deficit areas. This could overcome both abiotic and biotic stresses in pulse crops. To achieve the success of conservation agriculture in the long-run, cultivars need better water-use efficiency, improved soil health and pathogen resistance that would emerge from crop residues. Zero-tillage practices can have beneficial impacts on productivity as well as minimize environmental degradation, such as soil erosion in legume and cereal crop-growing areas (Das & Roychoudhury, 2014). Non-monetary inputs, like time of sowing and method of sowing, have a significant influence on crop growth, phenological development, insect-pests and weed dynamics and crop productivity. Delayed planting restricts vegetative growth and pod-bearing branches, decreases biological nitrogen fixation, leads to forced maturity and increases the incidence of pod-borer (Ali-Shtayeh et al., 1998). Off-season cultivation of pulses fetches higher income for the farmers owing to lower insect-pests and disease incidence. Powdery mildew, caused by *Erysiphe polygoni* DC., incidence can be reduced by adjusting the date of sowing with wider spacing in mung bean with resistant varieties (Pandey et al., 2017). Practices like suitable time of sowing, an optimum depth of sowing, spacing, seed rate and method of sowing depend on the cultivar, growing season, soil texture, seed size and planting system and are necessary for mitigating moisture and heat stress, weeds, insect-pests and diseases in pulses. In rainfed areas with erratic rainfall patterns, the seed rates for pulses are quite higher owing to less soil moisture availability and low temperature enough to harness the solar radiation to translate into higher crop yields in hill and mountain ecosystems (Choudhary et al., 2013). Bed planting and ridge and furrow systems are beneficial, particularly in areas having saline irrigation water. Bed planting, planting on ridge and use of mulches are helpful for water stress management in addition to their other added advantages. The supremacy of bed planting, ridge and furrow systems may avoid direct contact of seedlings with saline water, providing proper drainage and adequate aeration. This method is also successful in draining excess water from crop root-zone, reduces the incidence of insect-pests and diseases and results in 25–30% higher yield in kharif pulses over flat-bed planting (Das & Roychoudhury, 2014), besides significant water savings (16–20%) for a wide spectrum of legumes compared to traditional farmers' practice (Rana et al., 2016).

More than 70% of the area of pulses in India are cultivated under intercropping systems and are more productive, particularly under rainfed conditions. Pulses play a major role in intercropping systems with cereals, oilseeds and commercial crops for supplementing nitrogen requirement to the base crop, appreciable smothering effect on weeds, binding soil particles against soil degradation and many more benefits (Dass & Sudhishri, 2010). Sugarcane trash mulching at 8 t/ha is effective for the control of weeds, increasing yield, conservation of soil moisture and moderation of soil temperature in pigeon pea. Hence, mulching of preceding crop residues results in reducing the weed stress as well as concurrent insect-pests and disease stress in pulses. Pulses are distinctly diverse agricultural segments and are highly influenced by weed stress and need to shift to newer herbicide applications for better weed control. One of the newer herbicides, quizalafop-p-ethyl used at 0.5% is the safest and most beneficial for weed control in urd bean, as it had the least adverse effect on growth, yield and enzyme activities of the crop (Choudhary et al., 2013).

4.4 CONCLUSION

The abiotic and biotic stresses have multiplied in severity and incidence frequency and have threatened crop production. Nanomaterials have the potential to improve the abiotic stress tolerance of plants through a broad spectrum of actions. Some of the most pronounced actions of NPs include increasing water uptake, metabolism of starch reserves, stimulation of photosynthesis, alteration of phytohormone levels, modulation of oxidative stress or affecting nutrient absorption etc. However, most research has focussed NPs' impact on crop plants under only one type of stressful condition. Future research needs to focus on more realistic stress conditions in real scenarios like drought and heat stresses simultaneously. Moreover, the exact understanding of the molecular mechanisms which increase plant tolerance remains unclear. Therefore, further studies are needed to determine how NPs affect the antioxidant system of plant cells, thereby improving plant tolerance to various stresses. Such an understanding may aid to design NP-based strategies that help reduce stress and ensure sustainable agricultural production. Furthermore, the field application of many of the prepared new substances is still extremely limited due to changes in environmental conditions, soil types, plant species and most importantly, the physicochemical properties of the new metallic/nonmetallic substances. The limiting factors associated with field applications include toxicity and accumulation of NPs in crop plants. Future research on assessing the toxicological effects on model micro-organisms, flora and animals is critical to enable field applications of NPs. Last but not least, there is a dire need to set up policies and regulations to facilitate the adoption of nanotechnology-enabled stress tolerance in agricultural production. Furthermore, more research needs to be conducted to investigate how nanomaterials may affect plants under abiotic stresses from the viewpoint of source-sink regulation. It would be useful to study the effects of foliar-sprayed nanomaterials on the sink capacity of plants.

REFERENCES

Abbasi, A., Sufyan, M., Ashraf, H. J., uz Zaman, Q., Haq, I. U., Ahmad, Z., Saleem, R., Hashmi, M. R., Jaremko, M., Abdelsalam, N. R., & Ghareeb, R. Y. (2022). Determination of silicon accumulation in Non-Bt cotton (Gossypium hirsutum) plants and its impact on fecundity and biology of whitefly (Bemisia tabaci) under controlled conditions. *Sustainability (Switzerland), 14*(17). https://doi.org/10.3390/su141710996

Ahmad, Z., Barutçular, C., Zia Ur Rehman, M., Sabir Tariq, R. M., Afzal, M., Waraich, E. A., & Nawaz, H. (2022). Pod shattering in canola reduced by mitigating drought stress through silicon application and molecular approaches - A review. *Journal of Plant Nutrition, 46*(1), 101–128. https://doi.org/10.1080/01904167.2022.2027972

Ahmad, Z., Waraich, E. A., Iqbal, M. A., Barutcular, C., Alharby, H., Bamagoos, A., & Sabagh, A. E. (2021). Foliage applied silicon ameliorates drought stress through physio-morphological traits, osmoprotectants and antioxidant metabolism of camelina (*Camelina sativa* L.) genotypes. *Acta Scientiarum Polonorum Hortorum Cultus, 20*(4), 43–57. https://doi.org/10.24326/ASPHC.2021.4.4

Ahmed, T., Noman, M., Ijaz, M., Ali, S., Rizwan, M., Ijaz, U., & Li, B. (2021a). Current trends and future prospective in nanoremediation of heavy metals contaminated soils: A way forward towards sustainable agriculture. *Ecotoxicology and Environmental Safety, 227*, 112888. https://doi.org/10.1016/j.ecoenv.2021.112888

Ahmed, T., Noman, M., Manzoor, N., Shahid, M., Abdullah, M., Ali, L., Wang, G., Hashem, A., Al-Arjani, A. B. F., Alqarawi, A. A., Abd Allah, E. F., & Li, B. (2021b). Nanoparticle-based amelioration of drought stress and cadmium toxicity in rice via triggering the stress responsive genetic mechanisms and nutrient acquisition. *Ecotoxicology and Environmental Safety, 209*, 111829. https://doi.org/10.1016/j.ecoenv.2020.111829

Alabdallah, N. M., & Alzahrani, H. S. (2020). The potential mitigation effect of ZnO nanoparticles on [*Abelmoschus esculentus* L. Moench] metabolism under salt stress conditions. *Saudi Journal of Biological Sciences, 27*(11), 3132–3137. https://doi.org/10.1016/j.sjbs.2020.08.005

Alejandro, P. D. L., & Rubiales, D. (2009). Nanotechnology for parasitic plant control. *Pest Management Science, 65*(5), 540–545. https://doi.org/10.1002/ps.1732

Ali, S., Mehmood, A., & Khan, N. (2021). Uptake, translocation, and consequences of nanomaterials on plant growth and stress adaptation. *Journal of Nanomaterials, 2021*. https://doi.org/10.1155/2021/6677616

Ali, S., Rizwan, M., Hussain, A., Zia ur Rehman, M., Ali, B., Yousaf, B., Wijaya, L., Alyemeni, M. N., & Ahmad, P. (2019, April). Silicon nanoparticles enhanced the growth and reduced the cadmium accumulation in grains of wheat (*Triticum aestivum* L.). *Plant Physiology and Biochemistry, 140*, 1–8. https://doi.org/10.1016/j.plaphy.2019.04.041

Ali-Shtayeh, M. S., Yaghmour, R. M. R., Faidi, Y. R., Salem, K., & Al-Nuri, M. A. (1998). Antimicrobial activity of 20 plants used in folkloric medicine in the Palestinian area. *Journal of Ethnopharmacology, 60*(3), 265–271. https://doi.org/10.1016/S0378-8741(97)00153-0

Alkubaisi, N. A. O., Aref, N. M. M. A., & Hendi, A. A. (2014). Method of inhibiting plant virus using gold nanoparticles. *Science of the Total Environment.* United States Patent 9,198,434. https://doi.org/10.13140/RG.2.1.3600.5201

Arif, Y., Singh, P., Siddiqui, H., Bajguz, A., & Hayat, S. (2020, July). Salinity induced physiological and biochemical changes in plants: An omic approach towards salt stress tolerance. *Plant Physiology and Biochemistry, 156*, 64–77. https://doi.org/10.1016/j.plaphy.2020.08.042

Aurangzaib, M., Ahmad, Z., Jalil, M. I., Nawaz, F., Shaheen, M. R., Ahmad, M., Hussain, A., Ejaz, M. K., & Tabassum, M. A. (2022). Foliar spray of silicon confers drought tolerance in wheat (*Triticum aestivum* L.) by enhancing morpho-physiological and antioxidant potential. *Silicon*, *14*(9), 4793–4807. https://doi.org/10.1007/s12633-021-01271-5

Avestan, S., Ghasemnezhad, M., Esfahani, M., & Byrt, C. S. (2019). Application of nano-silicon dioxide improves salt stress tolerance in strawberry plants. *Agronomy*, *9*(5), 1–17. https://doi.org/10.3390/agronomy9050246

Azeez, N. A., Dash, S. S., Gummadi, S. N., & Deepa, V. S. (2021). Nano-remediation of toxic heavy metal contamination: Hexavalent chromium [Cr(VI)]. *Chemosphere*, *266*(Vi), 129204. https://doi.org/10.1016/j.chemosphere.2020.129204

Barros, C. H. N., Fulaz, S., Stanisic, D., & Tasic, L. (2018). Biogenic nanosilver against multidrug-resistant bacteria (MDRB). *Antibiotics*, *7*(3), 69. https://doi.org/10.3390/ANTIBIOTICS7030069

Choudhary, A. K., Thakur, S. K., & Suri, V. K. (2013). Technology transfer model on integrated nutrient management technology for sustainable crop production in high-value cash crops and vegetables in northwestern himalayas. *Communications in Soil Science and Plant Analysis*, *44*(11), 1684–1699. https://doi.org/10.1080/00103624.2013.783058

Cong, W., Miao, Y., Xu, L., Zhang, Y., Yuan, C., Wang, J., Zhuang, T., Lin, X., Jiang, L., Wang, N., Ma, J., Sanguinet, K. A., Liu, B., Rustgi, S., & Ou, X. (2019). Transgenerational memory of gene expression changes induced by heavy metal stress in rice (*Oryza sativa* L.). *BMC Plant Biology*, *19*(1), 1–14. https://doi.org/10.1186/s12870-019-1887-7

Das, K., & Roychoudhury, A. (2014, December). Reactive oxygen species (ROS) and response of antioxidants as ROS-scavengers during environmental stress in plants. *Frontiers in Environmental Science*, *2*, 53. https://doi.org/10.3389/FENVS.2014.00053/BIBTEX

Dass, A., & Sudhishri, S. (2010). Intercropping in fingermillet (*Eleusine coracana*) with pulses for enhanced productivity, resource conservation and soil fertility in uplands of Southern Orissa. *Indian Journal of Agronomy*, *55*(2), 89–94.

Dinesh-Kumar, S. P., Anandalakshmi, R., Marathe, R., Schiff, M., & Liu, Y. (2003). Virus-induced gene silencing. *Methods in Molecular Biology (Clifton, N.J.)*, *236*, 287–294. https://doi.org/10.1385/1-59259-413-1:287

El-Saadony, M. T., Saad, A. M., Najjar, A. A., Alzahrani, S. O., Alkhatib, F. M., Shafi, M. E., Selem, E., Desoky, E. S. M., Fouda, S. E. E., El-Tahan, A. M., & Hassan, M. A. A. (2021). The use of biological selenium nanoparticles to suppress *Triticum aestivum* L. crown and root rot diseases induced by Fusarium species and improve yield under drought and heat stress. *Saudi Journal of Biological Sciences*, *28*(8), 4461–4471. https://doi.org/10.1016/j.sjbs.2021.04.043

Elbeshehy, E. K. F., Elazzazy, A. M., & Aggelis, G. (2015, May). Silver nanoparticles synthesis mediated by new isolates of *Bacillus* spp., nanoparticle characterization and their activity against Bean Yellow Mosaic Virus and human pathogens. *Frontiers in Microbiology*, *6*. https://doi.org/10.3389/FMICB.2015.00453/PDF

Elsheery, N. I., Helaly, M. N., El-Hoseiny, H. M., & Alam-Eldein, S. M. (2020). Zinc oxide and silicone nanoparticles to improve the resistance mechanism and annual productivity of salt-stressed mango trees. *Agronomy*, *10*(4), 558. https://doi.org/10.3390/agronomy10040558

Farajzadeh Memari-Tabrizi, E., Yousefpour-Dokhanieh, A., & Babashpour-Asl, M. (2021). Foliar-applied silicon nanoparticles mitigate cadmium stress through physio-chemical changes to improve growth, antioxidant capacity, and essential oil profile of summer savory (*Satureja hortensis* L.). *Plant Physiology and Biochemistry*, *165*, 71–79. https://doi.org/10.1016/j.plaphy.2021.04.040

Farhangi-Abriz, S., & Torabian, S. (2018). Nano-silicon alters antioxidant activities of soybean seedlings under salt toxicity. *Protoplasma*, *255*(3), 953–962. https://doi.org/10.1007/s00709-017-1202-0

Fire, A., Xu, S., Montgomery, M. K., Kostas, S. A., Driver, S. E., & Mello, C. C. (1998). Potent and specific genetic interference by double-stranded RNA in Caenorhabditis elegans. *Nature*, *391*(6669), 806–811. https://doi.org/10.1038/35888

Giraldo, J. P., Landry, M. P., Faltermeier, S. M., McNicholas, T. P., Iverson, N. M., Boghossian, A. A., Reuel, N. F., Hilmer, A. J., Sen, F., Brew, J. A., & Strano, M. S. (2014). Plant nanobionics approach to augment photosynthesis and biochemical sensing. *Nature Materials*, *13*(4), 400–408. https://doi.org/10.1038/nmat3890

Haghighi, M., Abolghasemi, R., & da Silva, J. A. T. (2014). Low and high temperature stress affect the growth characteristics of tomato in hydroponic culture with Se and nano-Se amendment. *Scientia Horticulturae*, *178*, 231–240. https://doi.org/10.1016/j.scienta.2014.09.006

Hao, Y., Lv, R., Ma, C., Adeel, M., Zhao, Z., Rao, Y., & Rui, Y. (2021). Graphitic carbon nitride (g-C3N4) alleviates cadmium-induced phytotoxicity to rice (*Oryza sativa* L.). *Environmental Science and Pollution Research*, *28*(17), 21276–21284. https://doi.org/10.1007/s11356-020-12027-w

Hartikainen, H., Xue, T., & Piironen, V. (2000). Selenium as an anti-oxidant and pro-oxidant in ryegrass. *Plant and Soil*, *225*(1–2), 193–200. https://doi.org/10.1023/A:1026512921026

Hasanuzzaman, M., Nahar, K., & Fujita, M. (2014). Silicon and selenium: Two vital trace elements that confer abiotic stress tolerance to plants. In *Emerging technologies and management of crop stress tolerance: Biological techniques* (Vol. 1). Elsevier Inc. https://doi.org/10.1016/B978-0-12-800876-8.00016-3

Hezaveh, T. A., Pourakbar, L., Rahmani, F., & Alipour, H. (2019). Interactive effects of salinity and ZnO nanoparticles on physiological and molecular parameters of rapeseed (*Brassica napus* L.). *Communications in Soil Science and Plant Analysis*, *50*(6), 698–715. https://doi.org/10.1080/00103624.2019.1589481

Hnilickova, H., Kraus, K., Vachova, P., & Hnilicka, F. (2021). Salinity stress affects photosynthesis, malondialdehyde formation, and proline content in portulaca oleracea l. *Plants*, *10*(5), 845. https://doi.org/10.3390/plants10050845

Hussein, M. M., & Abou-Baker, N. H. (2018). The contribution of nano-zinc to alleviate salinity stress on cotton plants. *Royal Society Open Science*, *5*(8), 171809. https://doi.org/10.1098/rsos.171809

Iqbal, M. N., Rasheed, R., Ashraf, M. Y., Ashraf, M. A., & Hussain, I. (2018). Exogenously applied zinc and copper mitigate salinity effect in maize (*Zea mays* L.) by improving key physiological and biochemical attributes. *Environmental Science and Pollution Research*, *25*(24), 23883–23896. https://doi.org/10.1007/s11356-018-2383-6

Jeevanandam, J., Barhoum, A., Chan, Y. S., Dufresne, A., & Danquah, M. K. (2018). Review on nanoparticles and nanostructured materials: History, sources, toxicity and regulations. *Beilstein Journal of Nanotechnology*, *9*(1), 1050–1074. https://doi.org/10.3762/bjnano.9.98

Jin, Y., Liu, W., Li, X. L., Shen, S. G., Liang, S. X., Liu, C., & Shan, L. (2016). Nano-hydroxyapatite immobilized lead and enhanced plant growth of ryegrass in a contaminated soil. *Ecological Engineering*, *95*, 25–29. https://doi.org/10.1016/j.ecoleng.2016.06.071

Kareem, H. A., Saleem, M. F., Saleem, S., Rather, S. A., Wani, S. H., Siddiqui, M. H., Alamri, S., Kumar, R., Gaikwad, N. B., Guo, Z., Niu, J., & Wang, Q. (2022, February). Zinc oxide nanoparticles interplay with physiological and biochemical attributes in terminal heat stress alleviation in Mungbean (*Vigna radiata* L.). *Frontiers in Plant Science*, *13*, 1–15. https://doi.org/10.3389/fpls.2022.842349

Kashyap, P. L., Xiang, X., & Heiden, P. (2015). Chitosan nanoparticle based delivery systems for sustainable agriculture. *International Journal of Biological Macromolecules, 77*, 36–51. https://doi.org/10.1016/J.IJBIOMAC.2015.02.039

Khalid, M. F., Huda, S., Yong, M., Li, L., Li, L., Chen, Z. H., & Ahmed, T. (2022). Alleviation of drought and salt stress in vegetables: Crop responses and mitigation strategies. *Plant Growth Regulation, 99*(2), 177–194. https://doi.org/10.1007/s10725-022-00905-x

Khalid, M. F., Hussain, S., Ahmad, S., Ejaz, S., Zakir, I., Ali, M. A., Ahmed, N., & Anjum, M. A. (2019). Impacts of abiotic stresses on growth and development of plants. *Plant Tolerance to Environmental Stress, 2021*, 1–8. https://doi.org/10.1201/9780203705315-1

Khan, M. N. (2016). Nano-titanium dioxide (Nano-TiO_2) mitigates NaCl stress by enhancing antioxidative enzymes and accumulation of compatible solutes in tomato (*Lycopersicon esculentum* Mill.). *Journal of Plant Sciences, 11*(1), 1–11. https://doi.org/10.3923/jps.2016.1.11

Khan, Z. S., Rizwan, M., Hafeez, M., Ali, S., Adrees, M., Qayyum, M. F., Khalid, S., ur Rehman, M. Z., & Sarwar, M. A. (2019). Effects of silicon nanoparticles on growth and physiology of wheat in cadmium contaminated soil under different soil moisture levels. *Environmental Science and Pollution Research, 27*(5), 4958–4968. https://doi.org/10.1007/s11356-019-06673-y

Khan, Z. S., Rizwan, M., Hafeez, M., Ali, S., Javed, M. R., & Adrees, M. (2019). The accumulation of cadmium in wheat (*Triticum aestivum*) as influenced by zinc oxide nanoparticles and soil moisture conditions. *Environmental Science and Pollution Research, 26*(19), 19859–19870. https://doi.org/10.1007/s11356-019-05333-5

Kim, D., Shin, K., Kwon, S. G., & Hyeon, T. (2018). Synthesis and biomedical applications of multifunctional nanoparticles. *Advanced Materials, 30*(49), 1802309. https://doi.org/10.1002/ADMA.201802309

Kopittke, P. M., Lombi, E., Wang, P., Schjoerring, J. K., & Husted, S. (2019). Nanomaterials as fertilizers for improving plant mineral nutrition and environmental outcomes. *Environmental Science: Nano, 6*(12), 3513–3524. https://doi.org/10.1039/c9en00971j

Malerba, M., & Cerana, R. (2016). Chitosan effects on plant systems. *International Journal of Molecular Sciences, 17*, 996. https://doi.org/10.3390/IJMS17070996

Mirakhorli, T., Ardebili, Z. O., Ladan-Moghadam, A., & Danaee, E. (2021, September). Bulk and nanoparticles of zinc oxide exerted their beneficial effects by conferring modifications in transcription factors, histone deacetylase, carbon and nitrogen assimilation, antioxidant biomarkers, and secondary metabolism in soybean. *PLoS One, 16*(9), 1–10. https://doi.org/10.1371/journal.pone.0256905

Mittapally, S., Taranum, R., & Parveen, S. (2018). Metal ions as antibacterial agents. *Journal of Drug Delivery and Therapeutics, 8*(6), 411–419. https://doi.org/10.22270/JDDT.V8I6-S.2063

Mitter, N., Worrall, E. A., Robinson, K. E., Li, P., Jain, R. G., Taochy, C., Fletcher, S. J., Carroll, B. J., Lu, G. Q., & Xu, Z.P. (2017). Clay nanosheets for topical delivery of RNAi for sustained protection against plant viruses. *Nature Plants, 3*, 1–10. https://doi.org/10.1038/NPLANTS.2016.207

Munns, R., & Tester, M. (2008). Mechanisms of salinity tolerance. *Annual Review of Plant Biology, 59*, 651–681. https://doi.org/10.1146/annurev.arplant.59.032607.092911

Pandey, P., Irulappan, V., Bagavathiannan, M. V., & Senthil-Kumar, M. (2017). Impact of combined abiotic and biotic stresses on plant growth and avenues for crop improvement by exploiting physio-morphological traits. *Frontiers in Plant Science, 8*, 537. https://doi.org/10.3389/FPLS.2017.00537/BIBTEX

Pariona, N., Paraguay-Delgado, F., Basurto-Cereceda, S., Morales-Mendoza, J. E., Hermida-Montero, L. A., & Mtz-Enriquez, A. I. (2020). Shape-dependent antifungal activity of ZnO particles against phytopathogenic fungi. *Applied Nanoscience (Switzerland), 10*(2), 435–443. https://doi.org/10.1007/S13204-019-01127-W/METRICS

Paul, W., & Sharma, C. P. (2010). Fatty acid conjugated calcium phosphate nanoparticles for protein delivery. *International Journal of Applied Ceramic Technology*, *7*(2), 129–138. https://doi.org/10.1111/J.1744-7402.2009.02442.X

Pérez-Labrada, F., López-Vargas, E. R., Ortega-Ortiz, H., Cadenas-Pliego, G., Benavides-Mendoza, A., & Juárez-Maldonado, A. (2019). Responses of tomato plants under saline stress to. *Plants*, *8*(6), 1–17.

Prasad, P. V. V., Pisipati, S. R., Momčilović, I., & Ristic, Z. (2011). Independent and combined effects of high temperature and drought stress during grain filling on plant yield and chloroplast EF-Tu expression in spring wheat. *Journal of Agronomy and Crop Science*, *197*(6), 430–441. https://doi.org/10.1111/j.1439-037X.2011.00477.x

Qi, M., Liu, Y., & Li, T. (2013). Nano-TiO_2 improve the photosynthesis of tomato leaves under mild heat stress. *Biological Trace Element Research*, *156*(1–3), 323–328. https://doi.org/10.1007/s12011-013-9833-2

Rajput, V. D., Minkina, T., Kumari, A., Harish, Singh, V. K., Verma, K. K., Mandzhieva, S., Sushkova, S., Srivastava, S., & Keswani, C. (2021). Coping with the challenges of abiotic stress in plants: New dimensions in the field application of nanoparticles. *Plants*, *10*(6), 1–25. https://doi.org/10.3390/plants10061221

Rana, D. S., Dass, A., Rajanna, G. A., & Kaur, R. (2016). Biotic and abiotic stress management in pulses. *Indian Journal of Agronomy*, *61*, 238–248. https://www.researchgate.net/publication/311225970_Biotic_and_abiotic_stress_management_in_pulses

Rizwan, M., Ali, S., Ali, B., Adrees, M., Arshad, M., Hussain, A., Zia ur Rehman, M., & Waris, A. A. (2019). Zinc and iron oxide nanoparticles improved the plant growth and reduced the oxidative stress and cadmium concentration in wheat. *Chemosphere*, *214*, 269–277. https://doi.org/10.1016/j.chemosphere.2018.09.120

Roca, M., & Haes, A. J. (2008). Probing cells with noble metal nanoparticle aggregates. *Nanomedicine (London, England)*, *3*(4), 555–565. https://doi.org/10.2217/17435889.3.4.555

Rosi, N. L., Giljohann, D. A., Thaxton, C. S., Lytton-Jean, A. K. R., Han, M. S., & Mirkin, C. A. (2006). Oligonucleotide-modified gold nanoparticles for intracellular gene regulation. *Science (New York, N.Y.)*, *312*(5776), 1027–1030. https://doi.org/10.1126/SCIENCE.1125559

Sadeghi, M., Babaeian, E., Tuller, M., & Jones, S. B. (2017). The optical trapezoid model: A novel approach to remote sensing of soil moisture applied to Sentinel-2 and Landsat-8 observations. *Remote Sensing of Environment*, *198*, 52–68. https://doi.org/10.1016/J.RSE.2017.05.041

Saffie-Siebert, S., Guerriero, G., Maria Sutera, F., Torabi-Pour, N., Renaut, J., Hausman, J. F., Berni, R., Pennington, H. C., Welsh, M., Dehsorkhi, A., & Zancan, L. R. (2021). Phyto-courier, a silicon particle-based nanobiostimulant: Evidence from cannabis sativa exposed to salinity. *ACS Nano*, *15*(2), 3061–3069. https://doi.org/10.1021/acsnano.0c09488

Sardar, R., Ahmed, S., & Yasin, N. A. (2022). Titanium dioxide nanoparticles mitigate cadmium toxicity in *Coriandrum sativum* L. through modulating antioxidant system, stress markers and reducing cadmium uptake. *Environmental Pollution*, *292*(Part A), 34662592. https://doi.org/10.1016/j.envpol.2021.118373

Savicka, M., & Škute, N. (2010). Effects of high temperature on malondialdehyde content, superoxide production and growth changes in wheat seedlings (*Triticum aestivum* L.). *Ekologija*, *56*(1), 26–33. https://doi.org/10.2478/v10055-010-0004-x

Shahi, S., & Srivastava, M. (2018). Influence of foliar application of manganese on growth, pigment content, and nitrate reductase activity of *Vigna radiata* (L.) R. Wilczek under salinity. *Journal of Plant Nutrition*, *41*(11), 1397–1404. https://doi.org/10.1080/01904167.2018.1454470

Sharma, S., Singh, V. K., Kumar, A., & Mallubhotla, S. (2019). Effect of nanoparticles on oxidative damage and antioxidant defense system in plants. *Molecular Plant Abiotic Stress: Biology and Biotechnology*, 315–333. https://doi.org/10.1002/9781119463665.CH17

Siddiqi, K. S., & Husen, A. (2016). Engineered gold nanoparticles and plant adaptation potential. *Nanoscale Research Letters*, *11*(1), 1–10. https://doi.org/10.1186/s11671-016-1607-2

Siddiqui, H., Ahmed, K. B. M., Sami, F., & Hayat, S. (2020). Silicon nanoparticles and plants: Current knowledge and future perspectives, 129–142. https://doi.org/10.1007/978-3-030-33996-8_7

Singh, J., & Lee, B. K. (2016). Influence of nano-TiO_2 particles on the bioaccumulation of Cd in soybean plants (Glycine max): A possible mechanism for the removal of Cd from the contaminated soil. *Journal of Environmental Management*, *170*, 88–96. https://doi.org/10.1016/j.jenvman.2016.01.015

Singh, P., Arif, Y., Siddiqui, H., Sami, F., Zaidi, R., Azam, A., Alam, P., & Hayat, S. (2021). Nanoparticles enhances the salinity toxicity tolerance in *Linum usitatissimum* L. by modulating the antioxidative enzymes, photosynthetic efficiency, redox status and cellular damage. *Ecotoxicology and Environmental Safety*, *213*, 112020. https://doi.org/10.1016/j.ecoenv.2021.112020

Song, Y., Jiang, M., Zhang, H., & Li, R. (2021). Zinc oxide nanoparticles alleviate chilling stress in rice (*Oryza sativa* L.) by regulating antioxidative system and chilling response transcription factors. *Molecules*, *26*(8), 2196. https://doi.org/10.3390/molecules26082196

Tahjib-Ui-Arif, M., Sohag, A. A. M., Afrin, S., Bashar, K. K., Afrin, T., Sofi Uddin Mahamud, A. G. M., Polash, M. A. S., Hossain, M. T., Sohel, M. A. T., Brestic, M., & Murata, Y. (2019). Differential response of sugar beet to long-term mild to severe salinity in a soil–pot culture. *Agriculture (Switzerland)*, *9*(10), 223. https://doi.org/10.3390/agriculture9100223

Tang, J. W., Lai, F. Y. L., Nymadawa, P., Deng, Y. M., Ratnamohan, M., Petric, M., Loh, T. P., Tee, N. W. S., Dwyer, D. E., Barr, I. G., & Wong, F. Y. W. (2010). Comparison of the incidence of influenza in relation to climate factors during 2000–2007 in five countries. *Journal of Medical Virology*, *82*(11), 1958–1965. https://doi.org/10.1002/JMV.21892

Thakur, S., Asthir, B., Kaur, G., Kalia, A., & Sharma, A. (2022). Zinc oxide and titanium dioxide nanoparticles influence heat stress tolerance mediated by antioxidant defense system in wheat. *Cereal Research Communications*, *50*(3), 385–396. https://doi.org/10.1007/s42976-021-00190-w

Thiruvengadam, M., Gurunathan, S., & Chung, I. M. (2015). Physiological, metabolic, and transcriptional effects of biologically-synthesized silver nanoparticles in turnip (*Brassica rapa* ssp. rapa L.). *Protoplasma*, *252*(4), 1031–1046. https://doi.org/10.1007/s00709-014-0738-5

Thomashow, M. F. (1999). Plant cold acclimation: Freezing tolerance genes and regulatory mechanisms. *Annual Review of Plant Biology*, *50*, 571–599. https://doi.org/10.1146/annurev.arplant.50.1.571

Tian, W., He, G., Qin, L., Li, D., Meng, L., Huang, Y., & He, T. (2021). Genome-wide analysis of the NRAMP gene family in potato (*Solanum tuberosum*): Identification, expression analysis and response to five heavy metals stress. *Ecotoxicology and Environmental Safety*, *208*, 111661. https://doi.org/10.1016/j.ecoenv.2020.111661

Tolaymat, T., Genaidy, A., Abdelraheem, W., Dionysiou, D., & Andersen, C. (2017). The effects of metallic engineered nanoparticles upon plant systems: An analytic examination of scientific evidence. *Science of the Total Environment*, *579*, 93–106. https://doi.org/10.1016/j.scitotenv.2016.10.229

Tripathi, D. K., Singh, V. P., Prasad, S. M., Chauhan, D. K., & Dubey, N. K. (2015). Silicon nanoparticles (SiNp) alleviate chromium (VI) phytotoxicity in Pisum sativum (L.) seedlings. *Plant Physiology and Biochemistry*, *96*(VI), 189–198. https://doi.org/10.1016/j.plaphy.2015.07.026

uz Zaman, Q., Abbasi, A., Tabassum, S., Ashraf, K., Ahmad, Z., Siddiqui, M. H., Alamri, S., Maqsood, S., & Sultan, K. (2022). Calcium induced growth, physio-biochemical, antioxidants, osmolytes adjustments and phytoconstituents status in spinach under heat stress. *South African Journal of Botany*, *149*, 701–711. https://doi.org/10.1016/j.sajb.2022.06.065

Vanwallendael, A., Soltani, A., Emery, N. C., Peixoto, M. M., Olsen, J., & Lowry, D. B. (2019). A molecular view of plant local adaptation: Incorporating stress-response networks. *Annual Review of Plant Biology*, *70*, 559–583. https://doi.org/10.1146/annurev-arplant-050718-100114

Vlot, A. C., Liu, P. P., Cameron, R. K., Park, S. W., Yang, Y., Kumar, D., Zhou, F., Padukkavidana, T., Gustafsson, C., Pichersky, E., & Klessig, D. F. (2008). Identification of likely orthologs of tobacco salicylic acid-binding protein 2 and their role in systemic acquired resistance in Arabidopsis thaliana. *Plant Journal*, *56*(3), 445–456. https://doi.org/10.1111/j.1365-313X.2008.03618.x

Wang, P., Lombi, E., Zhao, F. J., & Kopittke, P. M. (2016). Nanotechnology: A new opportunity in plant sciences. *Trends in Plant Science*, *21*(8), 699–712. https://doi.org/10.1016/j.tplants.2016.04.005

Wu, J., & Wang, T. (2020). Synergistic effect of zinc oxide nanoparticles and heat stress on the alleviation of transcriptional gene silencing in arabidopsis thaliana. *Bulletin of Environmental Contamination and Toxicology*, *104*(1), 49–56. https://doi.org/10.1007/s00128-019-02749-0

Yassen, A., Abdallah, E., Gaballah, M., & Zaghloul, S. (2017). Role of silicon dioxide nano fertilizer in mitigating salt stress on growth, yield and chemical composition of cucumber (*Cucumis sativus* L.). *International Journal of Agricultural Research*, *12*(3), 130–135. https://doi.org/10.3923/ijar.2017.130.135

Ye, Y., Cota-Ruiz, K., Hernández-Viezcas, J. A., Valdés, C., Medina-Velo, I. A., Turley, R. S., Peralta-Videa, J. R., & Gardea-Torresdey, J. L. (2020). Manganese nanoparticles control salinity-modulated molecular responses in *Capsicum annuum* L. through priming: A sustainable approach for agriculture. *ACS Sustainable Chemistry and Engineering*, *8*(3), 1427–1436. https://doi.org/10.1021/acssuschemeng.9b05615

Younis, A. A., Khattab, H., & Emam, M. M. (2020). Impacts of silicon and silicon nanoparticles on leaf ultrastructure and tapip1 and tanip2 gene expressions in heat stressed wheat seedlings. *Biologia Plantarum*, *64*(1), 343–352. https://doi.org/10.32615/bp.2020.030

Zare, E., Pourseyedi, S., Khatami, M., & Darezereshki, E. (2017). Simple biosynthesis of zinc oxide nanoparticles using nature's source, and it's in vitro bio-activity. *Journal of Molecular Structure*, *1146*, 96–103. https://doi.org/10.1016/J.MOLSTRUC.2017.05.118

Zhao, L., Huang, Y., Hu, J., Zhou, H., Adeleye, A. S., & Keller, A. A. (2016). 1H NMR and GC-MS based metabolomics reveal defense and detoxification mechanism of cucumber plant under nano-Cu stress. *Environmental Science and Technology*, *50*(4), 2000–2010. https://doi.org/10.1021/acs.est.5b05011

Zhou, D., Jin, S., Li, L., Wang, Y., & Weng, N. (2011). Quantifying the adsorption and uptake of CuO nanoparticles by wheat root based on chemical extractions. *Journal of Environmental Sciences*, *23*(11), 1852–1857. https://doi.org/10.1016/S1001-0742(10)60646-8

5 Nanoparticles in Soil–Water–Plant Inference

*Binaya Kumar Pattnaik, Chandan Sahu and Debojyoti Moulick**

5.1 INTRODUCTION

With every passing day, crops are experiencing a wide arrange of stressors, i.e. biotic/abiotic stresses either in combination or alone. Stresses are known to be the biggest obstacles to attaining sustainability in agriculture (Choudhury & Moulick, 2022; Choudhury et al., 2022; Ghosh et al., 2020a, 2020b, 2022a, 2022b; Moulick et al., 2021). In order to achieve sustainability in terms of agricultural production and to ameliorate stress-induced adversities, numerous strategies have been tried and tested (Choudhury et al., 2021a, 2021b; Hossain et al., 2021a; Hossain et al., 2021b, 2021c, 2022; Mazumder et al., 2022; Saha et al., 2019; Sahoo et al., 2019). The advancement of nanotechnology has initiated a number of applications in various fields. Such developments intrigued scientists around the globe to understand their properties and their probable implications. Nanotechnology-based agricultural practices can provide solutions to achieve sustainable agricultural targets. Nanomaterials have several potential agricultural uses to increase crop yield and improve soil health through nanofertilizers, nanopesticides, nanobiosensors and soil remediation by nanomaterials (Usman et al., 2020). This leads to its extensive application in agriculture, which put nanoparticles' application under scrutiny.

Nanoparticles (NPs) are described as particles having three dimensions of less than 100 nm, while nanomaterials have one dimension less than 100 nm, nano-objects such as carbon nanotubes have two dimensions less than 100 nm (Stone et al., 2010; SCENIHR, 2007). In general, when the size of particles is reduced to the nanoscale, they display drastically altered or improved characteristics compared to their bigger counterparts. This alteration in behaviour is the result of two primary factors: a substantial increase in surface area and the predominance of quantum effects. As the size of a particle decreases, a higher proportion of atoms and molecules are exposed at its surface. Consequently, their surface qualities become increasingly significant and even predominate, whereas their mass property becomes significantly less significant. Shin et al. (2016) highlighted the three layers of nanoparticle structural composition, which justify its significant characteristics; these are (i) the top layer, which may be modified

* Corresponding author.

DOI: 10.1201/9781003376446-5

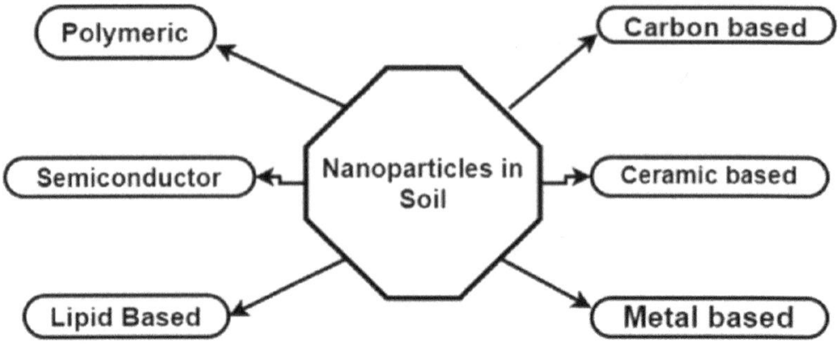

FIGURE 5.1 Types of nanoparticles in soil based on chemical composition

with chemicals like surfactants, polymers and metal ions, (ii) the shell layer, which in all respects is composed of chemically distinct material than the core and (iii) the core, which is effectively the centre of the NP. Based on the micro-structure, three different forms of nanoparticles of soil have been identified as nanowires, nanoplatelets and nanodots (Zhang, 2007). Nanomaterials can be categorized not only by size but also by composition like carbon, ceramic (metal oxides), metal, lipid-based, semiconductor and polymeric compounds (Paramo et al., 2020) (**Figure 5.1**). The surface chemistry of even the smallest nanoparticle will be different from that of the core substance. One of the reasons for the remarkable features of nanoparticles is their very high surface area to volume ratio (Christian et al., 2008). High reactivity and poor stability are concerns of the application of nanoparticles (Xu et al., 2018). Nanoparticles often make their way into agricultural soils through the use of untreated wastewater for irrigation or by the use of biosolids as fertilizer (Rawat et al., 2018). The presence of nanoparticles in sewage sludge (Wang et al., 2012b), wastewater effluents and landfill chelates indicates that nanoparticles do not remain bound to the products at the end of their life cycle. Various sources of nanoparticles in soil are shown in **Figure 5.2**.

Soil includes several solutes and colloidal components, including dissolved organic carbon, which may encapsulate NPs, influence their interactions with soil surfaces and eventually interfere with the absorption and other aspects of plant exposure to NPs (Coutris et al., 2012). Nanoparticles interact with plant molecules when they enter plant cells, either through endocytosis or other transport systems, and accumulate inside plant tissue, which eventually leads to alteration of morpho-anatomical features and many physiological activities of plants (Ahmed et al., 2021). The potential advantages of nanoparticles (NPs) are tremendous, and they are still being investigated. Investigation of NPs level in soil will not provide full disclosure of its bioavailability unless it covers further movement of NPs into the plant system. Therefore, the current study focusses on the intrinsic characteristics of nanoparticles, and its influences on the soil indigenous properties and vice-versa. A detailed investigation has been done to understand the NP-plant interaction through the evaluation of the mechanism, movement and factors affecting movements of NPs in plant cell. This study will also help the readers to assess the biosynthetic NPs and their impacts on plants.

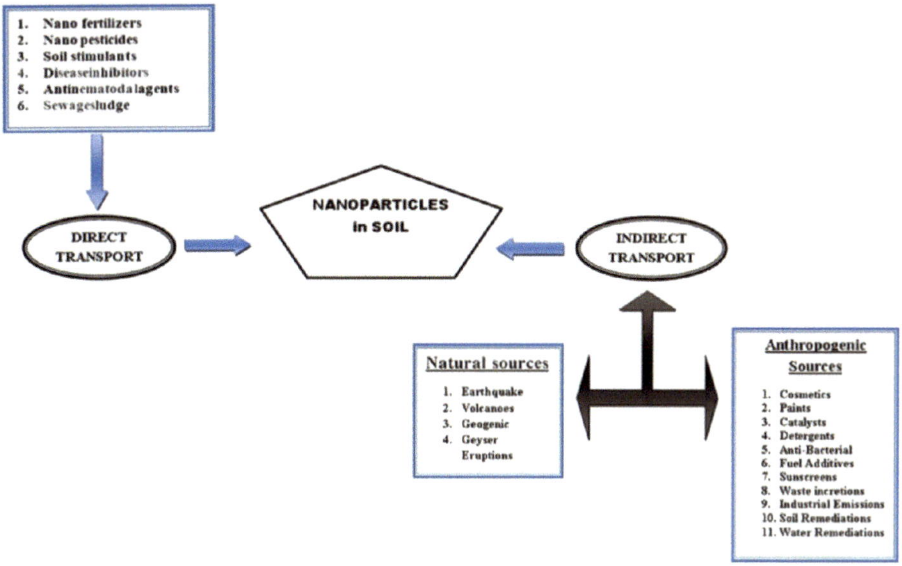

FIGURE 5.2 Sources of nanoparticles in soil

5.2 SOIL PROPERTIES AND NPs

5.2.1 Nanoporosity

In the soil ecosystem, nanoparticles interact with charged surfaces of clay particles and humic substances both in the solid and liquid phase, which augment the complexity of NPs in the context of their stability. Reactive charges surface in addition to the high specific surface area of NPs have a significant impact on soil physical, chemical and engineering properties of soil (Zhang, 2007). Nanoparticles with intraparticle space of nanosize can introduce nanoporosity in soil, which not only reduces the bulk density but also increases the sorption capabilities and surface area. At the nanoscale level, nanoplatelets have just one dimension. As with kaolinite, illite, smectite, chlorite and vermiculite, the majority of clay mineral particles are platy in shape. Nanoplatelets like smectite can host hydrated primary minerals like Na, K, Ca and Mg in the interlayer space apart from initiating nanoporosity. Smectitie surface charge can be attributed to negative permanent charges owing to isomorphism substitution on face surfaces and variable pH-dependent charges due to broken atomic bonds on edge surfaces. Soils rich in these nanoparticles often exhibit an aggregated structure with inter- and intra-aggregate space. Fe-oxide NPs prefer to interact with clay particles by cementation or electrostatic attraction methods to produce silt-sized aggregates (Cornell & Schwertmann, 2003). Through surface attraction and weaving, fibrous nanoparticles often initiate and accelerate the generation of three-dimensional networks of porous aggregates non-responsive to chemical treatments and resistant to dispersion. Soils high in nanoparticles with nanoporosity, such as imogolite, allophane and halloysite, tend to have a lower bulk density than the average granular soil which is attributed to the high porosity of micro-aggregates and intraparticle-nanoporosity. It is well-established that soils rich in smectite, halloysite and allophane have a high water-holding capacity, as indicated by the natural water content, liquid and plastic limitations.

5.2.2 Hydraulic Conductivities

Nanoparticles' charged surfaces (internal and exterior) are frequently coupled with hydrated cations and/or electrical double layers, and the huge surface area entails considerable interface contact. Hydrated cations contribute to oven-dried water content and Atterberg limitations (i.e., 105–110°C). Smectite clays have greater liquid and plastic limitations than kaolinites. Intra-particle water contributes to high natural water content and Atterberg limits (Zhang et al., 2004). On heating or drying (even moderate air-drying), these nanoparticles with nanoporosity often undergo irreversible loss of intra-particle water. Typically, soils containing nanoparticles have an aggregated structure with two distinct porosities: larger inter-aggregate pores and smaller intra-aggregate pores. In terms of variations in stress levels and pore water chemistry, this micro-structure has substantial effects on hydraulic conductivities.

5.2.3 Agglomeration/Aggregation

According to Stone et al. (2010), the primary NP characteristics that should be assessed in eco-toxicological investigations to characterize exposure include agglomeration/aggregation size, surface area and charge, dissolving rate, dispersibility and surface chemistry. These characteristics are the key features in the stability and mobilization of nanoparticles, which in turn influences NPs' bioavailability and toxicity (Ogunkunle et al., 2021; Tourinho et al., 2012). Aggregation is the interaction of primary particles by strong bonding, but agglomeration is the association of primary particles by weak bonding driven by Vander Waals forces (Jiang et al., 2009) (**Figure 5.3**). Physical factors, such as fluid motion, Brownian motion and gravity, as well as NP features such as surface properties and particle size, will have an impact on the agglomeration and aggregation of NPs in the environment (Farré et al., 2009). Particles can only aggregate if their centres come into contact with one another, hence the rate of aggregation is thought to be related to the likelihood of collision between particles (Rosická & Šembera, 2010). The stability of the electric double layer of nanoparticles shows an inverse relationship with the ionic strength of the medium. Particle association increases due to a reduction in the thickness of this electric double layer which resulted due to increased ionic strength (Badawy et al., 2010). Electrostatic repulsion between anions and the negatively charged surface of the nanoparticle determines the stability of the nanoparticles. Magnetic and hydrophobic interactions, as well as hydration force, greatly impact nanoparticle aggregation (Dwivedi et al., 2015). Identical nanoparticles aggregate to form homogeneous aggregation whereas heterogeneous aggregation is the association of nanoparticles having distinct characteristics and features (Ahmed et al., 2021).

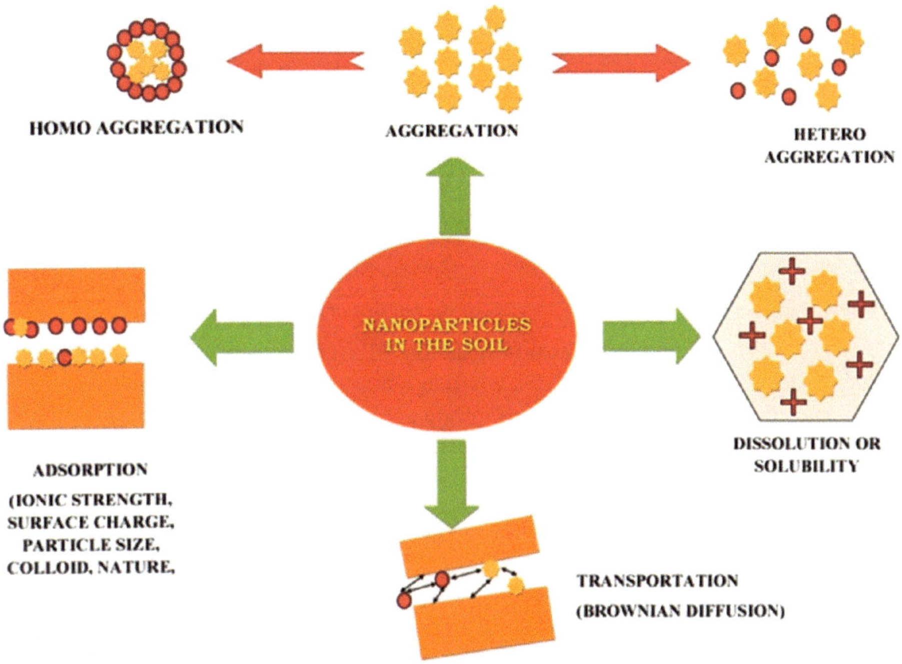

FIGURE 5.3 Aggregation and agglomeration of nanoparticles in soil

5.2.4 Surface Coating

The chemistry of the medium will affect the electrostatic surface charge of the particles, consequently influencing particle stability and rate of agglomeration/aggregation. Hydroxyl ions can charge the surface of the nanoparticle, which results in the adsorption of dissolved ligands and metal ions. Surface charging leads to the creation of an electrical double layer, which consists of the charged surface sites and a diffuse layer containing ions drawn from the solution to the particle surface by the charge. The electrical potential at the interface of the diffuse layer and the bulk solution (the zeta potential) is measurable, and its fluctuation with solution chemistry can effectively stand in for the variation in particle surface charge with solution chemistry. At high pH, the zeta potential of uncoated metal or metal oxide particles drops from positive values to negative values. As pH approaches the isoelectric point, also known as the "point of zero charge", where particle charge/zeta potential approaches zero, the aggregation rate will rise (Dunphy Guzman et al., 2006). The stability of the coating layers is very crucial for preserving the intrinsic characteristics of NPs. As per Derjaguin-Landau-Verwey-Overbeek theory, NPs' stability is dependent on the Vander Waals force (i.e attraction) and electrostatic force (i.e repulsion) (Ju-Nam & Lead, 2008; Stebounova et al., 2011).

5.2.5 Dissolution

In the absence of water-insoluble surface coating (which acts as kinetic constraints against dissolution), NPs undergo chemical disintegration due to unstable thermodynamic conditions. Under such a scenario, ions are separated from NPs and are dissolved in the solution by breaching through the protective electrical double layers of NPs (Borm et al., 2006). Shoults-Wilson et al. (2011) observed the release of 10 to 17% Ag-NPs as ionic species of Ag into the solution due to oxidative dissolution.

5.2.6 Soil pH

Soil pH conditions have a profound impact on NPs' behaviour. Alkaline condition in soil favours the aggregation of NPs, whereas a high rate of conversion of NPs to their ionic species was observed at low pH (Zhou et al., 2016). Aggregate formation under alkaline conditions may siege nanoparticle-specific properties and limit the capacity of dissolution (Dimkpa, 2018). Soil pH also influences the surface charge of the NPs in soil. In alkaline conditions, TiO_2 NP maintains its stability with a high negative charged surface, whereas a positive surface charge was observed in acidic conditions (Loosli & Stoll, 2017). Aggregation initiated by soil pH restricts the NP mobilization in soil, which influences NP fate in a biological system (Anderson et al., 2017).

5.2.7 Nanoparticles' Stability in Soil

Negatively charged organic matter form a conglomerate with nanoparticles, which restrict the bioavailability and uptake of the metal nanoparticle by plant cell (Unrine et al., 2008). Such conglomeration not only reduces aggregate formation but also restricts precipitation, thus enhancing the stability of the nanoparticle (Ju-Nam & Lead, 2008). At the pH point of zero charges (ZPC) or higher, the addition of Humic acids (HAs) significantly increased the colloidal stability of nanoparticles, but under acidic conditions, humic acid-containing NPs exhibited considerable aggregation. The hydrophobic character of the HA molecules had a significant effect on the aggregation of colloidal NPs, probably because of their conformational behaviour of HAs in a particular solution state (Ghosh et al., 2008). In an experimental study, Peyrot et al. (2014) observed that organic matter availability in the soil increased the soil pH (more than pH 5), which resulted in the release of ionic AgNPs which have more potential to be sorbed onto the organic matter surface due to their positive charge of Ag. Similarly, Hortin et al. (2020) also explained the role of dissolved organic matter in the dissolution of CuNPs in alkaline soil. In the case of flooded conditions in paddy soil, high soil organic matter in soil restricts the release of dissolved Ag through adsorption on soil particles, whereas low soil pH favours the increase in dissolved species of Ag (Li et al., 2017).

5.2.8 Heavy Metal Remediation

The size, surface area and reactivity of nanoparticles make them a promising candidate for use in the remediation of metal-contaminated soil. Processes like stabilization, conversion, redox reaction and adsorption give significant results in this regard. Adsorptive nanomaterials provide a large surface area, which favours the enhanced adsorption kinetics for heavy metals (Shan et al., 2009); on the other hand, reactive nanomaterials employ processes like photocatalysis, precipitation/dissolution, acid-base reaction, redox reaction and ion exchange (Wei et al., 2013).

Non-toxic zero-valent iron (ZVI) nanoparticles are abundantly available on earth. These nanoparticles utilize adsorption and reduction proceseses to counter heavy metals like cadmium (Cd), mercury (Hg) and nickel (Ni) (Li et al., 2006; Noubactep et al., 2012). At specific redox potential (Eh), ZVI reacts with heavy metals to immobilize or convert metals to non-toxic form (Cundy et al., 2008). A major limitation of its application is that it gets easily oxidized in the presence of air due to its highly reactive and enhanced surface area (Rabbani et al., 2016).

Iron sulphide (FeS) NPs combined with carboxymethyl cellulose (CMC) can be utilized in the immobilization of mercury (Hg) in the soil through adsorption, structural incorporation and precipitation (Gong et al., 2012). Iron phosphate nanoparticles trap the copper (Cu) in the soil through adsorption and precipitation. Such phosphate nanoparticles do not create any secondary complexes which can cause environmental hazards (Li & Zhang, 2007). Hydrous aluminosilicate NPs or allophanes are very much effective in the removal of Cu in soil. Allophanes utilize mechanisms like cation exchange reactions and specific compound complexation to alleviate Cu levels in soil (Yuan, 2004).

Almaroai et al. (2014) observed the enhanced adsorption and high binding capacity of Zirconium-iron oxide (Zr-FeO) NPs, which showed prominent results in the remediation study of arsenic (As) contaminated soil. Surface-modified carbon black NPs demonstrate better affinity to heavy metals in the soil environment (Zhou et al., 2010). Cheng et al. (2014) conducted a laboratory-based remediation experiment and observed that carbon black NPs reduced the concentration level of Cu and Zn in soil by complexation and adsorption mechanisms.

Nanoparticles modify the soil properties in various dimensions at the nanoscale level. Such a modified soil environment further influences its fate in the soil-plant system. Due to its ease of transport, compact nanosize, longer self-sustaining capacity and high efficiency, it is an appropriate choice for sustainable agriculture practices.

5.3 NP-PLANT INTERACTION

The NP-plant interaction (**Figure 5.4**) is guided basically by the route of NP entry into the plant. The NPs may be deposited in the aerial parts of the plants (primarily sourced from atmospheric deposition or foliar spray) or may become part of the soil system and enter through the underground route by root uptake. When entering through the aerial deposit, the size and morphology of the NPs play a pivotal role. Since the stomatal pore and guard cells are exposed to the NPs interaction, the physiological process of plants viz. the opening and closing of stomata, clogging of the stomatal pore, the chemistry of the NPs (especially reactivity), light intensity regulating opening and closing of stomata are other amendable criteria. The cell

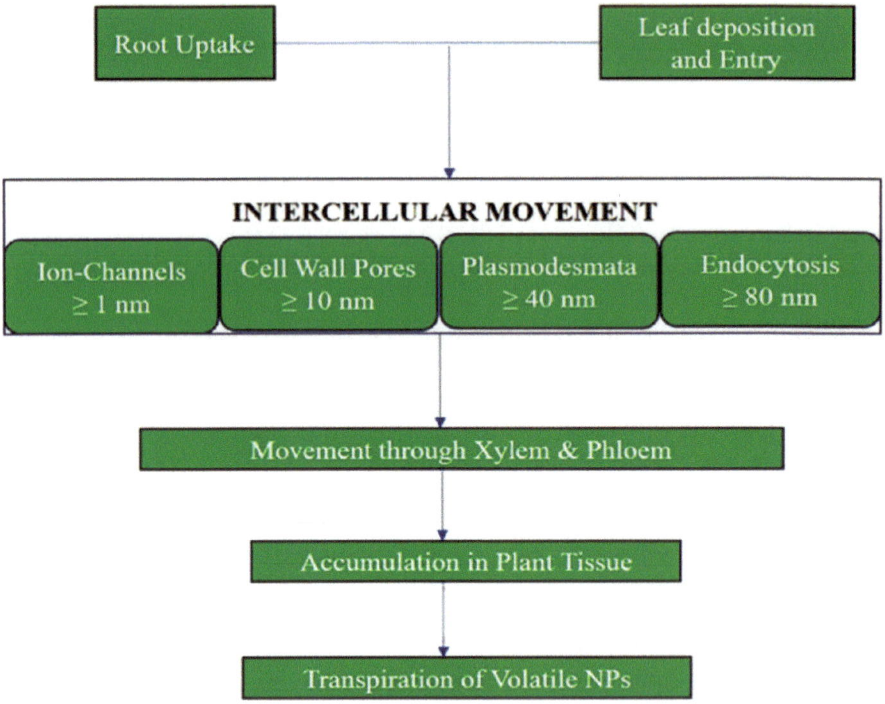

FIGURE 5.4 General schema of NP-plant interaction

wall, being negatively charged, has the ability to bind to positively charged NPs. Therefore, the charge of NPs also determines the NP-plant interaction.

When entering through the underground route, the adsorption of the NP by the roots is the first step for NP accumulation in soil. Some NPs are easily transformed in the soil with variable soil properties and soil chemistry. These transformations lead to the formation of various chemical complexes altering the initial NPs' chemistry. Such modifications may lead to synergistic or antagonistic effects on subsequent entry into the plant system. In addition, a certain type of NP has the tendency to form homogenous aggregates (with other molecules of the same NPs) or heterogeneous aggregates (with other NPs or soil surface). This aggregate formation may result in the reduction of their diffusion into the soil or hamper their reactivity potential. On subsequent entry into the plant body, NPs are translocated to different parts which are again dependent on multiple factors including the vesicles and hydrophobicity of the plant cell. Reports suggest that the effect of NPs (both positive and negative) on plants depends on several factors like soil properties, route of entry, NP properties, plant type and response (Perez-de-Luque, 2017). A proper understanding of these factors and the mechanism of translocation, nevertheless, can facilitate a better understanding of NP-plant interaction and benefit the cause of agriculture.

5.3.1 Factors Affecting NP-plant Interaction

5.3.1.1 Chemical Nature of NPs

NPs have a substantial effect on plants due to their massive surface area, shape, surface charge and charge: mass ratio. This NP chemistry determines the type of transporter complex formed with the membrane during NP entry into the plants. The solubility of NPs further determines their availability and interaction with the plant. For instance, ZnO has been reported to dissolve more quickly in acidic soil than in neutral or alkaline soil (Lv et al., 2022).

5.3.1.2 Size of NPs

The smaller-sized NPs with a diameter between 3 to 5 nm can penetrate the roots of the plants under osmotic pressure or cross the epidermal cells. Since the epidermal cells have small pores, they behave as a semi-permeable layer allowing the entry of small-sized NPs while restricting the larger ones. Some NPs have the tendency to bring new pores while making an entry into the epidermal cell. NPs entering aerially through cuticle are restricted to sizes within 5 nm while those entering through stomata may be greater than 10 nm. NPs between 10 and 50 nm have been reported to translocate through the symplastic route, whereas, NPs between 50 and 200 nm are translocated through the apoplastic route. The size of the cell wall pore regulates the entry of NPs into the plant cells. It has also been reported that NPs of 40–50 nm can penetrate the cell from the surface of the plant. It is therefore evident that the smaller-sized NPs find easy entry into the plant body via the root (soil) medium while the larger-sized NPs find an entry via the stomata, hydathodes and flower stigma through aerial deposits.

5.3.1.3 Reactivity of NPs

The reactivity of NPs orients the concentration of particles exposed or entering the plant cell. Positively charged NPs are better adhered to the negatively charged cell wall surface (through electrostatic attraction), thereby experiencing a restricted entry into the cell wall. Similarly, negatively charged NPs experience a possibility of repulsion by the cell wall. The mucilaginous layer secreted by the cells also has a defined role in attaching to the NPs by creating a negatively charged environment. Hydrophilicity and hydrophobicity of NPs are equally important parameters that guide the passage and entry through different plant barriers. The lipophilic NPs accumulate on the cell wall and become a part of the channelized movement through apoplast facilitated by pectins. On the other hand, the hydrophilic NPs bypass the polar channels and move through the cuticles.

5.3.1.4 Concentration of NP

The amount of NPs exposed or available for plant uptake is chiefly determined by the nature of the application (i.e. through the solution, spray, powdered solid etc.) and duration of exposure. NPs applied to the aerial plant parts through spray or mist have a greater chance of reaching the plant cells than those applied through the solution to the soil. The concentration of NPs in the plant is also dependent on the duration of the interaction of NPs and plants. Foliar sprayed NPs have displayed longer retention in the leaf surface of maize plants (Hu et al., 2020), thereby possibly increasing the duration of the interaction. However, there is a possibility of washout with the larger-sized NPs following a foliar spray or atmospheric deposit on the leaf surface. The dissolution of NPs, pH of the soil system, microbial inhabitation in soil and basic soil properties also guide the concentration of bioavailable fractions of NPs. Further, the type of plant, particular plant species, rhizosphere microbial community and the presence or absence of disease also regulate the availability, accumulation, transformation, uptake and translocation of NPs. This is achieved by manipulation of the plant metabolism and alteration in the physiological processes in plants. NP translocation rate fluctuates with changes in leaf transpiration and root pressure, and environmental factors involving humidity and temperature.

5.3.2 Mechanism of NPs Movement in Plants

The NP cell wall interaction is a complex process that may be case specific. Some species of NPs have proved to enlarge the pores with dynamic modifications in the cell wall while other species simply bind to the cell wall surface, increasing the electron concentration of the cell. This suggests that the size of NPs or cell wall pores are not the only guiding factors for NPs' entry into the plant cell. The NPs may find access to the plant cell via multiple mechanisms. The broad mechanism of NP movement in plants however accounts for two types of pathways viz. the apoplast pathway (pathway of intercellular movement) and the symplast pathway (pathway of connectivity between two cells). Based on these two pathways, the NPs may move via different transportation routes including the cell wall pore, cellular uptake (endocytosis), ion channels, xylem, phloem and plasmodesmata as described below.

5.3.2.1 Cell Wall Pore

The cell wall acts as a potent barrier for the delivery of the NPs. Therefore, the cell wall porosity, thickness and components play a big role in allowing the NPs' movement across them. When exposed to the plant, NPs cross the epidermis to enter into the epidermal cells. Some NPs tend to induce cell wall pores. The cell wall pores are made up of cellulose chains with a size between 5–10 nm. These pores allow the NPs to diffuse along with some protein molecules. Since these pores have a limited size, the NPs entering through the cell wall pores are restricted based on size permeability. However, NPs entering the cell wall reach the cytosol without being interrupted by any organelle.

5.3.2.2 Endocytosis

Endocytosis is defined as the process in which an object is engulfed by the cell membrane for cellular uptake. It is the most common method of NP entry into the plant cell. When NPs invade the close vicinity of the rhizosphere, they are wrapped onto the surface of the cell. Then the NP cell wall chemistry through the hydrophobic interactions, electrostatic interactions, ligand binding etc. facilitates the process of endocytosis. Intracellular vesicle formation with plasma membrane invagination is also an outcome of these interactions. The process of endocytosis may be accomplished by phagocytosis, pinocytosis, macro-pinocytosis etc. The foliar application also results in the internalization of NPs with some exceptions like the case of cerium oxide (Wu et al., 2017).

5.3.2.3 Ion Channels

An ion channel generally has a diameter in the range of 1 nm. This is the reason they have an important role in the balance of water concentration. Other ultra-molecules are also transported across this channel. Since 1 nm is a very small passage, the entry of NPs through this route is small. NPs below 1 nm are transported across the ion channels through the process of permeation. Ion channels have a structural resemblance to the cell wall pores. Although the role of ion channels in the transportation of NPs is not high, the down-regulation of water content through this pathway is of the utmost importance to plants.

5.3.2.4 The Upward-downward Movement

The intercellular translocation of NPs at variable rates is another important mechanism of movement. The NPs entering through aerial routes reach the plant cell and move towards phloem tissue which ultimately carry the NPs downward. Similarly, the NPs entering through the roots reach the xylem which then carries the NPs upwards to different parts of plants. This means NPs show the tendency of both root-leaf as well as leaf-root movements. Keeping the movement in view, if the NPs show better translocation through the phloem, then application through foliar spraying is preferred. On the contrary, application through soil is considered better for NPs, showcasing good translocation through the xylem. Besides the xylem and phloem, another specialized structure assisting the NP transportation inside plant cells is the plasmodesmata. However, transportation through the plasmodesmata is possible only when the NP is either present in the symplast or reaches the symplast.

5.3.3 Positive Impact of NPs on Plants

The positive impact of NPs on plants can be noticed at a concentration below the threshold limit of NPs in plant cells. One of the most fascinating aspects of NPs has been their use towards the enhancement of many plant physiological parameters, including seed germination potential and growth rate. Many NPs in recent times have been synthesized in the laboratory owing to their ability to reach the plant cell. Some positive impacts of NPs on plants include enhancement of nutrient availability, promotion of sprouting, increase in plant length (both root and shoot), production of anti-oxidants, promotion of enzyme activities (nitrate reductase, superoxide dismutase, catalase, glutamate dehydrogenase etc.), uptake and utilization of water, increase in dry weight, improvement in chlorophyll content, electron transportation, photochemical efficiency, increased leaf mass etc. Plants treated with NPs have shown increased efficiency to absorb and convert light energy reducing the ageing in chloroplasts. Si-rich NPs have been reported to act as a barrier through accumulation in the stelar tissues and protect the seedlings from rotting (Khan et al., 2019). Species of NPs that have been studied to induce stimulatory behaviour in plants include SiO_2, ZnO, Au-NP, Ag-NP and TiO_2.

Traditionally applied fertilizers are associated with low-nutrient-use efficiency. Fertilizers amended by nanotechnology help minimize nutrient loss while enhancing the nutrient-use efficiency of the plants (Rahman et al., 2021). The increased efficiency promotes plant growth and reduces environmental risks. With a high leaching rate, the probability of nitrate leaching into the nearby water bodies from the traditionally applied fertilizers is also high. This magnifies the chances of eutrophication (Cui et al., 2020). Further, the N-oxides generated from these fertilizers have a high magnitude of dumping greenhouse gases (of nitrogenous species) into the atmosphere. In addition, synthetic traditional fertilizers have long-term detrimental effects on soil biota, soil fertility and soil stability. The nitrate-origin intermediary products have shown the behaviour of carcinogenicity. Thus, nanotechnology ideally manipulates the chemical behaviour of synthetic fertilizers in favour of the plant-soil relationship that is delivered in the form of nanoemulsions and released slowly into the soil (Bidyarani et al., 2022). Generally, two types of nanoformulations are applied to improve the behaviour of traditional fertilizers i.e. (i) encapsulation, and (ii) thin coating (Pirzadah et al., 2020).

Studies involving the nanocoating on urea fertilizers showed a loading efficiency of 83%, simultaneously preventing premature nutrient release (Tarafder et al., 2020). Similarly, Cu-based nanofertilizer application has increased phenolic content, amino acid content, metabolite property and sugar concentration while displaying improvement in defence enzymes. The Si-based nanofertilizers, on the other hand, have enhanced seed germination, growth potential, growth rate, protection from UV radiation, increased nitrogen and phosphorus content etc. The Zn-based and Ti-based nanofertilizers have been reported to up-regulate superoxide dismutase, alleviate salt stress, increase pharmaceutical properties, improve the oil yield, increase indole and pigment concentration, restrict NP movement, increase lycopene and chlorophyll content, increase vegetative plant parts including branches size and number etc. Nanobiofertilizers have also exhibited positive effects in the rhizosphere zone promoting the rhizobacteria that assist in plant growth.

Another approach to enhancing agricultural productivity using nanotechnology is the use of traditional and nanofertilizers in a blended mode. The blending of nanofertilizers masks the adverse effect of traditional fertilizers. Some of the important advantages of using nanofertilizers are (i) pest resistance, (ii) improvement of secondary metabolite production, (iii) enhancement of soil water retention capacity, (iv) controlled nutrient delivery, (v) less nutrient drainage loss, (vi) disease control through anti-microbial and anti-fungal property, (vii) increase in nutrient self-life by protection against heat and UV damage, (viii) controlled and optimal application etc.

Nanopesticides can either be organic molecules or inorganic molecules (Alghuthaymi et al., 2021). While the former is a polymeric coating, the latter has a carrier material (Chaud et al., 2021). Out of these, the coated one is the more preferred option since it is target specific, sustainable and versatile use. Nanobiopesticides are biodegradable and have displayed the promise of enhancing plant growth by preventing pest infestations. Current nanoformulated biopesticides are also being used extensively in food packaging and food preservation (Koul, 2019).

5.3.4 Biosynthetic NPs

Biosynthetic NPs (also known as bionanomaterials synthesized using phyto-nanotechnology) are recently gaining attention due to their therapeutic uses. NPs also find use as a coating material for optoelectronic studies. Generally derived from the plant outputs and the microbes, biosynthetic NPs have proven their efficiency in anti-cancer, anti-bacterial, wound healing, diagnosing and imaging properties, gene therapy, drug delivery, biosensing etc. In addition, these NPs are eco-friendly, cost-effective, easily available and environmentally sustainable products. Besides, biosynthetic NPs are useful in treating various diseases that are inherent to plants. Reports suggest that rice plant disease has been controlled by destroying the responsible pathogenic bacteria with the help of NPs derived from endophytic bacteria (Ibrahim et al., 2019). However, plant-based extracts yield NPs in a short period, whereas, the employment of microbes for NP synthesis is a time-consuming affair. This is the reason that plant extracts are a more preferred and viable option in the present-day market.

The biosynthetic fabrication of NPs usually follows two types of approaches viz. the top-down approach and the bottom-up approach. While the top-down approach involves processes like ball milling, etching, lithography, laser technique, evaporation-condensation etc., the bottom-up approach is made up of biosynthesis, sol-gel method, vapour deposition, etc. The top-down technique involves chopping off the bulk materials to recover the NPs while the bottom-up technique involves stacking different-sized packing materials in various layers (ionic, atomic, molecular or aggregated layers) to generate nanosized particles. Out of the two available approaches, the top-down approach is adopted for the synthesis of NPs while the bottom-up approach is adopted for handling the NPs. Studies also reveal that during the synthesis of bio-nanomaterial, other volatile and toxic substances hazardous to the environment are also consumed thereby aiding in environmental remediation (Nguyen et al., 2022).

5.3.5 Toxicity Threats Associated with NPs

The toxicity threats posed by NPs largely rely on the material itself and the mode of their application. The direct effect of NPs in a plant is the induction of physiological, anatomical and morphological alterations on reaching the plant cell. The magnitude of the effect depends on the behaviour of the NPs and the plant's response to it. Some NPs' toxicity in plants may also include clogging of pores, damaging the barriers, reducing the nutrient uptake, competing with other nutrients during uptake, altering the hydraulic transfer, reducing the rate of photosynthesis and transpiration, NP-related ROS (reactive oxygen species) production and damaging the DNA structures. Besides, specific NPs may result in numerous other phytotoxic (both physiological and morphological) and cytotoxic effects in plants which are presented in Table 5.1, while the long-term detrimental effects in animals and plants are still not extensively known. However, some of the reported damages caused by the NPs in humans include plasma membrane damage, disruption of the nucleus, modifications of the cytoskeleton, toxicity of mitochondria, production of reactive oxygen species, intervention in signalling pathways etc. (Arora et al., 2022).

TABLE 5.1
Phytotoxic effect of various NPs

NPs species	Phytotoxic effect	Reference
Mo	Reduced plant growth	Aubert et al. (2010)
Ag	Reduced root and shoot biomass	Nair and Chung (2014)
$Ag^{+}CO$	Blockage of water channel; plant growth issues	Sanzari et al. (2019)
ZnO	Collapsed roots, epidermis, vacuoles; reduction in chlorophyll content; obstruction in translocation	Zhao et al. (2013)Mukherjee et al. (2014)Wang et al. (2013)
Al_2O_3	Reduction in enzyme activities	Poborilova et al. (2013)
TiO_2	Alteration in chlorophyll content	Servin et al. (2013)
CeO_2	Disorganized DNA	Lopez-Moreno et al. (2010)
Cu	Decreased transpiration rate	Musante and White (2012)
Au	Oxidative damage	Arora et al. (2012)

5.3.5.1 Accumulation of NPs

NP accumulation takes place in the destructed cells with the help of the mucilage entrapment in the rhizosphere. A negligible fraction of these accumulated NPs can however find an entry into the plant cell through translocation. In some instances, the nanofertilizers have displayed accumulation in the non-target cells, changing the gene expression and production of reactive nitrogen species. Although nanopesticide accumulation has not been sufficiently studied to date, the need for nanopesticide transformations, interaction, fate and persistence also demands immediate research explorations.

In some cases, NP accumulation has been reported to be affected due to the transpiration rate of plants (higher transpiration rate means lower NP accumulation), and charge interaction between cell wall NP and NP soil. On the other hand, NPs undergoing foliar application accumulate in the leaf and aerial parts of the plants. Some NPs have the tendency to dissolve into ions after accumulating in the leaf tissues. This is especially common in the application of ZnO. In most cases, NPs accumulate in plant tissues and organs (Varna et al., 2012) after entry into the cell. Accumulation of NPs not only leads to nutritional deficiencies of the food materials but also may enter the food chain posing a potential risk to humans and animals.

5.3.5.2 Inhibitory Action of NPs

NPs may show inhibitory action in the plant due to their overdose, over-accumulation or variable responses shown by plants of different genes. Certain NPs have shown inhibitory properties by decreasing the urease and catalase enzyme activities, preventing microbial growth by impacting their colony-forming units, reducing seed sprouting, thermogenic metabolism etc. The negative impact of NPs can even extend to the plant biomolecular level by interfering in the synthesis of chlorophyll, carotenoid, proline, soluble sugar etc. (Li et al., 2020). NPs also sometimes suppress the expression of aquaporins – the primary protein involved in the intercellular molecular movement. In addition, they may cause structural damage, interfering with plant metabolic processes and reducing biochemical concentration (sugars, acids, nutrients etc.).

5.4 CONCLUSION

Nanoparticles play a decisive role in soil property modification due to their influence at the nanoscale level and highly reactive surface area. Alleviating the effect of nontoxic NPs in the context of metal-contaminated soil can be very beneficial in sustainable agricultural practices. It is evident from the discussion that NPs have both regulatory and inhibitory roles in plants. While most of the inhibitory roles occur at a higher dose of NPs, the regulatory and synthesis-promoting impacts at lower doses may be beneficial for soil-plant relations. Therefore, the synthetic design of the NPs in relation to their shape, size, charge, reactivity, ease of plant entry, low accumulation potential etc. may help in gaining higher agricultural outputs. Further, the application of NPs through regulation, modification and channelization of NP-plant interaction can be a futuristic approach to bionanotechnology. Nevertheless, knowledge regarding the plant's relation with NPs is still in its early days. Future research in this line would aid significantly in achieving sustainable agricultural goals.

REFERENCES

Ahmed, B., Rizvi, A., Ali, K., Lee, J., Zaidi, A., Khan, M. S., & Musarrat, J. (2021). Nanoparticles in the soil–plant system: A review. *Environmental Chemistry Letters*, *19*(2), 1545–1609.

Alghuthaymi, M. A., Kalia, A., Bhardwaj, K., Bhardwaj, P., Abd-Elsalam, K. A., Valis, M., & Kuca, K. (2021). Nanohybrid antifungals for control of plant diseases: Current status and future perspectives. *Journal of Fungi*, *7*(1), 48.

Almaroai, Y. A., Meththika, V., Anushka, U. R., Sang, S. L., Xiaomin, D., Young, H. L., Jwa-Kyung, S., & Yong, S. O. (2014). Natural and synthesised iron-rich amendments for As and Pb immobilisation in agricultural soil. *Chemistry and Ecology*, *30*(3), 267–279.

Anderson, A., Joan, M., Paul, M., & David, B. (2017). Soil chemistry influences the phytotoxicity of metal oxide nanoparticles. *International Journal of Nanotechnology*, *14*, 15–21.

Arora, S., Gajiram, M., Koel, M., Sumit, S., & Dipak, M. (2022). A comprehensive overview of nanotechnology in sustainable agriculture. *Journal of Biotechnology*, *355*, 21–41.

Arora, S., Sharma, P., Kumar, S., Nayan, R., Khanna, P. K., & Zaidi, M. G. H. (2012). Gold-nanoparticle induced enhancement in growth and seed yield of *Brassica juncea*. *Plant Growth Regulation*, *66*(3), 303–310.

Aubert, T., Marie-Andrée, E., Stéphane, C., Fabien, G., & Francisco, C. (2010). *Phytotoxicity of nanosized molybdenum octahedral clusters*. XVII Congress of the FESPB. Valencia, Spain.

Badawy, A. M. E., Luxton, T. P., Silva, R. G., Scheckel, K. G., Suidan, M. T., & Tolaymat, T. M. (2010). Impact of environmental conditions (pH, ionic strength, and electrolyte type) on the surface charge and aggregation of silver nanoparticles suspensions. *Environmental Science & Technology*, *44*(4), 1260–1266.

Bidyarani, N., Vyas, G., Jaiswal, J., Deshmukh, S. K., & Kumar, U. (2022). Recent approaches in nanobioformulation for sustainable agriculture. In S. K. Deshmukh, M. Kochar, P. Kaur, & P. P. Singh (Eds.), *Nanotechnology in agriculture and environmental science* (pp. 166–182). CRC Press.

Borm, P., Klaessig, F. C., Landry, T. D., Moudgil, B., Pauluhn, J., Thomas, K., Trottier, R., & Wood, S. (2006). Research strategies for safety evaluation of nanomaterials, part V: Role of dissolution in biological fate and effects of nanoscale particles. *Toxicological Sciences*, *90*(1), 23–32.

Chaud, M., Souto, E. B., Zielinska, A., Severino, P., Batain, F., Oliveira-Junior, J., & Alves, T. (2021). Nanopesticides in agriculture: Benefits and challenge in agricultural productivity, toxicological risks to human health and environment. *Toxics*, *9*(6), 131.

Cheng, J., Yu-zhen, L., & Han-wei, W. (2014). Effects of surface-modified nano-scale carbon black on Cu and Zn fractionations in contaminated soil. *International Journal of Phytoremediation*, *16*(1), 86–94.

Choudhury, S., & Moulick, D. (2022). *Response of field crops to abiotic stress: Current status and future prospects* (pp. 1–332). CRC Press. https://doi.org/10.1201/9781003258063

Choudhury, S., Moulick, D., Ghosh, D., Soliman, M., Alkhedaide, A., Gaber, A., & Hossain, A. (2022). Drought-induced oxidative stress in pearl millet (*Cenchrus americanus* L.) at seedling stage: Survival mechanisms through alteration of morphophysiological and antioxidants activity. *Life*, *12*(8), 1171.

Choudhury, S., Moulick, D., & Mazumder, M. K. (2021a). Secondary metabolites protect against metal and metalloid stress in rice: An in silico investigation using dehydroascorbate reductase. *Acta Physiologiae Plantarum*, *43*(1), 1–10.

Choudhury, S., Sharma, P., Moulick, D., & Mazumder, M. K. (2021b). Unrevealing metabolomics for abiotic stress adaptation and tolerance in plants. *Journal of Crop Science and Biotechnology*, *24*(5), 479–493.

Christian, P., Von der Kammer, F., Baalousha, M., & Hofmann, T. (2008). Nanoparticles: Structure, properties, preparation and behaviour in environmental media. *Ecotoxicology*, *17*(5), 326–343.

Cornell, R. M., & Schwertmann, U. (2003). *The iron oxides: Structure, properties, reactions, occurrence and uses* (p. 664). Wiley-vch.

Coutris, C., Erik, J. J., & Deborah, H. O. (2012). Aging and soil organic matter content affect the fate of silver nanoparticles in soil. *Science of the Total Environment*, *420*, 327–333.

Cui, M., Zeng, L., Qin, W., & Feng, J. (2020). Measures for reducing nitrate leaching in orchards: A review. *Environmental Pollution*, *263*(Part B), 114553.

Cundy, A. B., Laurence, H., & Raymond, L. D. W. (2008). Use of iron-based technologies in contaminated land and groundwater remediation: A review. *Science of the Total Environment*, *400*(1), 42–51.

Dimkpa, C. O. (2018). Soil properties influence the response of terrestrial plants to metallic nanoparticles exposure. *Current Opinion in Environmental Science & Health*, *6*, 1–8.

Dunphy Guzman, K. A., Finnegan, M. P., & Banfield, J. F. (2006). Influence of surface potential on aggregation and transport of Titania nanoparticles. *Environmental Science & Technology*, *40*(24), 7688–7693.

Dwivedi, A. D., Dubey, S. P., Sillanpää, M., Kwon, Y. N., Lee, C., & Varma, R. S. (2015). Fate of engineered nanoparticles: Implications in the environment. *Coordination Chemistry Reviews*, *287*, 64–78.

Farré, M., Gajda-Schrantz, K., Kantiani, L., & Barceló, D. (2009). Ecotoxicity and analysis of nanomaterials in the aquatic environment. *Analytical and Bioanalytical Chemistry*, *393*(1), 81–95.

Ghosh, D., Brahmachari, K., Brestic, M., Ondrisik, P., Hossain, A., Skalicky, M., Sarkar, S., Moulick, D., Dinda, N. K., Das, A., & Pramanick, B. (2020b). Integrated weed and nutrient management improve yield, nutrient uptake and economics of maize in the rice-maize cropping system of Eastern India. *Agronomy*, *10*(12), 1906.

Ghosh, D., Brahmachari, K., Sarkar, S., Dinda, N. K., Das, A., & Moulick, D. (2022a). Impact of nutrient management in rice-maize-greengram cropping system and integrated weed management treatments on summer greengram productivity. *Indian Journal of Weed Science*, *54*(1), 25–30.

Ghosh, D., Brahmachari, K., Skalicky, M., Hossain, A., Sarkar, S., Dinda, N. K., Das, A., Pramanick, B., Moulick, D., Brestic, M., & Raza, M. A. (2020a). Nutrients supplementation through organic manures influence the growth of weeds and maize productivity. *Molecules*, *25*(21), 4924.

Ghosh, D., Brahmachari, K., Skalický, M., Roy, D., Das, A., Sarkar, S., Moulick, D., Brestič, M., Hejnak, V., Vachova, P., & Hassan, M. M. (2022b). The combination of organic and inorganic fertilizers influence the weed growth, productivity and soil fertility of monsoon rice. *PLoS One*, *17*(1), e0262586.

Ghosh, S., Mashayekhi, H., Pan, B., Bhowmik, P., & Xing, B. (2008). Colloidal behavior of aluminum oxide nanoparticles as affected by pH and natural organic matter. *Langmuir*, *24*(21), 12385–12391.

Gong, Y., Yuanyuan, L., Zhong, X., Dawn, K., & Dongye, Z. (2012). Immobilization of mercury in field soil and sediment using carboxymethyl cellulose stabilized iron sulfide nanoparticles. *Nanotechnology*, *23*(29), 294007.

Hortin, J. M., Anderson, A. J., Britt, D. W., Jacobson, A. R., & McLean, J. E. (2020). Copper oxide nanoparticle dissolution at alkaline pH is controlled by dissolved organic matter: Influence of soil-derived organic matter, wheat, bacteria, and nanoparticle coating. *Environmental Science: Nano*, *7*(9), 2618–2631.

Hossain, A., Ahmad, Z., Moulik, D., Maitra, S., Bhadra, P., Ahmad, A., Garai, S., Mondal, M., Roy, A., Sabagh, A. E., & Aftab, T. (2021b). Jasmonates and salicylates: Mechanisms, transport and signalling during abiotic stress in plants. In T. Aftab & M. Yusuf (Eds.), *Jasmonates and salicylates signaling in plants* (pp. 1–29). Springer.

Hossain, A., Ahmed, S., Ahmad, Z., Alam, M. J., Moulick, D., Saha, B., Samanta, S., Maitra, S., Bhadra, P., Bukhari, M. A., & Aftab, T. (2021c). Next-generation genetic engineering tools for abiotic stress tolerance in plants. In T. Aftab & K. R. Hakeem (Eds.), *Frontiers in plant-soil interaction* (pp. 153–197). Academic Press.

Hossain, A., Maitra, S., Pramanick, B., Bhutia, K. L., Ahmad, Z., Moulik, D., Syed, M. A., Shankar, T., Adeel, M., Hassan, M. M., & Aftab, T. (2022). Wild relatives of plants as sources for the development of abiotic stress tolerance in plants. In T. Aftab & A. Roychoudhury (Eds.), *Plant perspectives to global climate changes* (pp. 471–518). Academic Press.

Hossain, A., Pramanick, B., Bhutia, K. L., Ahmad, Z., Moulick, D., Maitra, S., Ahmad, A., & Aftab, T. (2021a). Emerging roles of osmoprotectant glycine betaine against salt-induced oxidative stress in plants: A major outlook of maize (*Zea mays* L.). In T. Aftab & K. R. Hakeem (Eds.), *Frontiers in plant-soil interaction* (pp. 567–587). Academic Press.

Hu, P., Jing, A., Maquela, M. F., Honghong, W., Zhaohu, L., Xiaoli, T., & Juan, P. G. (2020). Nanoparticle charge and size control foliar delivery efficiency to plant cells and organelles. *ACS Nano*, *14*(7), 7970–7986.

Ibrahim, E., Hatem, F., Muchen, Z., Yang, Z., Wen, Q., Chengqi, Y., Bin, L., Jianchu, M., & Jianping, C. (2019). Biosynthesis of silver nanoparticles using endophytic bacteria and their role in inhibition of rice pathogenic bacteria and plant growth promotion. *RSC Advances*, *9*(50), 29293–29299.

Jiang, J., Oberdörster, G., & Biswas, P. (2009). Characterization of size, surface charge, and agglomeration state of nanoparticle dispersions for toxicological studies. *Journal of Nanoparticle Research*, *11*(1), 77–89.

Ju-Nam, Y., & Lead, J. R. (2008). Manufactured nanoparticles: An overview of their chemistry, interactions and potential environmental implications. *Science of the Total Environment, 400*(1), 396–414.

Khan, M. R., Vojtech, A., Tanveer, F. R., Baohong, Z., Faheem, A., Izabela, J., Ye, Z., Mingying, Y., & Chuanbin, M. (2019). Nanoparticle–plant interactions: Two-way traffic. *Small, 15*(37), 1901794.

Koul, O. (2019). Nanobiopesticides: An introduction. In O. Koul (Ed.), *Nano-Biopesticides today and future perspectives* (pp. 1–15). Academic Press.

Li, M., Wang, P., Dang, F., & Zhou, D. M. (2017). The transformation and fate of silver nanoparticles in paddy soil: Effects of soil organic matter and redox conditions. *Environmental Science Nano, 4*(4), 919–928.

Li, X., Elliott, D. W., & Zhang, W. (2006). Zero-valent iron nanoparticles for abatement of environmental pollutants: Materials and engineering aspects. *Critical Reviews in Solid State and Materials Sciences, 31*(4), 111–122.

Li, X., & Zhang, W. (2007). Sequestration of metal cations with zerovalent iron nanoparticles: A study with high resolution X-ray photoelectron spectroscopy (HR-XPS). *The Journal of Physical Chemistry C, 111*(19), 6939–6946.

Li, Z., Ruijing, L., Qingfei, L., Junguo, Z., & Guangyin, W. (2020). Physiological response of cucumber (*Cucumis Sativus* L.) leaves to polystyrene nanoplastics pollution. *Chemosphere, 255*, 127041.

Loosli, F., & Stoll, S. (2017). Effect of surfactants, PH and water hardness on the surface properties and agglomeration behavior of engineered TiO_2 nanoparticles. *Environmental Science: Nano, 4*(1), 203–211.

López-Moreno, M. L., de la Rosa, G., Hernández-Viezcas, J. Á., Castillo-Michel, H., Botez, C. E., Peralta-Videa, J. R., & Gardea-Torresdey, J. L. (2010). Evidence of the differential biotransformation and genotoxicity of ZnO and CeO_2 nanoparticles on soybean (*Glycine Max*) plants. *Environmental Science & Technology, 44*(19), 7315–7320.

Lv, W., Geng, H., Zhou, B., Chen, H., Yuan, R., Chuanxin, M. A., Liu, R., Xing, B., & Wang, F. (2022). The behavior, transport, and positive regulation mechanism of ZnO nanoparticles in a plant-soil-microbe environment. *Environmental Pollution, 315*, 120368.

Mazumder, M. K., Sharma, P., Moulick, D., Tata, S. K., & Choudhury, S. (2022). Salicylic acid ameliorates zinc and chromium-induced stress responses in wheat seedlings: A biochemical and computational analysis. *Cereal Research Communications, 50*(3), 407–418.

Moulick, D., Samanta, S., Sarkar, S., Mukherjee, A., Pattnaik, B. K., Saha, S., Awasthi, J. P., Bhowmick, S., Ghosh, D., Samal, A. C., & Mahanta, S. (2021). Arsenic contamination, impact and mitigation strategies in rice agro-environment: An inclusive insight. *Science of the Total Environment, 800*, 149477.

Mukherjee, A., Peralta-Videa, J. R., Bandyopadhyay, S., Rico, C. M., Zhao, L., & Gardea-Torresdey, J. L. (2014). Physiological effects of nanoparticulate ZnO in green peas (*Pisum sativum* L.) cultivated in soil. *Metallomics, 6*(1), 132–138.

Musante, C., & White, J. C. (2012). Toxicity of silver and copper to *Cucurbita pepo*: Differential effects of nano and bulk-size particles. *Environmental Toxicology, 27*(9), 510–517.

Nair, P. M. G., & Chung, I. M. (2014). Physiological and molecular level effects of silver nanoparticles exposure in rice (*Oryza sativa* L.) seedlings. *Chemosphere, 112*, 105–113.

Nguyen, N. T. T., Nguyen, L. M., Nguyen, T. T. T., Nguyen, T. T., Nguyen, D. T. C., & Tran, T. V. (2022). Formation, antimicrobial activity, and biomedical performance of plant-based nanoparticles: A review. *Environmental Chemistry Letters, 20*(4), 2531–2571.

Noubactep, C., & Caré, S., & Crane, R. (2012). Nanoscale metallic iron for environmental remediation: Prospects and limitations. *Water, Air, & Soil Pollution, 223*, 1363–1382.

Ogunkunle, C. O., Oyedeji, S., Okoro, H. K., & Adimula, V. (2021). Interaction of nanoparticles with soil. In A. Amrane, D. Mohan, T. A. Nguyen, A. A. Assadi, & G. Yasin (Eds.), *Nanomaterials for soil remediation* (pp. 101–132). Elsevier.

Paramo, L. A., Feregrino-Pérez, A. A., Guevara, R., Mendoza, S., & Esquivel, K. (2020). Nanoparticles in agroindustry: Applications, toxicity, challenges, and trends. *Nanomaterials*, *10*(9), 1654.

Pérez-de-Luque, A. (2017). Interaction of nanomaterials with plants: What do we need for real applications in agriculture? *Frontiers in Environmental Science*, *5*, 12.

Peyrot, C., Wilkinson, K. J., Desrosiers, M., & Sauvé, S. (2014). Effects of silver nanoparticles on soil enzyme activities with and without added organic matter. *Environmental Toxicology and Chemistry*, *33*(1), 115–125.

Pirzadah, B., Pirzadah, T. B., Jan, A., & Hakeem, K. R. (2020). Nanofertilizers: A way forward for green economy. In K. R. Hakeem & T. B. Pirzadah (Eds.), *Nanobiotechnology in agriculture* (pp. 99–112). Springer International Publishing.

Poborilova, Z., Opatrilova, R., & Babula, P. (2013). Toxicity of aluminium oxide nanoparticles demonstrated using a BY-2 plant cell suspension culture model. *Environmental and Experimental Botany*, *91*, 1–11.

Rabbani, M. M., Ahmed, I., & Park, S. J. (2016). Application of nanotechnology to remediate contaminated soils. In H. Hasegawa, I. M. Rahman, & M. A. Rahman (Eds.), *Environmental remediation technologies for metal-contaminated soils* (pp. 219–229). Springer.

Rahman, M. H., Haque, K. M. S., & Khan, M. Z. H. (2021). A review on application of controlled released fertilizers influencing the sustainable agricultural production: A cleaner production process. *Environmental Technology & Innovation*, *23*, 101697.

Rawat, S., Pullagurala, V. L. R., Adisa, I. O., Wang, Y., Peralta-Videa, J. R., & Gardea-Torresdey, J. L. (2018). Factors affecting fate and transport of engineered nanomaterials in terrestrial environments. *Current Opinion in Environmental Science & Health*, *6*, 47–53.

Rosická, D., & Šembera, J. (2010). Assessment of influence of magnetic forces on aggregation of zero-valent iron nanoparticles. *Nanoscale Research Letters*, *6*(1), 10.

Saha, B., Chowardhara, B., Kar, S., Devi, S. S., Awasthi, J. P., Moulick, D., Tanti, B., &, Panda, S. K. (2019). Advances in heavy metal-induced stress alleviation with respect to exogenous amendments in crop plants. In M. Hasanuzzaman & V. Fotopoulos (Eds.), *Priming and pretreatment of seeds and seedlings: Implication in plant stress tolerance and enhancing productivity in crop plants* (pp. 313–332). Springer.

Sahoo, S., Borgohain, P., Saha, B., Moulick, D., Tanti, B., & Panda, S. K. (2019). Seed priming and seedling pre-treatment induced tolerance to drought and salt stress: Recent advances. In M. Hasanuzzaman & V. Fotopoulos (Eds.), *Priming and pretreatment of seeds and seedlings: Implication in plant stress tolerance and enhancing productivity in crop plants* (pp. 253–263). Springer.

Sanzari, I., Leone, A., & Ambrosone, A. (2019). Nanotechnology in plant science: To make a long story short. *Frontiers in Bioengineering and Biotechnology*, *7*, 120.

SCENIHR (Scientific Committee on Emerging and Newly Identified Health Risks). (2007). *Opinion on the scientific aspects of the existing and proposed definitions relating to products of Nanoscience and nanotechnologies*. European Commission.

Servin, A. D., Morales, M. I., Castillo-Michel, H., Hernandez-Viezcas, J. A., Munoz, B., Zhao, L., Nunez, J. E., Peralta-Videa, J. R., &, Gardea-Torresdey, J. L. (2013). Synchrotron verification of TiO_2 accumulation in cucumber fruit: A possible pathway of TiO_2 nanoparticle transfer from soil into the food chain. *Environmental Science & Technology*, *47*(20), 11592–11598.

Shan, G., Yan, S., Tyagi, R. D., Surampalli, R. Y., & Zhang, T. C. (2009). Applications of nanomaterials in environmental science and engineering. *Practice Periodical of Hazardous, Toxic, and Radioactive Waste Management, 13*(2), 110–119.

Shin, W. K., Cho, J., Kannan, A. G., Lee, Y. S., & Kim, D. W. (2016). Cross-linked composite gel polymer electrolyte using mesoporous methacrylate-functionalized SiO_2 nanoparticles for lithium-ion polymer batteries. *Scientific Reports, 6*(1), 26332.

Shoults-Wilson, W. A., Reinsch, B. C., Tsyusko, O. V., Bertsch, P. M., Lowry, G. V., & Unrine, J. M. (2011). Role of particle size and soil type in toxicity of silver nanoparticles to earthworms. *Soil Science Society of America Journal, 75*(2), 365–377.

Stebounova, L. V., Guio, E., & Grassian, V. H. (2011). Silver nanoparticles in simulated biological media: A study of aggregation, sedimentation, and dissolution. *Journal of Nanoparticle Research, 13*(1), 233–244.

Stone, V., Nowack, B., Baun, A., van den Brink, N., von der Kammer, F., Dusinska, M., Handy, R., Hankin, S., Hassellöv, M., Joner, E., & Fernandes, T. F. (2010). Nanomaterials for environmental studies: Classification, reference material issues, and strategies for physico-chemical characterisation. *Science of the Total Environment, 408*(7), 1745–1754.

Tarafder, C., Daizy, M., Alam, M. M., Ali, M. R., Islam, M. J., Islam, R., Ahommed, M. S., Aly Saad Aly, M., & Khan, M. Z. H. (2020). Formulation of a hybrid nanofertilizer for slow and sustainable release of micronutrients. *ACS Omega, 5*(37), 23960–23966.

Tourinho, P. S., Van Gestel, C. A. M., Lofts, S., Svendsen, C., Soares, A. M. V. M., & Loureiro, S. (2012). Metal-based nanoparticles in soil: Fate, behavior, and effects on soil invertebrates. *Environmental Toxicology and Chemistry, 31*(8), 1679–1692.

Unrine, J., Bertsch, P., & Hunyadi, S. (2008). Bioavailability, trophic transfer, and toxicity of manufactured metal and metal oxide nanoparticles in terrestrial environments. In V. H. Grassian (Ed.), *Nanoscience and nanotechnology: Environmental and health impacts* (Vol. 20, p. 345). Wiley.

Usman, M., Farooq, M., Wakeel, A., Nawaz, A., Cheema, S. A., Rehman, H., Ashraf, I., & Sanaullah, M. (2020). Nanotechnology in agriculture: Current status, challenges and future opportunities. *Science of the Total Environment, 721*, 137778.

Varna, M., Ratajczak, P., Ferreira, I., Leboeuf, C., Bousquet, G., & Janin, A. (2012). In vivo distribution of inorganic nanoparticles in preclinical models. *Journal of Biomaterials and Nanobiotechnology, 3*, 269–279.

Wang, P., Menzies, N. W., Lombi, E., McKenna, B. A., Johannessen, B., Glover, C. J., Kappen, P., & Kopittke, P. M. (2013). Fate of ZnO nanoparticles in soils and cowpea (*Vigna unguiculata*). *Environmental Science & Technology, 47*(23), 13822–13830.

Wei, Q., Yang, D., Fan, M., & Harris, H. G. (2013). Applications of nanomaterial-based membranes in pollution control. *Critical Reviews in Environmental Science and Technology, 43*(22), 2389–2438.

Wu, H., Tito, N., & Giraldo, J. P. (2017). Anionic cerium oxide nanoparticles protect plant photosynthesis from abiotic stress by scavenging reactive oxygen species. *ACS Nano, 11*(11), 11283–11297.

Xu, L., Liang, H. W., Yang, Y., & Yu, S. H. (2018). Stability and reactivity: Positive and negative aspects for nanoparticle processing. *Chemical Reviews, 118*(7), 3209–3250.

Yuan, G. (2004). Natural and modified nanomaterials as sorbents of environmental contaminants. *Journal of Environmental Science and Health, 39*(10), 2661–2670.

Zhang, G. (2007). Soil nanoparticles and their influence on engineering properties of soils. In D. J. DeGroot, C. Vipulanandan, J. A. Yamamuro, V. N. Kaliakin, P. V. Lade, M. Zeghal, U. E. Shamy, N. Lu, & C. R. Song (Eds.), *Advances in measurement and modeling of soil behavior* (pp. 1–13). American Society of Civil Engineers.

Zhang, G., Germaine, J. T., Whittle, A. J., & Ladd, C. C. (2004). Index properties of a highly weathered old alluvium. *Geotechnique*, *54*(7), 441–451.

Zhao, L., Sun, Y., Hernandez-Viezcas, J. A., Servin, A. D., Hong, J., Niu, G., Peralta-Videa, J. R., Duarte-Gardea, M., & Gardea-Torresdey, J. L. (2013). Influence of CeO_2 and ZnO nanoparticles on cucumber physiological markers and bioaccumulation of Ce and Zn: A life cycle study. *Journal of Agricultural and Food Chemistry*, *61*(49), 11945–11951.

Zhou, D. M., Wang, Y. J., Wang, H. W., Wang, S. Q., & Cheng, J. M. (2010). Surface-modified nanoscale carbon black used as sorbents for Cu (II) and Cd (II). *Journal of Hazardous Materials*, *174*(1–3), 34–39.

Zhou, W., Liu, Y. L., Stallworth, A. M., Ye, C., & Lenhart, J. J. (2016). Effects of pH, electrolyte, humic acid, and light exposure on the long-term fate of silver nanoparticles. *Environmental Science & Technology*, *50*(22), 12214–12224.

6 Nanopriming
A Comprehensive Perspective for Regulating Seed Germination of Crops Under Stress Conditions

Harsita Nayak, Satyabrata Mangaraj, Shriram Ratan Pradhan, Rabindra Kumar Paikaray and Akbar Hossain**

6.1 INTRODUCTION

At present, climate change and global warming have emerged as a potential threat to agriculture, which limits crop growing season and causes land deterioration, leading to drastic yield reduction or sometimes complete crop failure. As the world population is growing at an alarming rate, the demand for food is also increasing day by day, worsening the scenario. The global population is expected to rise from seven billion to 9–10 billion by 2050, necessitating an increase in global food production by 60–110 per cent (Godoy et al., 2021). However, under changing climatic conditions, crops are exposed to several abiotic stresses *viz.* drought, flood, cold, heat and salinity stress which affect the stand establishment and causes severe agricultural losses. The crop yield loss due to abiotic stresses accounts for more than 50%, out of which drought and salinity stresses are predominant, causing approximately 17% and 20% crop yield loss, respectively worldwide (Rehman et al., 2005; Ashraf et al., 2008; Ningombam et al., 2021). The root's ability to absorb water and nutrients is hampered under drought and salinity stress. Salinity stress also affects the morphological, physiological and biochemical responses of plants, causing significant changes in superoxide dismutase (SOD), antioxidant enzymes, growth regulators, lipid peroxidase, total chlorophyll content and root and shoot weight of plants (Siringam et al., 2011; Yadav et al., 2019).

Seeds are the most basic agricultural input and their germination is the prime and most sensitive stage of a crop life cycle. Quicker seed germination and uniform seedling emergence directly affect the crop stand establishment,

* Corresponding author.

 DOI: 10.1201/9781003376446-6

which ultimately depends on the quality of the seed *viz.* seed viability and vigour (Mahakham et al., 2017). Besides, the micro-climate of the seed such as temperature, moisture, light and salt concentration significantly affects seed germination and subsequent seedling development. Under extreme environmental conditions, seeds fail to germinate properly, which ultimately affects the crop yield. It is reported that seed germination as well as seedling growth and development in many crops are severely affected by drought and salt stress (AbdElgawad et al., 2016). Furthermore, the production of reactive oxygen species (ROS) as a result of low temperature or chilling stress also hinders the germination of seeds (Farooq et al., 2017). To address these issues, numerous technologies such as conventional breeding, genetic engineering, mutation breeding, polyploidy, genome editing and seed priming have been developed to hasten the seedling emergence and to offer tolerance to crops against adverse environmental conditions. Among these, pre-sowing treatments such as seed priming may play a pivotal role in successful crop establishment when the crop is grown under stress conditions. Better stand establishment reduces weed infestation by negatively affecting weed dry weight due to the smothering effect of crops (Chavan et al., 2018). The presence of an optimum plant stand helps minimize soil erosion as well as water and nutrient losses due to increased crop canopy density as compared to lower plant density (Al-Kaisi, 2001). Seed priming also increases crop yield by improving tolerance to environmental stress and enhancing the quality and germinability of weak and damaged seeds. Seed priming boosts crop performance and increases potential yield by facilitating germination even in a hostile environment (Ibrahim et al., 2016Marthandan et al., 2020).

In recent decades, nanotechnology has revolutionized several sectors of agriculture *viz.* precision farming, crop improvement, plant nutrition, plant protection, food processing and soil remediation by supplying useful nanoassets such as nanosensors, nanofertilizers, nanoherbicides, nanopesticides etc. (Park et al., 2006; Chinnamuthu & Kokiladevi, 2007; Tarafdar et al., 2012; Verma et al., 2012; Kandhol et al., 2022). Very recently, the application of nanotechnology has also started as an innovative seed priming tool in modern agriculture. The basic feature of nanopriming is the use of engineered nanomaterials (ENMs) manufactured with particle sizes in the range of 1–100 nm and a high surface area to volume ratio which provides ENMs very unique and invaluable physical properties compared to the respective conventional bulk materials (Gomez et al., 2021). Seed priming using NPs has been shown as more promising than conventional priming in order to achieve an attainable yield target (Abasi-Kalami et al., 2021). Nanopriming with metal-based or plant-synthesized NPs can activate specific metabolic processes that are typically active during the early stages of germination, which accelerates the rate of emergence and subsequent growth, yield and quality of the crop.

6.2 ROLES OF SEED PRIMING AND ITS TYPES

Seed priming is a pre-sowing seed treatment which permits seeds to be hydrated and go through the first stage of germination while preventing radical protuberance through the seed coat (McDonald, 2000). Earlier, the concept of seed priming involved the imbibition of seeds followed by drying just before sowing to get better plant vigour. Whereas, in the modern agricultural era, seed priming is classified into several methods such as minerals, growth regulators, salt solution, UV radiation, extreme temperatures, micro-organisms and nanomaterials.

6.2.1 Hydropriming

Hydropriming is the most primitive and low-cost priming method which involves soaking seeds in pure water, followed by drying to initial moisture content before sowing. The absence of the external addition of synthetic chemicals as a priming agent makes this method both economical as well as eco-friendly. Many studies have already confirmed the beneficial effect of hydropriming on seedling emergence and development on *Oryza sativa* (Matsushima & Sakagami, 2013; Ibrahim et al., 2013), *Vigna radiata* (Jisha & Puthur, 2018), *Vicia faba* (Damalas et al., 2019), *Momordica charantia* (Adhikari et al., 2021) and *Helianthus annuus* (Catiempo et al., 2021). However, a major limitation of this method is uncontrolled seed hydration which results in asynchronous activation of pre-germinative metabolic events within seeds, thereby lacking concomitant seedling emergence (Mc Donald, 2000).

6.2.2 Haloprimig/Osmopriming

Halopriming, also referred to as osmopriming, is a pre-sowing seed treatment which involves soaking seeds in a low osmotic aerated solution under varying time duration and water potential. Low water potential leads to the restricted entry of water inside the seed which results in slow seed imbibition and activation of the initial phases of seed germination. Polyethyl glycol (PEG) is the most commonly used osmotic agent. However, inorganic salts such as KCl, NaCl, KNO_3, $MgSO_4$, $CaCl_2$ and KH_2PO_4, sorbitol and mannitol are also used for the preparation of osmotic solution (Yacoubi et al., 2013).

6.2.3 Solid Matrix Priming

Solid matrix priming, otherwise known as matrix conditioning, is a controlled seed hydration technique which involves the mixing of moistened seeds with a solid organic carrier. In this method, the seed and solid carrier are used to compete for water, thereby reaching the moisture level required for radical protrusion. It has emerged as a replacement for osmopriming due to the high cost of osmotic agents and technical issues (Paparella et al., 2015). Carriers having specific physic-chemical properties such as high surface area and moisture holding capacity, low matrix potential and reduced solubility. No toxic effects on seeds and good binding properties are suitable for solid matrix priming. Vermiculite, expanded calcined clay, pat moss, sand, charcoal, agro lig and synthetic calcium silicates are some of the excellent solid carriers which have been used for matrix conditioning (Hacisallihagulu, 2007; Di Girolamo et al., 2012; Paparella et al., 2015).

6.2.4 Hormonal Priming

Seed priming with hormones or plant growth regulators solution is known as hormopriming. Hormones *viz.* auxin, cytokinin, kinetins, gibberellic acid, salicylic acid, ethylene and abscisic acid used in the process of hormopriming directly influence seed metabolism. Seeds primed with kinetin not only exhibited maximum growth and yield of wheat but also indicated enhanced salt tolerance (Iqbal et al., 2006). Similarly, increased grain yield and salt tolerance were noticed when spring wheat was primed with GA_3 solutions (Iqbal et al., 2013).

6.2.5 Biopriming

Biopriming is an ecological approach which integrates seed imbibition with inoculums of beneficial micro-organisms as priming agents in order to enhance seed germination rate, seedling vigour, growth and yield, tolerance to biotic and abiotic stress and protect crops from seed and soil-borne pathogens (Rakshit et al., 2015; Mahmood et al., 2016). Beneficial micro-organisms used for biopriming are plant growth–promoting rhizobacteria (PGPR) such as *Azatobacter*, *Azospirillum*, *Pseudomonas* and *Trichoderma* increase secretion of plant growth hormones which induce drought tolerance (Reddy, 2012). Seed priming with bacteria also produces siderophores, thereby improving iron uptake through the process of chelation and increasing the production of ACC deaminase, which lessens ethylene production leading to enhanced root and shoot development (Li et al., 2000; Dobbelaera et al., 2003). Biopriming facilitates better colonization, proliferation and establishment of biocontrol agents on the surface of the seed. which consequently helps increase seedling vigour and development of systemic resistance to various stresses (Singh et al., 2020). Seed biopriming of pearl millet with *Pseudomonas fluorescence* was reported to show enhanced germination, seedling vigour and triggered resistance against downy mildew (Raj et al., 2004).

6.3 SIGNIFICANCE OF NANOPRIMING IN MODERN AGRICULTURE

Nanopriming has started substituting conventional seed priming with water and other bulky priming agents due to its enhanced ability for seed germination, particularly under environmental stress conditions. Salinity, drought, high/low temperature stress and heavy metal contamination are the major abiotic environmental threats that affect plant growth and metabolism negatively. Seeds pretreated with NPs attenuate the severity of salinity stress by activating the antioxidant system, ionic regulation and biosynthesis of osmolytes. The presence of a photocatalytic mechanism in NPs leads to the conversion of the radiant energy of the sun into chemical energy, thereby stimulating plant photosynthetic activity. Studies showed that metal oxide nanoparticles enhance defence mechanisms in plants by stimulating superoxide dismutase (SOD), ascorbate peroxidase (APX) and catalase synthesis, mitigating several environmental stresses (Singh et al., 2021). Nanopriming tweaks the plasma membrane integrity by causing greater availability of stable lipids; hence it plays a key role in inducing a plant's tolerance to biotic and abiotic stresses (Sahebi et al., 2015). Khan et al. (2020), experimenting with seed priming with Ag NPs, noticed enhanced water uptake, proline content, reduced oxidative damage due to enhancement in antioxidant activity, increased K^+ and lowered Na^+ and Na^+/K^+ ratio, and increased flavonoid and phenolic contents in pearl millet.

The minuscule size and greater reactive surface area of NPs increase their bioavailability for plant uptake and minimize losses, leaving a promising effect on the soil microbial community and their surroundings. The NPs present in the soil either get immobilized in soil organic matter as humic acid or reacts with ions to form complex salts, leaving no residual toxicity in the soil and not having any damaging effect on soil microbionts. The application of NPs combined with plant growth–promoting rhizobacteria (PGPR) has been shown to increase plant growth under salinity stress due to an increase in photosynthetic and antioxidant activity, alleviating salt-induced stress (Yasmin et al., 2021). Many works of literature have also mentioned the antimicrobial properties of nanomaterials which directly affects the microbial community. Nanoparticles enter the cells of fungi and deform the cell configuration causing the death of many plant pathogenic fungi such as *Fusarium verticilloides*, *Pyricularia grisea* and *Xanthomonas oryzae* in rice (Ahuja et al., 2019; Sathiyabama & Muthukumar, 2020). Zinc-coated chitosan NPs were also reported to exhibit prolonged antifungal properties on *Curvularia lunata* and maize due to their controlled release, besides supplying essential plant elements (Choudhary et al., 2019). Nanoparticles such as oxides of zinc and magnesium have shown their promising effects in killing deleterious micro-organisms, thus being used as food preservatives (Shah & Belazorova, 2009). Similarly, nanoparticles of copper and iron oxides, when reacting with peroxide present in the environment, trigger the production of free radicals which are poisonous to microbes.

6.4 A NOVEL PRIMING APPROACH: NANOPRIMING AND ITS MECHANISM

Nanoparticle-mediated seed priming/nanopriming has emerged as a new dimension to smart farming, which deals with the priming of seeds using nanoparticles. It employs the use of organic and inorganic materials having dimensions in the range of 1–100 nm and having excellent physical and chemical properties than conventional bulk materials. The NPs possess some distinctive features such as a high surface to volume ratio and high surface energy, which make them enabled to boost enzymatic activities, supply materials of interest and absorb substances of interest. The effectiveness of nanopriming depends upon the nature and physicochemical characteristics of nanomaterials *viz.* size, zeta potential, concentrations used and duration of exposure. The prime objective of nanopriming is to increase the performance of crops as well as soil and sustenance of agriculture and the natural ecosystem by applying minimum inputs.

Seed nanopriming differs significantly from conventional seed priming since it primarily uses water (hydropriming) or solutions containing nutrients, hormones, salts and microbial inoculums as priming agents that adsorb on the seed as a seed coat. However, the media used in seed nanopriming are in the form of suspensions or nanoformulations, which may or may not be absorbed by the seed. When NPs are taken up by the seed, a significant fraction of nanomaterials remain on the seed surface as coatings and these coatings function as bactericides or fungicides preventing pathogenic infection in the field or during storage (Gross et al., 2020; Ahuja et al., 2020). Moreover, the formation of a convoluted pathway during NP absorption discourages incessant water uptake while lowering gas permeability, resulting in more stable and safe seed storage. The mechanisms underlying nanopriming-induced seed germination are the development of nanopores for improved water uptake and activation of aquaporin gene expression, production of ROS/antioxidant and hydroxyl radicals for cell wall loosening, endosperm weakening, generation of nanocatalyst for greater starch hydrolysis, rapid embryo growth and rapid root-shoot development (Shelar et al., 2021). The increase in soluble sugar content in seeds due to nanopriming leads to an increase in amylase activity, thereby enhancing water uptake in response to decreased intrinsic osmotic water potential. Nanopretreated seeds usually generate more ROS than non-primed or seeds primed with other media, indicating the significance of ROS in fostering seed germination. It is well-established fact that ROS such as H_2O_2, and OH^- regulate radial growth, cellular reconfiguration and the degeneration of endosperm and testa (Muller et al., 2009).

Nanopriming commonly employs the use of metallic NPs such as oxides of silica, zinc, selenium, titanium and iron etc., bio-synthesized metallic NPs *viz.*, extracts of the plant, algae, fungi, bacteria and polymeric nanocompounds which monitor the gene expression responsible for the synthesis of growth regulators, thus hastening plant morphological and metabolic attributes and ultimately stimulating plant growth. Priming using nanomaterials enhances seedling emergence, quality, yields, storage and environmental stress tolerance as compared to unprimed seeds or seeds primed using other agents (Pereira et al., 2021). Some NPs may undergo the process of re-oxidation and induce negative effects such as germination retardation and seedling phytotoxicity (Feizi et al., 2013; Falco et al., 2020). Ndeh et al. (2017) reported the inhibitory effect of gold NPs on the shoot and root length of rice.

6.5 FUTURE PERSPECTIVE

Though seed nanopriming helps develop homeostasis in seeds to sustain under different types of biotic and abiotic stresses *viz.* drought, salinity, high temperature, chilling, heavy metals stress etc., it is necessary to evolve some techniques that produce nanomaterials (NMs) in huge quantities at minimum expense. Furthermore, the exact mechanisms that help mitigate abiotic stress in the seed are not well understood yet; it needs to be explored at molecular as well as hormonal levels to develop stress tolerance through genetic manipulation. The crosstalk of nanoparticles with the phytohormones and beneficial soil microbes is a new area, which needs to be focused upon. Also, little information is available on the interaction of nanoparticles with the natural flora and fauna and their synergistic and/or antagonistic effect on plant growth and stress management. Moreover, there are variations in seed coat and the cell anatomy of seeds depending upon the crop species. The NPs penetrate the intracellular spaces of seeds through parenchymatous tissues. The diameter of the parenchymatous tissue and intracellular spaces may vary among and within the crop species, indicating differential penetrability and thereby variations in doses. Therefore, intensive laboratory studies based on dose optimization and zeta potential adjustment of NPs are highly recommended for their use at a larger scale in field trials. It is supposed that the seed which has fewer parenchyma and intracellular spaces may need a lower dose while other species having multiple parenchyma and intracellular spaces may require higher doses of NMs to improve seed tolerance to a different strain. Subsequently, proper designing of NMs and a better understanding of seed coat anatomy could be helpful to reduce the negative impact of NMs. Besides, all these areas, the nutritional quality of food grains and their impact on human health should also be considered in the due course of time for the commercial use of NMs in future.

6.6 CONCLUSION

In the current scenario, most of the developed countries have adopted intensive modern agricultural production systems to produce sufficient food to feed the burgeoning world population. Food production increased tremendously by introducing high-yielding varieties (HYVs) and irrigation facilities, providing sufficient nutrition to the plant by involving extensive uses of chemical fertilizers and agrochemicals as plant protection measures. But, indiscriminate use of agrochemicals has led to the degradation of natural resources like soil, water and vegetation. Despite all these, agriculture nowadays faces many other challenges. Climate change and global warming looming large and the impact is adversely affecting the food production system. Thus, it is inevitable to develop innovative and eco-friendly strategies that can cope with these constraints in order to promote food production systems sustainability. Nanotechnology seems a better approach and is a user-friendly and economically viable option to achieve these goals. Nanotechnology used in agriculture like nanofertilizers, nanoherbicides etc. shows promising results in crop production. Seed priming through nanoparticles could be a simple, cost-effective and eco-friendly approach that helps in the mitigation of biotic and abiotic stress that impedes seed germination and emergence in the field. Seed nanobiofortification aims at minimizing environmental pollution caused by the runoff of fertilizers and other agrochemicals to the ecosystem. Nanotechnology in general, and seed nanopriming in particular, will provide safe and healthy food and environments to the farmers and consumers.

REFERENCES

Abbassi, K. M., Moameri, M., Asgari, B. L., & Astatkie, T. (2021). Influence of nano-priming on seed germination and plant growth of forage and medicinal plants. *Plant Growth Regulation*, *93*(1), 13–28.

Abdel-Aziz, H. (2019). Effect of priming with chitosan nanoparticles on germination, seedling growth and antioxidant enzymes of broad beans. *Catrina: International Journal of Environmental Science*, *18*(1), 81–90.

AbdElgawad, H., Zinta, G., Hegab, M. M., Pandey, R., Asard, H., & Abuelsoud, W. (2016). High salinity induces different oxidative stress and antioxidant responses in maize seedlings organs. *Frontiers in Plant Science, 7*, 276.

Acharya, P., Jayaprakasha, G. K., Crosby, K. M., Jifon, J. L., & Patil, B. S. (2020). Nanoparticle-mediated seed priming improves germination, growth, yield, and quality of watermelons (*Citrullus lanatus*) at multi-locations in Texas. *Scientific Reports, 10*(1), 1–16.

Adhikari, B., Dhital, P. R., Ranabhat, S., & Poudel, H. (2021). Effect of seed hydro-priming durations on germination and seedling growth of bitter gourd (*Momordica charantia*). *PloS One, 16*(8), e0255258.

Afzal, S., Sharma, D., & Singh, N. K. (2021). Eco-friendly synthesis of phytochemical-capped iron oxide nanoparticles as nano-priming agent for boosting seed germination in rice (*Oryza sativa* L.). *Environmental Science Pollution Research, 28*(30), 40275–40287.

Ahmed, H. S., Ahmed, M. F., Shoala, T., & Salah, M. (2018). Impact of single or fractionated radiation and selenium nano-particles on acid lime (*Citrus aurantifolia* L.) seed germination ability and seedlings growth. *Advances in Agriculture and Environmental Science, 1*, 91–100.

Ahuja, R., Sidhu, A., & Bala, A. (2019). Synthesis and evaluation of iron (ii) sulfide aqua nanoparticles (FeS-NPs) against Fusarium verticillioides causing sheath rot and seed discoloration of rice. *European Journal of Plant Pathology, 155*(1), 163–171.

Ahuja, R., Sidhu, A., & Bala, A. (2020). Effect of iron sulfide nanopriming in reducing *Drechslera* seed rot and seedling mortality in rice. *Agricultural Research Journal, 57*, 60–65.

Al-Kaisi, M. M. (2001). *Soil erosion, crop productivity and cultural practices*. Iowa State University, University Extension.

Ashraf, M., Athar, H. R., Harris, P. J. C., & Kwon, T. R. (2008). Some prospective strategies for improving crop salt tolerance. *Advances in Agronomy, 97*, 45–110.

Azimi, R., Borzelabad, M. J., Feizi, H., & Azimi, A. (2014). Interaction of SiO_2 nanoparticles with seed prechilling on germination and early seedling growth of tall wheat grass (*Agropyron elongatum* L.). *Polish Journal of Chemical Technology, 16*(3), 45–65.

Bayat, M., Zargar, M., Murtazova, K. M. S., Nakhaev, M. R., & Shkurkin, S. I. (2022). Ameliorating seed germination and seedling growth of nano-primed wheat and flax seeds using seven biogenic metal-based nanoparticles. *Agronomy, 12*(4), 811.

Catiempo, R., Photchanachai, S., Bayogan, E. R., & Wongs-Aree, C. (2021). Impact of hydropriming on germination and seedling establishment of sunflower seeds at elevated temperature. *Plant, Soil and Environment, 67*(9), 491–498.

Chavan, A. S., Raj, V. C., & Waghmare, P. K. (2018). Residues studies of pendimethalin and other herbicides in pigeon pea field through bioassay technique. *International Journal of Current Microbiology and Applied Sciences, 6*, 1866–1873.

Chinnamuthu, C. R., & Kokiladevi, E. (2007). Weed management through nanoherbicides. *Application of Nanotechnology in Agriculture, 10*, 978–981.

Choudhary, R. C., Kumaraswamy, R. V., Kumari, S., Sharma, S. S., Pal, A., Raliya, R., Biswas, P., & Saharan, V. (2019). Zinc encapsulated chitosan nanoparticle to promote maize crop yield. *International Journal of Biological Macromolecules, 127*, 126–135.

Damalas, C. A., Koutroubas, S. D., & Fotiadis, S. (2019). Hydro-priming effects on seed germination and field performance of faba bean in spring sowing. *Agriculture, 9*(9), 201.

Das, I., Sonowal, A., Gogoi, B., Sharma, B., Patowary, K., & Borah, D. (2021). Citrus maxima (Burm.) Merr. fruit juice and peel extract mediated synthesis of silver nanoparticles (Ag NPs) and their applications as antimicrobial agents and plant growth enhancers. *Kuwait Journal of Science*. https://doi.org/10.48129/kjs.19889.

Di Girolamo, G., & Barbanti, L. (2012). Treatment conditions and biochemical processes influencing seed priming effectiveness. *Italian Journal of Agronomy, 7*(2), e25.

Dobbelaere, S., Vanderleyden, J., & Okon, Y. (2003). Plant growth-promoting effects of diazotrophs in the rhizosphere. *Critical Reviews in Plant Sciences*, *22*(2), 107–149.

Falco, W. F., Scherer, M. D., Oliveira, S. L., Wender, H., Colbeck, I. T., & Caires, A. R. (2020). Phytotoxicity of silver nanoparticles on Vicia faba: Evaluation of particle size effects on photosynthetic performance and leaf gas exchange. *Science of the Total Environment*, *701*, 134816.

Farooq, M., Hussain, M., Nawaz, A., Lee, D. J., Alghamdi, S. S., & Siddique, K. H. (2017). Seed priming improves chilling tolerance in chickpea by modulating germination metabolism, trehalose accumulation and carbon assimilation. *Plant Physiology and Biochemistry*, *111*, 274–283.

Feizi, H., Kamali, M., Jafari, L., & Moghaddam, P. R. (2013). Phytotoxicity and stimulatory impacts of nanosized and bulk titanium dioxide on fennel (*Foeniculum vulgare* Mill). *Chemosphere*, *91*(4), 506–511.

Galaktionova, L. V., Korotkova, A. M., Voskobulova, N. I., Lebedev, S. V., Terehova, N. A., & Vershinina, I. A. (2020). *Evaluation of the effect of SiO2 and Fe3O4 nanoparticles on Pisum sativum seeds in laboratory and field experiments*. https://doi.org/10.1101/2020.08.31.275859

Garza-Alonso, C. A., Gonzalez-Garcia, Y., Cadenas-Pliego, G., Olivares-Saenz, E., Trejo-TéLlez, L. I., & Benavides-Mendoza, A. (2021). Seed priming with ZnO nanoparticles promotes early growth and bioactive compounds of Moringa oleifera. *Notulae Botanicae Horti Agrobotanici Cluj-Napoca*, *49*(4), 12546.

Gholami, S., Dehaghi, M. A., Rezazadeh, A., & Naji, A. M. (2022). Seed germination and physiological responses of quinoa to selenium priming under drought stress. *Bragantia*, *81*, e0722.

Girolamo, D. G., & Barbanti, L. (2012). Treatment conditions and biochemical processes influencing seed priming effectiveness. *Italian Journal of Agronomy*, *7*(2), e25.

Godoy, F., Olivos-Hernández, K., Stange, C., & Handford, M. (2021). Abiotic stress in crop species: Improving tolerance by applying plant metabolites. *Plants*, *10*(2), 186.

Gomez, A., Narayan, M., Zhao, L., Jia, X., Bernal, R. A., Lopez-Moreno, M. L., & Peralta-Videa, J. R. (2021). Effects of nano-enabled agricultural strategies on food quality: Current knowledge and future research needs. *Journal of Hazardous Materials*, *401*, 123385.

Gross, M. S., Bean, T. G., Hladik, M. L., Rattner, B. A., & Kuivila, K. M. (2020). Uptake, metabolism, and elimination of fungicides from coated wheat seeds in Japanese quail (*Coturnix japonica*). *Journal of Agriculture and Food Chemistry*, *68*, 1514–1524.

Gupta, N., Singh, P. M., Sagar, V., Pandya, A., Chinnappa, M., Kumar, R., & Bahadur, A. (2022). Seed priming with ZnO and Fe_3O_4 nanoparticles alleviate the lead toxicity in *Basella alba* L. through reduced lead uptake and regulation of ROS. *Plants*, *11*(17), 2227.

Hacisalihoglu, G. (2007). Germination characteristics of three warm-season turfgrasses subjected to matriconditioning and aging. *HortTechnology*, *17*(4), 480–485.

Ibrahim, E. A. (2016). Seed priming to alleviate salinity stress in germinating seeds. *Journal of Plant Physiology*, *192*, 38–46.

Ibrahim, N. D., Bhadmus, Z., & Singh, A. (2013). Hydro-priming and re-drying effects on germination, emergence and growth of upland rice (*Oryza sativa* L.). *Nigeria Journal of Basic Applied Science*, *21*, 157–164.

Iqbal, M., & Ashraf, M. (2013). Gibberellic acid mediated induction of salt tolerance in wheat plants: Growth, ionic partitioning, photosynthesis, yield and hormonal homeostasis. *Environmental and Experimental Botany*, *86*, 76–85.

Iqbal, M., Ashraf, M., & Jamil, A. (2006). Seed enhancement with cytokinins: Changes in growth and grain yield in salt stressed wheat plants. *Plant Growth Regulation*, *50*(1), 29–39.

Itroutwar, P. D., Kasivelu, G., Raguraman, V., Malaichamy, K., & Sevathapandian, S. K. (2020). Effects of biogenic zinc oxide nanoparticles on seed germination and seedling vigor of maize (*Zea mays*). *Biocatalysis and Agricultural Biotechnology*, *29*, 101778.

Jisha, K. C., & Puthur, J. T. (2018). Seed halopriming outdo hydropriming in enhancing seedling vigor and osmotic stress tolerance potential of rice varieties. *Journal of Crop Science and Biotechnology*, *17*, 209–219.

Kandhol, N., Singh, V. P., Peralta-Videa, J., Corpas, F. J., & Tripathi, D. K. (2022). Silica nanoparticles: The rising star in plant disease protection. *Trends in Plant Science*, *27*(1), 7–9.

Khan, I., Raza, M. A., Awan, S. A., Shah, G. A., Rizwan, M., Ali, B., & Huang, L. (2020). Amelioration of salt induced toxicity in pearl millet by seed priming with silver nanoparticles (AgNPs): The oxidative damage, antioxidant enzymes and ions uptake are major determinants of salt tolerant capacity. *Plant Physiology and Biochemistry*, *156*, 221–232.

Li, J., Ovakim, D. H., Charles, T. C., & Glick, B. R. (2000). An ACC deaminase minus mutant of Enterobacter cloacae UW4No longer promotes root elongation. *Current Microbiology*, *41*(2), 101–105.

Mahajan, P., Dhoke, S. K., & Khanna, A. S. (2011). Effect of nano-ZnO particle suspension on growth of mung (*Vigna radiata*) and gram (*Cicer arietinum*) seedlings using plant agar method. *Journal of Nanotechnology*. https://doi.org/10.1155/2011/696535

Mahakham, W., Sarmah, A. K., Maensiri, S., & Theerakulpisut, P. (2017). Nanopriming technology for enhancing germination and starch metabolism of aged rice seeds using phytosynthesized silver nanoparticles. *Scientific Reports*, *7*, 1–21.

Mahmood, A., Turgay, O. C., Farooq, M., & Hayat, R. (2016). Seed biopriming with plant growth promoting rhizobacteria: A review. *FEMS Microbiology Ecology*, *92*(8), fiw112.

Marthandan, V., Geetha, R., Kumutha, K., Renganathan, V. G., Karthikeyan, A., & Ramalingam, J. (2020). Seed priming: A feasible strategy to enhance drought tolerance in crop plants. *International Journal of Molecular Science*, *21*, 8258.

Matsushima, K. I., & Sakagami, J. I. (2013). Effects of seed hydropriming on germination and seedling vigor during emergence of rice under different soil moisture conditions. *American Journal of Plant Sciences*, *4*, 1584–1593.

McDonald, M. B. (2000). Seed priming. In M. Black & J. D. Bewley (Eds.), *Seed technology and its biological basis* (pp. 287–325). Sheffield Academic Press.

Muller, K., Linkies, A., Vreeburg, R. A., Fry, S. C., Krieger-Liszkay, A., & Leubner-Metzger, G. (2009). In vivo cell wall loosening by hydroxyl radicals during cress seed germination and elongation growth. *Plant Physiology*, *150*(4), 1855–1865.

Nasibi, F., Aminian, F., Mohammadinejad, G., & Hassanshahian, M. (2022). Seed priming with selenium nanoparticle and plant growth promoting rhizobacteria improve seedling development of foxtail millet (*Setaria italica*) under salinity stress. *Research Square*.

Ndeh, N. T., Maensiri, S., & Maensiri, D. (2017). The effect of green synthesized gold nanoparticles on rice germination and roots. *Advances in Natural Science in Nanoscience and* Nanotechnology, *8*(3), 035008.

Ningombam, B., Devi, A. S., & Darvhankar, M. (2021). Effects of abiotic stresses on crop yield: A review. *Pharma Innovation*, *10*(5), 418–422.

Paparella, S., Araújo, S. S., Rossi, G., Wijayasinghe, M., Carbonera, D., & Balestrazzi, A. (2015). Seed priming: State of the art and new perspectives. *Plant Cell Reports*, *34*(8), 1281–1293.

Park, H. J., Kim, S. H., Kim, H. J., & Choi, S. H. (2006). A new composition of nanosized silica-silver for control of various plant diseases. *The Plant Pathology Journal*, *22*(3), 295–302.

Pereira, A. E. S., Oliveira, H. C., Fraceto, L. F., & Santaella, C. (2021). Nanotechnology potential in seed priming for sustainable agriculture. *Nanomaterials*, *11*(2), 267.

Rahimi, D., Kartoolinejad, D., Nourmohammadi, K., & Naghdi, R. (2016). Increasing drought resistance of Alnus subcordata CA Mey. Seeds using a nano priming technique with multi-walled carbon nanotubes. *Journal of Forest Science*, *62*(6), 269–278.

Raj, S. N., Shetty, N. P., & Shetty, H. S. (2004). Seed bio-priming with Pseudomonas fluorescens isolates enhances growth of pearl millet plants and induces resistance against downy mildew. *International Journal of Pest Management*, *50*(1), 41–48.

Rakshit, A., Sunita, K., Pal, S., Singh, A., & Singh, H. B. (2015). Bio-priming mediated nutrient use efficiency of crop species. In A. Rakshit, H. B. Singh & A. Sen (Eds.), *Nutrient use efficiency: From basics to advances* (pp. 181–191). Springer.

Reddy, P. P. (2012). Bio-priming of seeds. In P. P. Reddy (Ed.), *Recent advances in crop protection* (pp. 83–90). Springer.

Rehman, S., Harris, P. J. C., & Ashraf, M. (2005). Stress environments and their impact on crop production. In M. Ashraf & P. J. C. Harris (Eds.), *Abiotic stresses: Plant resistance through breeding and molecular approaches* (pp. 3–18). Haworth Press.

Sahebi, M., Hanafi, M. M., Siti Nor Akmar, A. M., Rafii, Y., Azizi, P., & Shabanimofrad, M. (2015). Importance of silicon and mechanisms of biosilica formation in plants. *BioMed Research International*. https://doi.org/10.1155/2015/396010

Sathiyabama, M., & Muthukumar, S. (2020). Chitosan guar nanoparticle preparation and its in vitro antimicrobial activity towards phytopathogens of rice. *International Journal of Biological Macromolecules*, *153*, 297–304.

Shah, V., & Belozerova, I. (2009). Influence of metal nanoparticles on the soil microbial community and germination of lettuce seeds. *Water, Air, & Soil Pollution*, *197*(1), 143–148.

Shelar, A., Singh, A. V., Maharjan, R. S., Laux, P., Luch, A., Gemmati, D., & Patil, R. (2021). Sustainable agriculture through multidisciplinary seed nanopriming: Prospects of opportunities and challenges. *Cells*, *10*(9), 2428.

Singh, D. P., Singh, V., Shukla, R., Sahu, P., Prabha, R., Gupta, A., & Gupta, V. K. (2020). Stage-dependent concomitant microbial fortification improves soil nutrient status, plant growth, antioxidative defense system and gene expression in rice. *Microbiological Research*, *239*, 126538.

Singh, N., Bhuker, A., & Jeevanadam, J. (2021). Effects of metal nanoparticle-mediated treatment on seed quality parameters of different crops. *Naunyn-Schmiedeberg's Archives of Pharmacology*, *394*(6), 1067–1089.

Siringam, K., Juntawong, N., Cha-um, S., & Kirdmanee, C. (2011). Salt stress induced ion accumulation, ion homeostasis, membrane injury and sugar contents in salt-sensitive rice (*Oryza sativa* L. spp. indica) roots under isoosmotic conditions. *African Journal of Biotechnology*, *10*(8), 1340–1346.

Tarafdar, J. C., Raliya, R., & Rathore, I. (2012). Microbial synthesis of phosphorous nanoparticle from tri-calcium phosphate using *Aspergillus tubingensis* TFR-5. *Journal of Bionanoscience*, *6*(2), 84–89.

Verma, A. K., Singh, V. P., & Vikas, P. (2012). Application of nanotechnology as a tool in animal products processing and marketing: An overview. *American Journal of Food and Technology*, *10*, 389–398.

Yacoubi, R., Job, C., Belghazi, M., Chaibi, W., & Job, D. (2013). Proteomic analysis of the enhancement of seed vigour in osmoprimed alfalfa seeds germinated under salinity stress. *Seed Science Research*, *23*, 99–110.

Yadav, S. P., Bharadwaj, R., Nayak, H., Mahto, R., Singh, R. K., & Prasad, S. K. (2019). Impact of salt stress on growth, productivity and physicochemical properties of plants: A review. *International Journal of Chemical Studies*, *7*(2), 1793–1798.

Yasmin, H., Mazher, J., Azmat, A., Nosheen, A., Naz, R., Hassan, M. N., & Ahmad, P. (2021). Combined application of zinc oxide nanoparticles and biofertilizer to induce salt resistance in safflower by regulating ion homeostasis and antioxidant defence responses. *Ecotoxicology and Environmental Safety*, *218*, 112262.

7 Perspectives of Nanoparticles as Priming Agents for Amelioration of Abiotic Stresses in Crops

Zahoor Ahmad, Ejaz Ahmad Waraich, Muhammad Aamir Iqbal, Muhammad Ashar Ayub, Amir Aziz, Amber Raza, Sundas Sana, Jawaria Jameel, Hira Nawaz, Muhammad Ahmad, Sarfraz Ahmed and Shehar Bano*

7.1 INTRODUCTION

Climate change and world food security are the two prime challenges that humans are dealing with in the twenty-first century (Ahmad et al., 2017; Liaqat et al., 2022). Abiotic stresses (AS) such as drought (DS), heat (HS), chilling (CS), salinity (SS), heavy metals (HMs) toxicity, water logging (WL), eroded soils (ES) and nutrients imbalance (NI) are continuously getting worse due to climate change (Ahmad et al., 2018a, 2018b, 2018c). The increased quantities of greenhouse gases, particularly carbon dioxide (CO_2), cause global warming and a variety of abiotic stresses (Vaughan et al., 2018). Plant health and production are impacted by AS such as extreme HS and CS conditions (Saeed et al., 2022), and the presence of HMs in the soil (Naz et al., 2022; Ali et al., 2022). Numerous abiotic stressors constantly pose a threat to the plants' ability to grow, survive and produce. When plants experience stress for the first time, they are able to collect molecular data and retain it as a type of memory, enabling them to respond skillfully and quickly to subsequent shocks. The priming-induced stress response is the term used to describe this process (Nair et al., 2022).

Nanotechnology-based solutions are being used more and more frequently in various human endeavours, such as crop production and (agro) ecosystems (Ondrasek et al., 2021). For instance, the use of specific nanobased technologies and various methods which include nanofiltration are in practice nowadays for stress tolerance in crop

* Corresponding author.

DOI: 10.1201/9781003376446-7

plants (Zhao et al., 2019). The priming of seeds is essential for managing stressful situations. Seed priming is essential for the plant to have better memory and to adapt tolerance against various AS. During seed imbibition, a process known as "seed priming" is used before sowing (Sen & Puthur, 2020). Three stages of seed development are involved in imbibition: fast water intake, a saturation of water absorbing capacity and water intake, along with the beginning of new cell formation and expansion of the cell. It indicates that the second stage is crucial for seed priming (SP) (Cheng et al., 2017; Arshad et al., 2020). Recent climate estimates notified that a rise in global temperatures, disturbance in rainfall patterns and CO_2 emissions could have a significant impact on how much main crops are produced worldwide (Imran et al., 2018). HS is the term used to describe an increase in temperature that lasts longer than necessary and causes irreversible harm to growth and development (Li et al., 2018). By decreasing the weight and number of grain plants, HS also disrupts the grain filling (Patriyawaty et al., 2018). Gradual rise in temperature has a detrimental effect on shoot growth by reducing the length of the first internode and results in plant death at early stages of life (Ahmad et al., 2019). As a result, growing season temperatures and their fluctuation will rise, changing the phenology, physiological function and production of plants (Dusenge et al., 2019). Heat-shock proteins (HSPs), molecular chaperones that help protect cells from heat-induced disruptions and assist with the refolding of heat-inactivated proteins, are produced rapidly as a result of HS (Yadav et al., 2020).

Life on Earth depends on water, which is a crucial component of cell protoplasm. However, dramatic climate changes can reduce water availability, leading to a variety of problems, such as the current water-scarce regimes in several biome regions (Wahab et al., 2022). The loss of crops brought on by DS has significantly increased in recent years as a result of climate change and altered rainfall patterns. To counteract the negative effects of drought stress, naturally drought-tolerant plants mobilize a variety of metabolites and low molecular weight proteins. The DS alone has caused a $30 billion global crop yield loss over the past ten years (Gupta et al., 2020). It is a multifaceted stress that negatively affects the rate of carbon assimilation, reduces turgor pressure and modifies photosynthesis, respiration, carbohydrate metabolism and nutrient uptake in the leaves of plants. In many regions of the world, salinity has emerged as a prime issue that has adversely affected agricultural soils. Because salt interferes with several fundamental metabolic processes, it has a detrimental impact on new plant emergence and developmental patterns (Hameed et al., 2021). Higher plants, which are particularly adaptable to bear extreme salinity, have evolved a variety of AS tolerance mechanisms. These mechanisms include salt detoxification, osmotic adjustment regulation, maintenance of cation and anion equilibrium, also scavenging activity of reactive oxygen species and manufacturing suitable solutes (Ye et al., 2019). It is important to note that various xenobiotics are released into the environment as a result of human activities. The HMs and metalloids are known as trace elements due to their presence in trace amounts (less than 10 ppm) in different mediums. These metals have surpassed various known pollutants like synthetic plant protection products and oxides of carbon and sulphur (Sytar et al., 2019). The CS disturbs the balance of macro- and micro-nutrients in the topsoil and promotes the uptake of metallic ions by plants. This may intensify the issue of heavy metal accumulation (Zeiner et al., 2022). In addition, HMs harm many plant species and other living things. Increased ROS formation is brought on by either direct or indirect involvement in redox reactions (less reactive metals) in concentrations above the maximum tolerable amount, which also affects plant functional and biochemical processes (Rehman et al., 2020; Ahmed et al., 2022). These physiological processes include the biosynthesis of chlorophyll, increased photosynthetic ability, changes in transpiration rate, increased water uptake and nutrient balance and enzyme actions responsible for different metabolic activities (Liu et al., 2022).

Nearly every subcellular compartment produces the four primary ROS. These are hydrogen peroxide, superoxide, singlet oxygen and hydroxyl radical, primarily as waste products of regular plant metabolism (Castro et al., 2021). All of these cellular alterations lead to morphological modifications, including shoot and root dwarfisms, the appearance of necrotic spots and chlorotic blemishes and the reduction in the number of leaves and size leading to early ageing of the plant (Muszyska et al., 2021).

7.2 ROLE OF NPS DURING HEAT-INDUCED OXIDATIVE STRESS IN PLANTS

A new scientific revolution has been sparked by the developing science of nanotechnology, which has been ingrained in all facets of daily life. The fundamental unit of nanotechnology is a particle created at the nanoscale, with at least a one-dimension size of less than 100 nm (Kulkarni et al., 2022). Crop plants may become stressed from prolonged high temperatures. Plants can die from drought and excessive heat, and they become much more vulnerable to biotic and abiotic dangers as a result (Onyekachi et al., 2019). The agricultural sector's challenges will only worsen as global warming lengthens and intensifies heat waves (Sun et al., 2019). In response to the drying and dehydration process, stress-related responses such as heat-shock proteins and antioxidant systems are triggered, enhancing cross-resistance against various abiotic stresses (Salam et al., 2022). When NPs are administered to plants in low concentrations, they have antioxidative capabilities; however, when NPs are treated in large quantities, oxidative stress results. Plants under heat stress produce molecular chaperones and heat-shock proteins. In addition to being engaged in heat stress resistance, a heat-shock protein helps other proteins keep their stability under stressful circumstances.

7.2.1 Physio-biochemical Aspects

Nanopriming has been shown to affect plants' stability, growth and physiology in addition to controlling seed germination by shielding them from abiotic stress (Iqbal et al., 2020). This intervention contributes to enhancing plant resilience to abiotic stress, influencing nutritional uptake, modifying metabolic processes, regulating cellular antioxidants, optimizing photosynthetic efficiency, and impacting molecular pathways(Hassan et al., 2018). The drying and hydration processes trigger stress-related reactions, such as heat-shock proteins and antioxidant systems, which strengthen cross-resistance against various abiotic stresses. Numerous researchers were conducted to understand how NPs' application impacts plant physiology and growth under abiotic stress. In contrast to non-primed seeds, SiO_2-treated Purna HI 1544 (a wheat cultivar) had more proline and catalase (Rai-Kalal et al., 2021). This increased proline and catalase led to better maintenance of biochemical balance and photosynthetic parameters and increased tolerance against drought stress. The application of selenium NPs at 100g/ml concentration improves plant growth, photosynthetic rate and gas exchange at high temperatures (El-Saadony et al., 2021). Another study recorded similar results when ZnO NPs were applied in mungbean, enhancing the chlorophyll activity, gas exchange and enzyme balance in mungbean. Furthermore, it also enhanced the agronomic and yield characteristics i.e., pod size number and total grain yield (Kareem et al., 2022). Similarly, TiO_2 and ZnO application enhanced the root and shoot characteristics and antioxidant defence mechanisms in wheat under heat stress (Thakur et al., 2022). The application of NP aids in maintaining ion concentration in plants, which in turn promotes membrane stability and osmotic control. Production is also guaranteed by avoiding enzymatic oxidative stress in stressful situations. Under conditions of heat and cold stress, the production of enzymatic anti-oxidizers, such as SOD, POD, CAT and APX, is promoted by the application of NPs. However, the concentration of NPs' application is crucial since at greater concentrations, NPs' toxicity can be seriously detrimental (Rajput et al., 2021). The HS can have various deleterious impacts on a plant's growth, development, physiological processes and yield. Oxidative stress occurs when plants experience an excessive amount of ROS generation because of high temperatures (Mittler et al., 2022). Various metabolic processes require different enzymes, and they respond differently to temperature. Enzymes may detach from metabolic pathways in reaction to HS, which leads to the buildup of ROS, especially singlet oxygen ($^1O_2$), superoxide radicals (O_2), hydrogen peroxide (H_2O_2) and hydroxyl radicals (Debnath et al., 2022).

7.2.2 Molecular Aspect

The regulation of plant stress response, which is controlled by gene expression and subsequently affects the production of enzymes and proteins, has a direct impact on crop productivity. Tolerance to HS and CS is regulated by several genes, transcription factors (TFs) and proteins. Plant survival is directly impacted by the activation of different TFs, whether it be up-regulation or down-regulation. ZnO-NPs applied topically to rice leaves increased the antioxidative system's chilling-induced gene expression (Song et al., 2021). The TFs i.e., EREB, R2R3MYB, HSF-34, WRKY1, MAPK1, HDA3 and CAT proteins were transactionally elevated by ZnO-NPs in soybean seedlings, which led to an increase in photosynthetic pigments, proline concentration, antioxidant enzyme activity and plant yields (Mirakhorli et al., 2021). Brassinosteroids production, oxidative stress response and root development genes were discovered to be regulated by (50 nm) Cu-based NPs in transcriptome research (Ali et al., 2019). Accumulation of secondary metabolites (acetyl glucosamine, phenyl lactate and 4-aminobutyrate) involved in cell signalling and immune responses was calculated under Cu-NPs with a size range of 40 nm. The degradation of flavonoids, fatty acids, riboflavin and amino acids in metabolites involved in production and defence responses has been previously reported (Zhao et al., 2016). Similarly, TaPIP1 and TaNIP2 aquaporin genes were shown to be overexpressed in wheat seedlings when silicon was applied rather than Si NPs, increasing the relative water content (RWC) in wheat (Younis et al., 2020). The HS-induced reduction of transcriptional gene silencing (TGS)- beta-glucuronidase (GUS) gene was greatly improved in *Arabidopsis thaliana* seedlings grown in ZnO-NPs and subjected to HS (37°C) (Wu and Wang, 2020). As a result of increased rubisco activase (RCA) mRNA levels and activity brought on by nano-anatase, rubisco carboxylation and rates of photosynthetic carbon production were improved. The administration of lanthanum oxide (La_2O_3) to the root tips of maize seedlings altered the expression of genes for aquaporins such tonoplast intrinsic proteins (TIPs), plasma membrane intrinsic proteins (PIPs), basic intrinsic proteins (SIPs) and Nod26-like intrinsic proteins (NIPs) (Yue et al., 2017). Numerous genes and TFs that are induced or regulated by cold and heat stress are influenced differently by various NPs. The ability of plants to withstand stress is increased by the overexpression of HS and CS-regulating genes and TFs, which allows for physiological, molecular and biochemical changes in plants (Khalid et al., 2022).

7.3 ROLE OF NPS DURING DROUGHT-INDUCED OXIDATIVE STRESS IN PLANTS

The NPs enter the plant body through the roots and leaves, and after entering the plant body, NPs induce biochemical, morphological, molecular and physiological changes in the crops (Khan et al., 2019). These changes significantly affect plant growth depending on the concentration, size and application method of the NPs. Moreover, the size, chemical nature and reactivity of NPs can substantially impact plants. The available evidence indicates that NPs mitigate the adverse effects of DS and appreciably improve plant growth and development (Ahmed et al., 2021).

7.3.1 Physio-biochemical Aspects

The water shortage in the soil is not always the cause of the physiological drought (Gupta et al., 2020), rather it occurs when a plant cannot get enough water (Malinowska et al., 2020). Physiological, biochemical, anatomical, morphological and long- and short-term developmental and growth-related adaptable techniques might be involved (Chaudhry & Sidhu, 2022). Reducing leaf relative water content, turgor loss and stomatal closure are the frequent consequences of drought stress in barley (*Hordeum vulgare* L.) (Istanbuli et al., 2020). During drought, leaf wilting and abscission reduce water loss via transpiration (Wu et al., 2022). When there is DS, cell enlargement in higher plants is hampered by the interruption of xylem water flow. When DS is minimal to nonexistent, stomatal closure, cell membrane structural damage and plant metabolic disturbances occur (Kim et al., 2017). The results suggested by Tůmová et al. (2018) concluded that numerous internal and external conditions govern internal plant water interactions in maize such as the stomatal resistance, relative water contents (RWC), rate of transpiration, leaf temperature of wheat crop (Triticum aestivum L.), leaf water potential and the canopy temperature just above the plant (Badr & Brüggemann, 2020). Leaf RWC is a crucial controller of physiological processes in plants. RWC reduction is the first symptom of the drought stress response (Hussain et al., 2018). The relative water content of leaves strongly correlates simultaneously with leaf tissue growth rate and rate of transpiration (Kapoor et al., 2020). Lower RWC reduces leaf water potential, causing stomata to contract. Transpiration is the primary mechanism governing leaf temperature; increasing stomatal resistance minimizes the transpiration rate in rice leaves due to ABA content and increases leaf temperature (Wu et al., 2022). In wheat, the leaf's relative water content increases throughout development and decreases as dry matter accumulates as the leaf ages (Alghory, 2019). Water-stressed wheat and rice plants contained less water content as compared to those wheat and rice plants that were grown under controlled conditions. A decline in relative water content induced a drop in water content and osmotic potential under stress regimes. In wheat, the state of reduced leaf turgor pressure disrupts plant metabolic functions. Under drought conditions, crop development is impeded by a lower soil water potential, and the resulting lowered plant osmotic potential leads to low nutrient absorption (Yan et al., 2018).

The DS causes plants to be adapted accordingly by regulating their stomata movement, adjusting their osmotic balance and mounting an antioxidant defence (Kamran et al., 2020). However, a protracted period of high-intensity, limited water conditions might slow plant development, alter the morphological structure and biomass distribution pattern in tomato crops (*Solanum lycopersicum* L.) or cause mortality (Chai et al., 2016). Drought stress significantly influences the photosynthetic system and its pigments, such as chlorophyll a, b and carotenoids (Zia-Ul-Haq et al., 2021). Drought stress also impacts complex systems such as photosystems I and II. Drought stress significantly influences plant starch production by affecting the Calvin cycle and enzyme activity (ribulose phosphate). The first sign of a plant's drought stress response is closing its stomata. When drought stress becomes more severe during the day, stomata progressively close in sugar beet (*Beta vulgaris* L.) (Zia-Ul-Haq et al., 2021). Stomata are entirely closed in extreme drought stress conditions. Still, full closure varies among plant species depending on their specific tolerances to drought conditions, as shown in pea crops (*Pisum sativum* L.) (Hashmat et al., 2021). As a result, plant species tolerance influences the stomatal mechanism, which regulates carbon fixation rates, photosynthesis and water usage efficiency. When stomata restrict CO_2 uptake into the leaves, more electrons are available to produce active oxygen species (Yang et al., 2021). When physiological processes at the stomata are reduced by environmental conditions that increase transpiration rates, then the pH of the leaf sap is elevated; Foyer, (2018) observed reductions of photosynthesis, ROS production regulations decreased and stomatal conductance under drought stress could be recovered by following re-watering (Yang et al., 2021).

Accumulating biochemicals such as proline, protein, sugar and glycine betaine (GB) improve crop production by scavenging ROS-generated oxidative stress (Perveen & Hussain, 2021). Moreover, physiological processes including cellular respiration, rate of photosynthesis, mineral nutrition, enzymatic activities and redox (oxidation/reduction) homeostasis are influenced by drought stress regimes.

Likewise, biochemicals, including membrane lipo-proteins, DNA and cellular protein content, deteriorate under water-limited conditions (Khan et al., 2021). Plants withstand drought stress regimes by developing various biochemical, structural and molecular strategies, including the accumulation of certain osmolytes such as proline, proteins, sugars and glycine betaine. Applying salicylic acid improved drought stress tolerance by upholding redox potential and activating proline biosynthesis (Abdelaal et al., 2020) compatible solutes such as proteins, proline, glycine betaine, phenolic compounds, soluble sugars and organic acids accumulated chiefly in the cytoplasm in response to limited water availability by scavenging ROS, improving the water potential and protecting biological molecules from lipid peroxidation (Raza et al., 2019). Plant cells collect soluble chemicals during drought stress and increase cytoplasm viscosity. Under some situations, the content of these unique chemicals may become toxic, causing issues with enzyme development and the entire photosynthetic process (Mahmood et al., 2021). The rate of regeneration of ribulose-1, 5-bisphosphate, the maximum rate of ribulose-1, 5-carboxylate, NADP-malic enzyme, phosphoenolpyruvate carboxylase, rubisco, fructose-1, 6-bisphosphatase and orthophosphate-Di kinase pyruvate are all reduced as a result of the rapid decrease in "dry" photosynthesis (Wu et al., 2021,). Non-cyclic electron transport is similarly lowered to satisfy the needs of decreased NADPH synthesis, ATP production and ROS production. Different cultivars may respond and adapt differently to drought stress. According to transcriptome studies, drought-tolerant and sensitive wheat genotypes may use distinct molecular processes to deal with drought stress. Differential expression of numerous drought-inducible genes involved in regulation, cell defence and cellular component remodelling is one of the most noticeable changes (Yang et al., 2021). According to transcriptome research, drought-tolerant and sensitive wheat (*Triticum aestivum* L.) genotypes may use molecular methods to cope with drought stress (Kapoor et al., 2020). One of the most noticeable changes is the differential expression of several drought-inducible genes involved in cell defence regulation and cellular component remodelling (Iqbal, 2018). While many of these genes are activated in drought-sensitive wheat (*Triticum aestivum L.*) genotypes and contribute to limiting drought impacts and perception, many of these genes are expressed constitutively in tolerant genotypes (Rauf et al., 2007)

Furthermore, signal transduction and hormone-dependent regulation mechanisms change among *Triticum aestivum L.* genotypes (Ghatak et al., 2017). Drought stress-tolerant genotypes perceive drought quickly and activate signal transduction pathways that trigger downstream components, helping them withstand drought stress (Kaur et al., 2021). When there is a lack of water, chemicals and metabolites including proline, glycine betaine (GB) and soluble sugar accumulate in the cytoplasm, assisting in osmotic adjustment and preparing the plant to cope with the adverse effects of oxidative stress in *Triticum aestivum L.* (Dhanda et al., 2004). These metabolites are significant because their distinct biochemical processes promote plant tolerance drought signalling results in crosstalk between various biological molecules and metabolites. Proline is an essential metabolite that accumulates in higher amounts in water-stressed environments (Pourghasemian et al., 2020).

7.3.2 Molecular Aspect

The deficiency of water causes oxidative stress by increasing the accumulation of MDA, H_2O_2 and ROS (Sutulienė et al., 2022). The NPs possess an excellent potential to improve antioxidant activity and mitigate the detrimental impacts of DS. The application of NPs (ZnO, Se and Si) significantly improved APX, CAT and SOD activity, which reduced drought-induced oxidative damage (Ali et al., 2021; Sutulienė et al., 2022). In the same context, the NPs also increase the relative abundance of Cu/Zn-SOD, APX and CAT under DS, which noticeably scavenge the ROS. The application of NPs (ZnO) also improves non-enzymatic antioxidant activities (phenolic compounds, AsA) that work in coordination with antioxidant enzymes (APX, CAT and SOD) (El-Zohri et al., 2021). NPs (SiO_2) significantly increased non-enzymatic activities by increasing TPC, DPPH and FRAPS activity in plants grown under DS (Sutulienė et al., 2022). The NPs trigger the accumulation of antioxidant genes, osmolytes, nutrients and amino acids, which increases antioxidant activities and thus protects the plants from oxidative stress (Mittal & Kakkar, 2020).

The use of NPs (TiO_2) also increases nitrate reductase activity and accumulation of osmolytes. NP-mediated increases in NR activity lead to NO formation being induced, which stimulates proline and glycine betaine (GB) synthesis, and protects the plants from oxidative damage (Khan et al., 2019). NPs also significantly increase the accumulation of glucose, fructose, trehalose and sucrose, which increase antioxidant activities and, as a result, improve drought tolerance (Heikal et al., 2022). The NPs mediated improvements in antioxidant activities mitigate the adverse effects of DS on plants by scavenging the ROS.

Applying Zn-NPs (100 ppm) improved the length and width of the stem and vascular cylinder, and increased the thickness of the cortex and vascular cylinder (Semida et al., 2021). The plant processes, including physiological, biochemical and anatomical mechanisms, are responsible for the performance of plants, and these functions are linked to a plant's internal anatomy. Application of NPs improved anatomical parameters, RWC, membrane stability and nutrient status. Water deficiency decreases vessel diameter and vascular length and width. A foliar spray of Zn-NPs significantly increased the thickness of the epidermis, the size and width of the vascular bundle and the diameter of the vessel, which improved drought tolerance (Al-Dhalimi & Al-Ajeel, 2020). There is limited information available in the literature regarding the role of NPs on plant ultra-structural changes under DS. Therefore, more studies are urgently required to investigate the role of NPs in these aspects under DS.

In photosynthesis, CO_2 and H_2O within the chloroplast of plant cells produce sugars and O_2 as a byproduct in the presence of light. Chlorophyll is an essential component of chloroplasts required for photosynthesis (Duursma et al., 2019). Chlorophyll pigments are essential for photosynthesis and are affected by water-stressed conditions during stomatal closure and openings in *Nicotiana tabacum L.* (Hu et al., 2018). Plants must capture light and use it during the photosynthesis process. Under drought stress, the chlorophyll concentration is dramatically reduced due to increased oxidative stress, degeneration or photo-oxidation of chlorophyll pigments (Allakhverdiev

2020). Drought sensitivity in (*Triticum aestivum L.*) was predominantly connected with reductions in stomatal conductance, which decreased the delivery of carbon dioxides to chloroplasts and, consequently, reduced net photosynthesis (Soares et al., 2019). The results determined that drought stress affected plant growth and development by lowering the rate of photosynthesis (Ferrara et al., 2011). The major factors responsible for slowing photosynthesis might be stomatal closure (reduced stomatal CO_2 fixation), non-stomatal (decreased photosynthesis activity in mesophyll tissues) or both (Li et al., 2021). The high sensitivity of connecting photosystems II (PS-II) following limiting tensions induced by external variables motivates drought stress in harming these systems, which are reaction locations. Methods of chlorophyll fluorescence revealed a hazard and suggested that manufacturing operations were not balanced (Pourghasemian et al., 2020).

7.4 ROLE OF NPS DURING SALINITY-INDUCED OXIDATIVE STRESS IN PLANTS

Soil salinization is a serious threat to crop productivity and is predicted to increase in the face of global climate change (Gustafsson et al., 2013). Estimations show that as much as 12 billion US$ will be lost globally, each year, due to the reduction in agricultural production from salt-affected areas. The situation is the worst in arid and semi-arid regions, which are deficient in water and face high temperatures, resulting in more water loss from plants due to higher evapotranspiration rates, aggravating the effects of salinity (Flowers et al., 2010).

7.4.1 Physio-biochemical Aspect

Salinity stress induces a multitude of responses in plants including morphological, physiological, biochemical and molecular changes (Ambede et al., 2012). It causes ionic imbalance, which results in ionic toxicity, osmotic stress and the generation of reactive oxygen species (Chawla et al., 2013). For instance, the accumulation of Na^+ under salinity stress competes with K^+ binding in proteins, causing inhibition of protein synthesis and metabolic enzymes (Pardo & Quintero, 2002). High concentrations of NaCl outside the roots reduce the water potential, making it more difficult for plants to extract water, and this results in osmotic stress. In leaves, high salt levels cause stomatal closure, and impairment of electron transport and the photosynthetic apparatus, leading to reduced photosynthesis and productivity (Deinlein et al., 2014). To cope with osmotic stress, salt-stressed plants tend to accumulate compatible solutes such as proline (Pro) and glycine betaine (Qureshi et al., 2013), which decreases the cytoplasmic osmotic potential, enabling water absorption (Puniran-Hartley et al., 2014).

High salinity also induces the formation of ROS within plant cells, and its over-accumulation results in oxidative damage of membrane lipids, proteins and nucleic acids. To scavenge high ROS levels, an efficient system of non-enzymatic and enzymatic antioxidants is involved (Gill & Tuteja, 2010). Enzymatic antioxidants include superoxide dismutase (SOD), peroxidase (POX), catalase (CAT) as well as the enzymes of the ascorbate (ASC) – glutathione (GSH) cycle [GSH reductase (GR), ASC peroxidase (APX), monodehydroascorbate dehydrogenase (MDHAR), and dehydroascorbate reductase (DHAR)] that detoxify ROS (Gill & Tuteja, 2010; Chawla et al., 2013). Up-regulation of antioxidants has been observed in salt-tolerant cultivars of tomato, pea, Jatropha and Calendula, suggesting a pertinent role of antioxidants in alleviating salt stress-induced oxidative damage. So far, no salinity-specific ROS or salinity-specific antioxidants have been identified in plants in general, the same species of ROS are produced under different abiotic and some biotic stresses, in part as byproducts of the stress and to serve as signals to nearby tissues as well (reviewed by Apel & Hirt, 2004). However, some salinity-specific physiological and biochemical mechanisms are identified. For instance, the uptake, compartmentalization and/or excretion of ions from specialized cells under salinity stress is essential for normal growth (Viehweger, 2014).

7.4.2 Molecular Aspect

The byproducts of cellular regulatory mechanisms are metabolites. These metabolites are secreted in response to the external stimuli faced by the organism. More than 200,000 metabolites are secreted by plants and these metabolites are divided into two classes; these are primary and secondary metabolites for plant growth and development (Dixon, 2001). Primary metabolites are essential and include carbohydrates, fatty acids, vitamins, amino acids and organic acids (Fiehn, 2002). Polyketides, alkaloids, terpenoids and phenylpropanoids are secondary metabolites synthesized from primary metabolites and are required by plants for adaptation and defence responses (Hounsome et al., 2008). Throughout the plant kingdom, primary metabolites are common in all plants and conserved in their structure, while on the other hand, plant secondary compounds may vary in their chemical composition and are species-specific. In xenobiotic plants, the modifications in plant physiology induced by engineered nanoparticles are monitored by molecular events. These molecular events also reflect the metabolites that participate in biological pathways, for instance, the citric acid cycle, glycolysis, gluconeogenesis, amino acid and secondary metabolite biosynthesis, nitrogen and fatty acid metabolism. In order to defend against or adapt to various abiotic stresses, plant roots excrete metabolites as signalling molecules. Plants also alter soil chemistry and biochemical pathways to enhance nutrient bioavailability (Mhlongo et al., 2018).

7.5 ROLE OF NPS DURING HEAVY METALS-INDUCED OXIDATIVE STRESS IN PLANTS

Heavy metals are being increased in natural soil in response to various environmental threats. Commercialization and rapidly growing industries are adding a lot of heavy metals as waste to the soil. Plants growing in these heavy metal-affected soil face oxidative stress. To counter the adverse effect of these heavy metals on plants, various nanoparticles have been designed. These nanoparticles can effectively alleviate the stress imposed by these heavy metals. These heavy metals not only disturb the normal physiochemical processes of plants but also trigger the normal molecular mechanism of plants. Here, we will shortly discuss the various aspect of plants regarding the role of nanoparticles on oxidative stress induced by heavy metals.

7.5.1 Physio-biochemical Aspect

Heavy metals are being added to the soil and water at a much faster rate due to rapid industrialization, mining and commercialization. These are highly available to plants and other living organisms due to their bioavailability, thus they interfere with the normal physicochemical aspects of plants leading to oxidative damage which includes the production of reactive oxygen species (ROS). Reactive oxygen species are comprised of super oxide dismutase, peroxidase, hydrogen peroxide (H_2O_2), hypochlorous acid (HOCl) and hydroxyl radicals which cause oxidative stress also known as secondary stress (Ahmed et al., 2022). Oxidative stress will continue to rise with the increase in the concentration of ROS. ROS are molecules and radicals having oxygen atoms which are chemically activated due to the lack of electrons and, thus, cause oxidative stress. Heavy metals such as zinc, lead, nickel and cadmium are non-redox metals while chromium and copper are redox-active metals. Non-redox metals can produce a single atom of oxygen (O^-) which then induces the production of superoxide (O_2^-) while redox metals induce the production of a hydroxyl group (OH). These charged atoms then interfere with the normal structure and function of the cell leading to immature growth leading to cell or plant death (Xu et al., 2022; Ahmed et al., 2021).

Thus, they may affect the following plant parts and processes. They may affect the rate of photosynthesis, can cause poor seed germination, loss of coloured pigments (i.e. chlorophyll *a*, *b*, c and d along with carotenoids), altered level of plant hormones and altered root structure leading to overall stunted plant growth (Sharma et al., 2012). These chemically active radicals react with the normal structure of the lipid membrane, causing the leakage of the membrane and also disturbing the normal redox balance of the cell. Oxidative stress may attack the protein structure, damaging the active functional domain of amino acids, altering enzyme function, peptide chain fragmentation, degradation of enzymes and damaging cross-links between large protein structures, thus leading to altered protein structure and function. Accumulation of ROS may damage the structure and function of DNA and RNA, thus causing mutations, transversion and deletions (Pinho & Ladeiro, 2012). Oxidative stress has a strong impact on the process of photosynthesis, all pigments and respiration.

Nanoparticles have commercially gained importance due to their unique properties regarding their structure, physiology, electrochemical properties and application methods. Metal-based and green nanoparticles are being used for the improvement of plant growth against an increased concentration of heavy metals in the soil. Research has supported the role of nanoparticles in the adsorption or removal of heavy metals from the soil, making soil suitable for the successful growth of plants. Many results have shown an increase in seed germination, plant growth, root length, yield production and overall plant biomass of various crops including wheat, radish, cucumber and *Brassica* growing in HMs-polluted soil after NPs application (Sebastian et al., 2018; Praveen et al., 2017).

Some studies have shown the beneficial role of nanoparticles as they improved the lipid peroxidation (MDA), superoxide dismutase (SOD), peroxidase (POD) and salicylic acid content in plants treated with engineered nanoparticles against heavy metals as compared to those directly treated with heavy metal contamination. Also, nanoparticles are believed to play the role of antioxidants in plant defence mechanisms. Due to their small surface area and high affinity for heavy metals, nanoparticles are able to attach the heavy metals to their surface reducing their availability to plant mechanisms thus preventing them to interfere with normal plant processes. Research has indicated the role of nanoparticles in improving photosynthesis by reducing the production of ROS. Nano-SiO_2 and TiO_2 in a combined mixture have been used in research for improving the rate of seed germination and plant growth as they treat the ROS, thus helping to improve the plant defence mechanism. Engineered nanoparticles have been known to have the ability to exchange electrons with available biomolecules in cell systems, thus developing favourable interactions. According to some studies, among all engineered nanoparticles, silver nanoparticles have been recognized as the best candidate for electron exchange with CO_3^+ and Fe_2^+ (Yan et al., 2022; Ahmed et al., 2022).

However, after an immense increase in the demand for nanoparticles for reducing heavy metal stress, some authors have suggested that nanoparticles themselves are involved in the generation of oxidative stress. Researchers have conceptualized that engineered metallic nanoparticles release their metallic component in the cell system, thus generating oxidative stress leading to damage to the overall cell processes. After entering into the plant cell system, nanoparticles release their attached metal ion into the cell cytoplasm, resulting in interaction with functional groups of proteins, i.e. carboxylic and sulfhydryl groups, which leads to a change in the protein activity. Nanoparticles promote wilting and necrosis by adsorbing organic and inorganic ions and molecules to their surface and creating indirect toxicity. Some have reported that zinc, iron, silicone, cadmium, titanium oxide, copper oxide and other metal-based nanoparticles interfere with normal photosynthetic machinery. They interfere with the normal functioning of photosystem I and II leading to the inhibition of electron transfer through the electron transport chain. Nanoparticles are also found to have toxicity effects on algae. Some studies have suggested a decrease in plant seed germination, growth rate, root length and shoot overall biomass after the application of nanoparticles. According to some other reports, silver-based nanoparticles have been declared as detrimental due to the potential hazards of producing oxidative stress (Ahmed et al., 2022).

7.5.2 Molecular Aspect

Reactive oxygen species are produced in response to stress but they also serve as a secondary messenger to activate the stress responses. Gene expression is usually activated with increased production of reactive oxygen species in organelles i.e. peroxisomes, mitochondria and chloroplasts. Oxidative stress disturbs the mitotic divisions. Also, cell proliferation is inhibited, and signalling pathways along with biochemical protein structures are damaged. ROS are reported to interact with the normal DNA structure leading to damage, deletion, transition and mutation in DNA structure (Ali et al., 2021). These mutations have been confirmed through various modern techniques including AFLP (amplified fragment length polymorphism), RFLP (restriction fragment length) and micro-satellites. Total gene expression analysis, transcriptional analysis and transcriptase analysis have suggested that nanoparticles selectively activate gene expressions in response to nanoparticles-mediated toxicity. Carcinogenesis, genotoxicity, fibrosis, cytokinesis and inflammation are reported due to the influence of nanoparticles. Redox-sensitive mitogen-activated protein kinase (MAPK) and nuclear factor kappa-light-chain enhancer of activated B cells (NF-κB) and nuclear factor erythroid 2-related factor 2 (Nrf2) are involve in infammatory response against oxidative stress (Ye et al., 2020; Asgari et al., 2018).

7.6 CONCLUSIONS

Abiotic stresses have emerged as serious threats to modern, intensive and commercial-oriented farming systems which reduce crop yield and nutritional quality. Among abiotic stresses, drought, heat stress and salinity have become the most pronounced challenges which must be countered effectively in order to ensure the food security of a rapidly increasing population. The heat stress is more deleterious at sensitive phases of development, early plant set up, flowering phase and the gametogenesis process, while plants respond through biosynthesis of reactive oxygen species. Likewise, drought stress causes the production of reactive oxygen species (ROS), which imposes oxidative damage and changes cellular membranes, creating an ionic imbalance in crop plants. Similarly, salinity stress lessens photosynthesis, damages chlorophyll levels, lowers stomatal conductance and slows seed germination. Nanotechnology has evolved into the synthesis of nanoparticles that hold bright perspectives in ameliorating the adverse effects of abiotic stresses through improvement in morphological traits, physiological functioning and molecular adaptations. However, there are certain concerns and reservations regarding nanoparticle efficacy and their underlying mechanisms that await further research in order to ensure crop productivity under unfavourable environmental conditions.

REFERENCES

Abdelaal, K. A., Attia, K. A., Alamery, S. F., El-Afry, M. M., Ghazy, A. I., Tantawy, D. S., & Hafez, Y. M. (2020). Exogenous application of proline and salicylic acid can mitigate the injurious impacts of drought stress on barley plants associated with physiological and histological characters. *Sustainability*, *12*(5), 1736.

Ahmad, A. F., Abbas, Z., Abrass, H. A., & You, K. Y. (2019). Optimum design of a microstrip ring resonator sensor to determine the moisture content in oil palm fruits and seeds. *BioResources*, *14*(1), 1819–1837.

Ahmad, Z., Ejaz, A. W., Sajjad, A., Shazia, A., Tanveer, A., Wajid, M., Osama, B. A. H., Terence, T., Maryke, L., & Muhammad, R. (2018a). Physiological responses of wheat to drought stress and its mitigation approaches. *Acta Physiologiae Plantarum*. https://doi.org/10.1007/s11738-018-2651-6

Ahmad, Z., Shazia, A., Ejaz, A. W., Muhammad, A. A., Tanveer, A., Rana, M. S. T., Rashid, A., & Muhammad, A. I. (2018b). Growth, physiology, and biochemical activities of plant responses with foliar potassium application under drought stress–A review. *Journal of Plant Nutrition*. https://doi.org/10.1080/01904167.2018.1459688

Ahmad, Z., Shazia, A., Muhammad, A. I., & Hafiz, S. R. (2018c). Foliar applied potassium enhances fibre quality, water foliar applied potassium enhances fibre quality, water relations and yield of cotton. *Journal of Agricultural Research*, *56*(1), 17–25.

Ahmad, Z., Waraich, E. A., Ahmad, R., & Shahbaz, M. (2017). Modulation in water relations, chlorophyll contents and antioxidants activity of maize by foliar phosphorus application under drought stress. *Pakistan Journal of Botany*, *49*(1), 11–19.

Ahmed, S., Muhammad, I., Zahoor, A., Muhammad, A. I., Arkadiusz, A., Gospodarstwa, W., Warszawie, W. B., Inzynierii, S., Ayman, E. S., & Hesham, F. A. (2022). Foliar application of silicon-based nanoparticles improve the adaptability of maize (Zea Mays L.) in cadmium contaminated soils. *Environmental Science and Pollution Research*, 0123456789. https://doi.org/10.1007/s11356-023-25189-0

Ahmed, T., Muhammad, N., Natasha, M., Muhammad, S., Muhammad, A., Liaqat, A., & Gang, W. (2021). Nanoparticle-based amelioration of drought stress and cadmium toxicity in rice via triggering the stress responsive genetic mechanisms and nutrient acquisition. *Ecotoxicology and Environmental Safety*, *209*, 111829. https://doi.org/10.1016/j.ecoenv.2020.111829

Ahmed, T., Noman, M., Manzoor, N., Shahid, M., Hussaini, K. M., Rizwan, M., & Li, B. (2021). Green magnesium oxide nanoparticles-based modulation of cellular oxidative repair mechanisms to reduce arsenic uptake and translocation in rice (*Oryza sativa* L.) plants. *Environmental Pollution*, *288*, 117785.

Al-Antary, T. A., Kahlel, A., Ghidan, A., & Asoufi, H. (2020). Effects of nanotechnology liquid fertilizers on fruit set and pods of broad bean (*Vicia faba* L.). *Fresenius Environmental Bulletin*, *29*(6), 4794–4798.

Al-Dhalimi, A. M., & Al-Ajeel, S. A. H. (2020). Effect of plant regulators, zinc nanoparticles and irrigation intervals on leaf content of endogenous hormones and nutrients in sunflower (*Helianthus annuus* L.). *Plant Archives*, *20*(1), 2720–2725.

Alghory, A. (2019). Evaluation of crop water stress index and leaf water potential for deficit irrigation management of sprinkler-irrigated wheat. *Irrigation Science*, *37*(1), 61–77.

Ali, B., Hafeez, A., Javed, M. A., Afridi, M. S., Abbasi, H. A., Qayyum, A., & Selim, S. (2022). Role of endophytic bacteria in salinity stress amelioration by physiological and molecular mechanisms of defense: A comprehensive review. *South African Journal of Botany*, *151*, 33–46.

Ali, S., Asif, M., & Naeem, K. (2021). Uptake, translocation, and consequences of nanomaterials on plant growth and stress adaptation. *Journal of Nanomaterials*. https://doi.org/10.1155/2021/6677616

Ali, S., Rizwan, M., Hussain, A., Rehman, M. Z., Ali, B., Yousaf, B., Wijaya, L., Alyemeni, M. N., & Ahmad, P. (2019). Silicon nanoparticles enhanced the growth and reduced the cadmium accumulation in grains of wheat (*Triticum aestivum* L.). *Plant Physiology and Biochemistry*, *140*, 1–8.

Allakhverdiev, S. I. (2020). Optimising photosynthesis for environmental fitness. *Functional Plant Biology*, *47*(11), iii–vii.

Ambede, J. G., Netondo, G. W., Mwai, G. N., & Musyimi, D. M. (2012). NaCl salinity affects germination, growth, physiology, and biochemistry of bambara groundnut. *Brazilian Journal of Plant Physiology*, *24*, 151–160.

Apel, K., & Hirt, H. (2004). Reactive oxygen species: Metabolism, oxidative stress, and signaling transduction. *Annual Review of Plant Biology*, *55*, 373.

Arshad, A., Ashraf, M., Sundari, R. S., Qamar, H., Wajid, M., & Hasan, M. U. (2020). Vulnerability assessment of urban expansion and modelling green spaces to build heat waves risk resiliency in Karachi. *International Journal of Disaster Risk Reduction*, *46*, 101468.

Asgari, F., Ahmad, M., Parissa, J., & Farzaneh, N. (2018). Effects of silicon nanoparticles on molecular, chemical, structural and ultrastructural characteristics of oat (*Avena Sativa* L.). *Plant Physiology and Biochemistry*, *127*, 152–160. https://doi.org/10.1016/j.plaphy.2018.03.021

Asgari, S., Moradi, H., & Afshari, H. (2018). Evaluation of some physiological and morphological characteristics of narcissus tazatta under BA treatment and nano-potassium fertilizer. *Journal of Chemical Health Risks*, *4*(4), 63–70.

Badr, A., & Brüggemann, W. (2020). Comparative analysis of drought stress response of maize genotypes using chlorophyll fluorescence measurements and leaf relative water content. *Photosynthetica*, *58*(2), 38–45.

Castro, B., Citterico, M., Kimura, S., Stevens, D. M., Wrzaczek, M., & Coaker, G. (2021). Stress-induced reactive oxygen species compartmentalization, perception and signalling. *Nature Plants*, *7*(4), 403–412.

Chai, Q., Gan, Y., Zhao, C., Xu, H. L., Waskom, R. M., Niu, Y., & Siddique, K. H. (2016). Regulated deficit irrigation for crop production under drought stress. A review. *Agronomy for Sustainable Development*, *36*(1), 1–21.

Chaudhry, S., & Sidhu, G. P. S. (2022). Climate change regulated abiotic stress mechanisms in plants: A comprehensive review. *Plant Cell Reports*, *41*(1), 1–31.

Chawla, S., Jain, S., & Jain, V. (2013). Salinity induced oxidative stress and antioxidant system in salt-tolerant and salt-sensitive cultivars of rice (*Oryza sativa* L.). *Journal of Plant Biochemistry and Biotechnology*, *22*(1), 27–34.

Cheng, J., Wang, L., Zeng, P., He, Y., Zhou, R., Zhang, H., & Wang, Z. (2017). Identification of genes involved in rice seed priming in the early imbibition stage. *Plant Biology*, *19*(1), 61–69.

Debnath, S., Chakraborty, M., & Kumawat, R. K. (2022). Physiological responses of plants under high temperature stress. *Biotica Research Today*, *4*(5), 347–349.

Deinlein, U., Stephan, A. B., Horie, T., Luo, W., Xu, G., & Schroeder, J. I. (2014). Plant salt-tolerance mechanisms. *Trends in Plant Science*, *19*(6), 371–379.

Dhanda, S. S., Sethi, G. S., & Behl, R. K. (2004). Indices of drought tolerance in wheat genotypes at early stages of plant growth. *Journal of Agronomy and Crop Science*, *190*(1), 6–12.

Dixon, R. A. (2001). Natural products and plant disease resistance. *Nature*, *411*(6839), 843–847.

Dusenge, M. E., Duarte, A. G., & Way, D. A. (2019). Plant carbon metabolism and climate change: Elevated CO_2 and temperature impacts on photosynthesis, photorespiration and respiration. *New Phytologist*, *221*(1), 32–49.

Duursma, R. A., Blackman, C. J., Lopéz, R., Martin-StPaul, N. K., Cochard, H., & Medlyn, B. E. (2019). On the minimum leaf conductance: Its role in models of plant water use, and ecological and environmental controls. *New Phytologist*, *221*(2), 693–705.

El-Saadony, M. T., Saad, A. M., Najjar, A. A., Alzahrani, S. O., Alkhatib, F. M., Shafi, M. E., Selem, E., Desoky, E.-S. M., Fouda, S. E., & El-Tahan, A. M. (2021). The use of biological selenium nanoparticles to suppress *Triticum aestivum* L. crown and root rot diseases induced by Fusarium species and improve yield under drought and heat stress. *Saudi Journal of Biological Sciences*, *28*(8), 4461–4471.

El-Zohri, M., Al-Wadaani, N. A., & Bafeel, S. O. (2021). Foliar sprayed green zinc oxide nanoparticles mitigate drought-induced oxidative stress in tomato. *Plants*, *10*(11). https://doi.org/10.3390/plants10112400

Ferrara, A., Lovelli, S., Di Tommaso, T., & Perniola, M. (2011). Flowering, growth and fruit setting in greenhouse bell pepper under water stress. *Journal of Agronomy, 10*(1), 12–19.

Fiehn, O. (2002). Metabolomics the link between genotypes and phenotypes. *Plant molecular biology*, 48, 155–171.

Flowers, T. J., Galal, H. K., & Bromham, L. (2010). Evolution of halophytes: Multiple origins of salt tolerance in land plants. *Functional Plant Biology, 37*(7), 604–612.

Foyer, C. H. (2018). Reactive oxygen species, oxidative signaling and the regulation of photosynthesis. *Environmental and Experimental Botany, 154*, 134–142.

Ghatak, A., Chaturvedi, P., & Weckwerth, W. (2017). Cereal crop proteomics: Systemic analysis of crop drought stress responses towards marker-assisted selection breeding. *Frontiers in Plant Science, 8*, 757.

Gill, S. S., & Tuteja, N. (2010). Reactive oxygen species and antioxidant machinery in abiotic stress tolerance in crop plants. *Plant Physiology and Biochemistry, 48*(12), 909–930.

Gupta, A., Rico-Medina, A., & Caño-Delgado, A. I. (2020). The physiology of plant responses to drought. *Science, 368*(6488), 266–269.

Gupta, A., Singh, S. K., Singh, M. K., Singh, V. K., Modi, A., Singh, P. K., & Kumar, A. (2020). Plant growth–promoting rhizobacteria and their functional role in salinity stress management. In Pardeep Singh, Ajay Kumar and Anwesha Borthakur*Abatement of environmental pollutants* (pp. 151–160). Elsevier.

Gustafsson, J., Cederberg, C., Sonesson, U., & Emanuelsson, A. (2013). *The methodology of the FAO study: Global food losses and food waste-extent, causes and prevention-FAO, 2011*The Swedish Institute for Food and Biotechnology.

Hameed, A., Ahmed, M. Z., Hussain, T., Aziz, I., Ahmad, N., Gul, B., & Nielsen, B. L. (2021). Effects of salinity stress on chloroplast structure and function. *Cells, 10*(8), 2023.

Hashmat, S., Shahid, M., Tanwir, K., Abbas, S., Ali, Q., Niazi, N. K., & Javed, M. T. (2021). Elucidating distinct oxidative stress management, nutrient acquisition and yield responses of Pisum sativum L. fertigated with diluted and treated wastewater. *Agricultural Water Management, 247*, 106720.

Hassan, N. S., Salah El Din, T. A., Hendawey, M. H., Borai, I. H., & Mahdi, A. A. (2018). Magnetite and zinc oxide nanoparticles alleviated heat stress in wheat plants. *Current Nanomaterials, 3*(1), 32–43.

Heikal, Y. M., El-Esawi, M. A., & Galilah, D. A. (2022). Morpho-anatomical, biochemical and molecular genetic responses of canola (*Brassica napus* L.) to sulphur application. *Environmental and Experimental Botany, 194*, 104739.

Hounsome, N., Hounsome, B., Tomos, D., & Edwards-Jones, G. (2008). Plant metabolites and nutritional quality of vegetables. *Journal of Food Science, 73*(4), R48–R65.

Hu, W., Tian, S. B., Di, Q., Duan, S. H., & Dai, K. (2018). Effects of exogenous calcium on mesophyll cell ultrastructure, gas exchange, and photosystem II in tobacco (*Nicotiana tabacum Linn.*) under drought stress. *Photosynthetica, 56*(4), 1204–1211.

Hussain, S., Khalid, M. F., Saqib, M., Ahmad, S., Zafar, W., Rao, M. J., & Anjum, M. A. (2018). Drought tolerance in citrus rootstocks is associated with better antioxidant defense mechanism. *Acta Physiologiae Plantarum, 40*(8), 1–10.

Imran, M. A., Ali, A., Ashfaq, M., Hassan, S., Culas, R., & Ma, C. (2018). Impact of Climate Smart Agriculture (CSA) practices on cotton production and livelihood of farmers in Punjab, Pakistan. *Sustainability, 10*(6), 2101.

Iqbal, M. J. (2018). Role of osmolytes and antioxidant enzymes for drought tolerance in wheat. *Global Wheat Production, 51*. 51-66. http://dx.doi.org/10.5772/intechopen.75926

Iqbal, M. S., Singh, A. K., Singh, S. P., & Ansari, M. I. (2020). Nanoparticles and plant interaction with respect to stress response. In Indu Bhushan, Vivek Kumar Singh, Durgesh Kumar Tripathi*Nanomaterials and environmental biotechnology* (pp. 1–15). Springer.

Istanbuli, T., Baum, M., Touchan, H., & Hamwieh, A. (2020). Evaluation of morpho-physiological traits under drought stress conditions in barley (*Hordeum vulgare* L.). *Photosynthetica*, *58*(4), 1059–1067.
Kamran, K., Kemmerling, B., Shutaywi, M., & Mashwani, Z. U. R. (2020). Nano zinc elicited biochemical characterization, nutritional assessment, antioxidant enzymes and fatty acid profiling of rapeseed. *Plos One*, *15*(11), e0241568.
Kapoor, D., Bhardwaj, S., Landi, M., Sharma, A., Ramakrishnan, M., & Sharma, A. (2020). The impact of drought in plant metabolism: How to exploit tolerance mechanisms to increase crop production. *Applied Sciences*, *10*(16), 5692.
Kareem, H. A., Saleem, M. F., Saleem, S., Rather, S. A., Wani, S. H., Siddiqui, M. H., Alamri, S., Kumar, R., Gaikwad, N. B., & Guo, Z. (2022). Zinc oxide nanoparticles interplay with physiological and biochemical attributes in terminal heat stress alleviation in Mungbean (*Vigna radiata* L.). *Frontiers in Plant Science*, *13*, 842349.
Kaur, H., Chowrasia, S., Gaur, V. S., & Mondal, T. K. (2021). Allantoin: Emerging role in plant abiotic stress tolerance. *Plant Molecular Biology Reporter*, *39*(3), 648–661.
Khalid, M. F., Iqbal Khan, R., Jawaid, M. Z., Shafqat, W., Hussain, S., Ahmed, T., Rizwan, M., Ercisli, S., Pop, O. L., & Alina Marc, R. (2022). Nanoparticles: The plant saviour under abiotic stresses. *Nanomaterials*, *12*(21), 3915.
Khan, I., Saeed, K., & Khan, I. (2019). Nanoparticles: Properties, applications and toxicities. *Arabian Journal of Chemistry*, *12*(7), 908–931.
Khan, S., Basit, A., Hafeez, M. B., Irshad, S., Bashir, S., Bashir, S., & Li, Y. (2021). Moringa leaf extract improves biochemical attributes, yield and grain quality of rice (*Oryza sativa* L.) under drought stress. *Plos One*, *16*(7), e0254452.
Kim, K. H., Kabir, E., & Jahan, S. A. (2017). Exposure to pesticides and the associated human health effects. *Science of the Total Environment*, *575*, 525–535.
Kulkarni, M. B., Ayachit, N. H., & Aminabhavi, T. M. (2022). Recent advancements in nano-biosensors: Current trends, challenges, applications, and future scope. *Biosensors*, *12*(10), 892.
Li, B., Gao, K., Ren, H., & Tang, W. (2018). Molecular mechanisms governing plant responses to high temperatures. *Journal of Integrative Plant Biology*, *60*(9), 757–779.
Li, S., Liu, J., Liu, H., Qiu, R., Gao, Y., & Duan, A. (2021). Role of hydraulic signal and ABA in decrease of leaf stomatal and mesophyll conductance in soil drought-stressed tomato. *Frontiers in Plant Science*, *12*, 653186.
Liaqat, W., Barutçular, C., Farooq, M., Ahmad, H., Jan, M., Ahmad, Z., & Li, M. (2022). Climate change in relation to agriculture: A review. *Spanish Journal of Agricultural Research*, *20*(2). e03R01. https://doi.org/10.5424/sjar/2022202-17742
Liu, D., Gao, Z., Li, J., Yao, Q., Tan, W., Xing, W., & Lu, Z. (2022). Effects of cadmium stress on the morphology, physiology, cellular ultrastructure, and BvHIPP24 gene expression of sugar beet (Beta vulgaris L.). *International Journal of Phytoremediation*, *30*, 1–11.
Mahmood, T., Rana, R. M., Ahmar, S., Saeed, S., Gulzar, A., Khan, M. A., & Du, X. (2021). Effect of drought stress on capsaicin and antioxidant contents in pepper genotypes at reproductive stage. *Plants*, *10*(7), 1286.
Malinowska, M., Donnison, I., & Robson, P. (2020). Morphological and physiological traits that explain yield response to drought stress in Miscanthus. *Agronomy*, *10*(8), 1194.
Mhlongo, M. I., Piater, L. A., Madala, N. E., Labuschagne, N., & Dubery, I. A. (2018). The chemistry of plant–microbe interactions in the rhizosphere and the potential for metabolomics to reveal signaling related to defense priming and induced systemic resistance. *Frontiers in Plant Science*, *9*, 112.
Mirakhorli, T., Ardebili, Z. O., Ladan-Moghadam, A., & Danaee, E. (2021). Bulk and nanoparticles of zinc oxide exerted their beneficial effects by conferring modifications in transcription factors, histone deacetylase, carbon and nitrogen assimilation, antioxidant biomarkers, and secondary metabolism in soybean. *Plos One*, *16*(9), e0256905.

Mittal, A., & Kakkar, R. (2020). The effect of solvent polarity on the antioxidant potential of echinatin, a retrochalcone, towards various ROS: A DFT thermodynamic study. *Free Radical Research, 54*(10), 777–786.

Mittler, R., Zandalinas, S. I., Fichman, Y., & Van Breusegem, F. (2022). Reactive oxygen species signalling in plant stress responses. *Nature Reviews Molecular Cell Biology, 23*(10), 663–679.

Muszyńska, E., Tokarz, K., Dziurka, M., Labudda, M., Dziurka, K., & Piwowarczyk, B. (2021). Photosynthetic apparatus efficiency, phenolic acid profiling and pattern of chosen phytohormones in metal-tolerant and intolerant Alyssum montanum ecotypes. *Scientific Reports, 11*, 4135.

Nair, A. U., Bhukya, D. P. N., Sunkar, R., Chavali, S., & Allu, A. D. (2022). Molecular basis of priming-induced acquired tolerance to multiple abiotic stresses in plants. *Journal of Experimental Botany, 73*(11), 3355–3371.

Naz, R., Khan, M. S., Hafeez, A., Fazil, M., Khan, M. N., Ali, B., & Ahmed, A. E. (2022). Assessment of phytoremediation potential of native plant species naturally growing in a heavy metal-polluted industrial soils. *Brazilian Journal of Biology, 84*. https://doi.org/10.1590/1519-6984.264473

Ondrasek, G., Kovačić, M. B., Carević, I., Štirmer, N., Stipičević, S., Udiković-Kolić, N., & Rengel, Z. (2021). Bioashes and their potential for reuse to sustain ecosystem services and underpin circular economy. *Renewable and Sustainable Energy Reviews, 151*, 111540.

Onyekachi, O. G., Boniface, O. O., Gemlack, N. F., & Nicholas, N. (2019). The effect of climate change on abiotic plant stress: A review. *Abiotic and Biotic Stress in Plants, 17*. http://dx.doi.org/10.5772/intechopen.82681

Pardo, J. M., & Quintero, F. J. (2002). Plants and sodium ions: Keeping company with the enemy. *Genome Biology, 3*(6), 1–4.

Patriyawaty, N. R., Rachaputi, R. C., & George, D. (2018). Physiological mechanisms underpinning tolerance to high temperature stress during reproductive phase in mungbean (*Vigna radiata* (L.) Wilczek). *Environmental and Experimental Botany, 150*, 188–197.

Perveen, S., & Hussain, S. A. (2021). Methionine-induced changes in growth, glycinebetaine, ascorbic acid, total soluble proteins and anthocyanin contents of two Zea mays L. varieties under salt stress. *JAPS: Journal of Animal & Plant Sciences, 31*(1).131-142

Perveen, S., Iqbal, M., Parveen, A., Akram, M. S., Shahbaz, M., Akber, S., & Mehboob, A. (2017). Exogenous triacontanol-mediated increase in phenolics, proline, activity of nitrate reductase, and shoot k+ confers salt tolerance in maize (*Zea mays* L.). *Brazilian Journal of Botany, 40*, 1–11.

Pinho, S., & Ladeiro, B. (2012). Phytotoxicity by lead as heavy metal focus on oxidative stress. *Journal of Botany. 2012*(369572), 10. doi:10.1155/2012/369572

Pourghasemian, N., Moradi, R., Naghizadeh, M., & Landberg, T. (2020). Mitigating drought stress in sesame by foliar application of salicylic acid, beeswax waste and licorice extract. *Agricultural Water Management, 231*, 105997.

Puniran-Hartley, N., Hartley, J., Shabala, L., & Shabala, S. (2014). Salinity-induced accumulation of organic osmolytes in barley and wheat leaves correlates with increased oxidative stress tolerance: In planta evidence for cross-tolerance. *Plant Physiology and Biochemistry, 83*, 32–39.

Qureshi, M. I., Abdin, M. Z., Ahmad, J., & Iqbal, M. (2013). Effect of long-term salinity on cellular antioxidants, compatible solute and fatty acid profile of Sweet Annie (*Artemisia annua* L.). *Phytochemistry, 95*, 215–223.

Rai-Kalal, P., Tomar, R. S., & Jajoo, A. (2021). Seed nanopriming by silicon oxide improves drought stress alleviation potential in wheat plants. *Functional Plant Biology, 48*(9), 905–915.

Rajput, V. D., Minkina, T., Kumari, A., Singh, V. K., Verma, K. K., Mandzhieva, S., Sushkova, S., Srivastava, S., & Keswani, C. (2021). Coping with the challenges of abiotic stress in plants: New dimensions in the field application of nanoparticles. *Plants*, *10*(6), 1221.

Rauf, M., Munir, M., ul Hassan, M., Ahmad, M., & Afzal, M. (2007). Performance of wheat genotypes under osmotic stress at germination and early seedling growth stage. *African Journal of Biotechnology*, *6*(8).971-975

Raza, A., Mehmood, S. S., Tabassum, J., & Batool, R. (2019). Targeting plant hormones to develop abiotic stress resistance in wheat. In Mirza Hasanuzzaman, Kamrun Nahar, Md. Amzad Hossain*Wheat production in changing environments* (pp. 557–577). Springer.

Rehman, M. Z., Mehwish, Z., Aisha, A., Waris, M. R., Shafaqat, A., Muhammad, S., Muhammad, A. A., & Zahoor, A. (2020). Residual effects of frequently available organic amendments on cadmium bioavailability and accumulation in wheat. *Chemosphere*, *244*, 125548. https://doi.org/10.1016/j.chemosphere.2019.125548

Saeed, S., Ullah, A., Ullah, S., Noor, J., Ali, B., Khan, M. N., & Alamri, S. (2022). Validating the impact of water potential and temperature on seed germination of wheat (*Triticum aestivum* L.) via hydrothermal time model. *Life*, *12*(7), 983.

Salam, A., Afridi, M. S., Javed, M. A., Saleem, A., Hafeez, A., Khan, A. R., Zeeshan, M., Ali, B., Azhar, W., & Ulhassan, Z. (2022). Nano-priming against abiotic stress: A way forward towards sustainable agriculture. *Sustainability*, *14*(22), 14880.

Sebastian, A., Nangia, A., & Prasad, M. N. V. (2018). A green synthetic route to phenolics fabricated magnetite nanoparticles from coconut husk extract: Implications to treat metal contaminated water and heavy metal stress in *Oryza sativa* L. *Journal of Cleaner Production*, *174*, 355–366.

Semida, W. M., Abdelkhalik, A., Mohamed, G. F., Abd El-Mageed, T. A., Abd El-Mageed, S. A., Rady, M. M., & Ali, E. F. (2021). Foliar application of zinc oxide nanoparticles promotes drought stress tolerance in eggplant (*Solanum melongena* L.). *Plants*, *10*(2), 421.

Sen, A., & Puthur, J. T. (2020). Seed priming-induced physiochemical and molecular events in plants coupled to abiotic stress tolerance: An overview. *Priming-Mediated Stress and Cross-Stress Tolerance in Crop Plants*, 303–316. https://doi.org/10.1016/B978-0-12-817892-8.00018-0

Sharma, P., Jha, A. B., Dubey, R. S., & Pessarakli, M. (2012). Reactive oxygen species, oxidative damage, and antioxidative defense mechanism in plants under stressful conditions. *Journal of Botany*, *2012*. https://doi.org/10.1155/2012/217037

Soares, J. C., Santos, C. S., Carvalho, S. M., Pintado, M. M., & Vasconcelos, M. W. (2019). Preserving the nutritional quality of crop plants under a changing climate: Importance and strategies. *Plant and Soil*, *443*(1), 1–26.

Song, Y., Jiang, M., Zhang, H., & Li, R. (2021). Zinc oxide nanoparticles alleviate chilling stress in rice (*Oryza Sativa* L.) by regulating antioxidative system and chilling response transcription factors. *Molecules*, *26*(8), 2196.

Sun, Q., Miao, C., Hanel, M., Borthwick, A. G., Duan, Q., Ji, D., & Li, H. (2019). Global heat stress on health, wildfires, and agricultural crops under different levels of climate warming. *Environment International*, *128*, 125–136.

Sutulienė, R., Ragelienė, L., Samuolienė, G., Brazaitytė, A., Urbutis, M., & Miliauskienė, J. (2022). The response of antioxidant system of drought-stressed green pea (*Pisum sativum* l.) affected by watering and foliar spray with silica nanoparticles. *Horticulturae*, *8*(1), 35.

Sytar, O., Kumari, P., Yadav, S., Brestic, M., & Rastogi, A. (2019). Phytohormone priming: Regulator for heavy metal stress in plants. *Journal of Plant Growth Regulation*, *38*(2), 739–752.

Thakur, S., Asthir, B., Kaur, G., Kalia, A., & Sharma, A. (2022). Zinc oxide and titanium dioxide nanoparticles influence heat stress tolerance mediated by antioxidant defense system in wheat. *Cereal Research Communications*, *50*(3), 385–396.

Tůmová, L., Tarkowská, D., Řehořová, K., Marková, H., Kočová, M., Rothová, O., Čečetka, P., & Holá, D. (2018). Drought-tolerant and drought-sensitive genotypes of maize (Zea mays L.) differ in contents of endogenous brassinosteroids and their drought-induced changes. *PLoS One*, *13*(5), e0197870. https://doi.org/10.1371/journal.pone.0197870

Vaughan, M. M., Block, A., Christensen, S. A., Allen, L. H., & Schmelz, E. A. (2018). The effects of climate change associated abiotic stresses on maize phytochemical defenses. *Phytochemistry Reviews*, *17*(1), 37–49.

Viehweger, K. (2014). How plants cope with heavy metals. *Botanical Studies*, *55*(1), 1–12.

Wahab, A., Abdi, G., Saleem, M. H., Ali, B., Ullah, S., Shah, W., & Marc, R. A. (2022). Plants' physio-biochemical and phyto-hormonal responses to alleviate the adverse effects of drought stress: A comprehensive review. *Plants*, *11*(13), 1620.

Wu, J., & Wang, T. (2020). Synergistic effect of zinc oxide nanoparticles and heat stress on the alleviation of transcriptional gene silencing in Arabidopsis thaliana. *Bulletin of Environmental Contamination and Toxicology*, *104*(1), 49–56.

Wu, J., Wang, J., Hui, W., Zhao, F., Wang, P., Su, C., & Gong, W. (2022). Physiology of plant responses to water stress and related genes: A review. *Forests*, *13*(2), 324.

Wu, P., Xiao, C., Cui, J., Hao, B., Zhang, W., Yang, Z., & Cui, H. (2021). Nitric oxide and its interaction with hydrogen peroxide enhance plant tolerance to low temperatures by improving the efficiency of the calvin cycle and the ascorbate–glutathione cycle in cucumber seedlings. *Journal of Plant Growth Regulation*, *40*(6), 2390–2408.

Xu, X., Peng, L., Shanghang, L., Qinghai, Z., & Dali, S. (2022). Distributions of heavy metals in rice and corn and their health risk assessment in guizhou province. *Bulletin of Environmental Contamination and Toxicology*, 0123456789. https://doi.org/10.1007/s00128-021-03407-0

Yadav, A., Singh, J., Ranjan, K., Kumar, P., Khanna, S., Gupta, M., & Sirohi, A. (2020). Heat shock proteins: Master players for heat-stress tolerance in plants during climate change. *Heat Stress Tolerance in Plants: Physiological, Molecular and Genetic Perspectives*, 189–211. https://doi.org/10.1002/9781119432401.ch9

Yan, F., Sun, Y., Xu, H., Yin, Y., Wang, H., Wang, C., & Ma, J. (2018). Effects of wheat straw mulch application and nitrogen management on rice root growth, dry matter accumulation and rice quality in soils of different fertility. *Paddy and Water Environment*, *16*(3), 507–518.

Yan, X., Jing, A., Yongchao, Y., Chengcheng, G., Baoyu, W., & Shuhe, W. (2022). Heavy metals uptake and translocation of typical wetland plants and their ecological effects on the coastal soil of a contaminated Bay in Northeast China. *Science of the Total Environment*, *803*, 149871. https://doi.org/10.1016/j.scitotenv.2021.149871

Yang, X., Lu, M., Wang, Y., Wang, Y., Liu, Z., & Chen, S. (2021). Response mechanism of plants to drought stress. *Horticulturae*, *7*, 50.

Yang, Y. J., Bi, M. H., Nie, Z. F., Jiang, H., Liu, X. D., Fang, X. W., & Brodribb, T. J. (2021). Evolution of stomatal closure to optimize water-use efficiency in response to dehydration in ferns and seed plants. *New Phytologist*, *230*(5), 2001–2010.

Ye, Y., Keni, C. R., José, A., Hernández, V., Carolina, V., Illya, A., Medina, V., Reagan, S., Turley, J. R., Peralta, V., Jorge, L., & Gardea, T. (2020). Manganese nanoparticles control salinity-modulated molecular responses in capsicum annuum L. through priming: A sustainable approach for agriculture. *ACS Sustainable Chemistry & Engineering*, *8*(3), 1427–1436. https://doi.org/10.1021/acssuschemeng.9b05615

Ye, Y., Medina-Velo, I. A., Cota-Ruiz, K., Moreno-Olivas, F., & Gardea-Torresdey, J. L. (2019). Can abiotic stresses in plants be alleviated by manganese nanoparticles or compounds? *Ecotoxicology and Environmental Safety*, *184*, 109671.

Younis, A., Khattab, H., & Emam, M. (2020). Impacts of silicon and silicon nanoparticles on leaf ultrastructure and TaPIP1 and TaNIP2 gene expressions in heat stressed wheat seedlings. *Biologia Plantarum*, *64*, 343–352.

Yue, L., Ma, C., Zhan, X., White, J. C., & Xing, B. (2017). Molecular mechanisms of maize seedling response to La 2 O 3 NP exposure: Water uptake, aquaporin gene expression and signal transduction. *Environmental Science: Nano*, *4*(4), 843–855.

Zeiner, M., Juranović Cindrić, I., Nemet, I., Franjković, K., & Salopek Sondi, B. (2022). Influence of soil salinity on selected element contents in different brassica species. *Molecules*, *27*(6), 1878.

Zhao, G., Zhao, Y., Lou, W., Su, J., Wei, S., Yang, X., & Shen, W. (2019). Nitrate reductase-dependent nitric oxide is crucial for multi-walled carbon nanotube-induced plant tolerance against salinity. *Nanoscale*, *11*(21), 10511–10523.

Zhao, L., Huang, Y., Hu, J., Zhou, H., Adeleye, A. S., & Keller, A. A. (2016). 1H NMR and GC-MS based metabolomics reveal defense and detoxification mechanism of cucumber plant under nano-Cu stress. *Environmental Science & Technology*, *50*(4), 2000–2010.

Zia-Ul-Haq, M., Dewanjee, S., & Riaz, M. (2021). *Carotenoids: Structure and function in the human body* (pp. 1–42). Springer.

8 Effective Use of Nanomaterials to Maintain Soil Fertility and Reduce Heavy Metal Toxicity in Soil

Biswajit Pramanick, Anurag Bera, Priyanka Saha, Anamika Barman, Shiv Vendra Singh, Rishav Kumar, Sagar Maitra and Akbar Hossain*

8.1 INTRODUCTION

Soil fertility in lots of Asian nations including India is declining due to different management vis-à-vis soil inherent characteristics viz. high pH, calcareous nature of the soil, improper fertilizer scheduling, presence of heavy metals in soils, etc. (Durgude et al., 2022; Laik et al., 2021; Pramanick et al., 2014). Such low soil fertility status ultimately impacts crop productivity on a global basis (Lakshmi et al., 2021; Pramanick et al., 2022). Maintaining soil fertility needs a comprehensive focus on soil management systems such as the adoption of sustainable plant nutrition approaches (Dey et al., 2021). Globally, the micro-nutrient fertilizer efficiency is less than 5% (Durgude et al., 2022). Moreover, the macro-nutrient efficiency is about 25–40% in developing countries (Pramanick et al., 2022). All these are serious matters of concern and demand for the smart delivery systems of nutrients that improve the use-efficiency and reduce soil stresses due to several external as well as internal factors. Nanotechnology offers all these in one go. The well-known features of nanotechnology and nanocomposites can help to enhance micro-nutrient use-efficiency by increasing the uptake and partitioning. Scientists throughout the world are constantly trying to prepare smart fertilizer sources through nanotechnology which can improve productivity by enhancing seed germination, shoot and root growth, abiotic stress tolerance, photosynthesis and photosynthate partitioning as well as crop quality. The mechanism of enhancing seed germination may include the up-regulation of aquaporin genes and the enhancement of a-amylase production, while shoot and root growth could be explained as the expression of cell division genes and water

* Corresponding author.

 DOI: 10.1201/9781003376446-8

channel genes (Ghosh & Bera, 2021). Scientists are exploring the various dimensions of nanotechnology such as the mitigation of environmental pollution, antibacterial nanoagents, nanofertilizers and nanomaterials for genetic engineering. In this chapter, a comprehensive approach has been made to illustrate the effective use of nanomaterials in maintaining the soil fertility status and reducing heavy metal stress.

8.2 APPLICATION OF NANOMATERIALS

Nanomaterials have emerged as a gift to humanity in the twenty-first century. It is currently widely employed in a variety of fields. Nanomaterials are used in many different disciplines, including medicine, industry, agriculture, engineering and many others.

8.2.1 Application in Medicine

Nanomaterials are nowadays used in fluorescent biological labelling (Bruchez et al., 1998), manufacturing drugs (Mah et al., 2000), gene delivery (Pantarotto et al., 2003), detection of pathogens (Edelstein et al., 2000) and for probing DNA structure (Mahtab et al., 1995). Not only these but nanomaterials are also found to help destroy tumours through heating (Matsuoka et al., 2004), separation and purification of biological cells and molecules (Molday & Mackenzie, 1982), cellular imaging (Roco et al., 2000), etc.

8.2.2 Application in Industrial Fields

In industry, several chemical reactions are carried out at a time. Nickel, lead, silver and platinum have all been utilized as unique metal catalysts in chemical reactions. In the case of gold, it can not be used due to its reactivity. But, in different experiments, it has been found that gold nanoparticles can be used as they is not reactive during such reactions (Haruta, 2003). In such a way, nanoparticles can be a viable option for several chemical reactions due to their small size and high efficacy as a catalyst (Saliba et al., 1998).

8.2.3 Application in Agriculture

In recent times, nanofertilizers have become a growing industry. A significant increase in yields has been recorded due to the foliar application of nanoparticles as fertilizers (El-Saadony et al., 2021). Currently, research is underway to develop nanocomposites to supply all the required essential nutrients in suitable proportions through smart delivery systems. Nanoherbicides (usually coated materials), nanofungicides and nano-insecticides (surface-modified hydrophobic nanosilica) are developed by employing nanomaterials. This newly developed chemical would be safe for plants and cause less environmental pollution.

8.2.4 Application in Engineering

In the engineering field, scientists are trying to utilize NPs to produce energy from the splitting of electrochemical water and photoelectrochemical (Ning et al., 2016).

The NPs (nanoparticles) can store energy in a variety of ways at the nanoscale. This is why these materials are also put to use in the storage of energy (Wang & Su 2014; Liu et al., 2015). The prospect of printed electronics being more cost-effective than more conventional silicon printing methods has sparked the industry's interest in the field in recent years. It is anticipated that electronics printed with different inks will flow swiftly. These inks are using carbon nanotubes, and organic and ceramic nanoparticles which are making them more effective (Kosmala et al., 2011). Manufacturing modern-day semiconductors, microchips, micro-wiring, etc. all rely on nanomaterials in the field of electrics and electronics.

From the above discussion, it is quite clear that nanoparticles have improved the quality of our lives immensely. Numerous scientific disciplines are finding promising uses for nanotechnology, which bodes well for its current and future applications.

8.3 RECOVERY OF HEAVY METAL(OID) CONTAMINATED SOIL VIA CARBON AND NON-CARBON-BASED NANOMATERIALS

Heavy metals are non-degradable compounds that will remain in the contaminated environment once they are introduced, making them one of the challenging soil contaminants. Large amounts of polluted soil can be treated in-situ (on-site) or ex-situ (taken and treated off-site); however, conventional techniques of treating contaminated soil are prohibitively expensive and very challenging. Therefore, the best strategy to safeguard the environment is to stop heavy metal contamination or to stop the spread of heavy metals in soil using an immobilization technique (Ma et al., 1993). A wide range of amendment agents has been utilized to modify heavy metal bioavailability and hinder their diffusion in soil by inducing diverse sorption processes such as adsorption to mineral surfaces, formation of stable complexes with organic ligands, surface precipitation and ion exchange. This is because the activity of heavy metals in soil is controlled by sorption-desorption reactions with other soil constituents (Singh et al., 2001). Unique size-dependent properties of nanomaterials include a highly specific surface area, rapid dissolution, high reactivity and strong adsorption and other characteristics include quantum confinement, superparamagnetism and localized surface plasmon resonance, which make it a logical choice for heavy metal removal (Gehrke et al., 2015). Nanomaterials are classified into zero-dimensional (0-D), one-dimensional (1-D), two- dimensional (2-D) and three-dimensional (3-D) nanoparticles. Zero-dimensional nanomaterials are those with all three dimensions less than 100 nm; fullerene and quantum dots are two examples (Han et al., 2016). One-dimensional nanomaterials are those with just one dimension more than 100 nm and two dimensions smaller than 100 nm, such as carbon and titanium nanotubes (Lee et al., 2014). Two-dimensional nanomaterials are those whose two dimensions are larger than 100 nm; graphene is a well-known example. Graphite and various nanomaterial composites are examples of three-dimensional materials, which are defined as three-dimensional materials with all dimensions greater than 100 nm (Singh, 2011). Among the various nanomaterials such as nano-adsorbents, nanomembranes and nanocatalysts, nano-adsorbents are mainly used for heavy metal removal from soil and water. Specialized nanoparticles are known as "nano-adsorbents" and serve as the best adsorbents because of their small dimensions, enormous surface area, catalytic potential, high reactivity, ease of separation and numerous active sites for interacting with various pollutants (Dubey et al., 2017). Carbon-based nano-adsorbents (carbon nanotubes, carbon nanoparticles and graphene) and metal-based nano-adsorbents are two primary categories of nano-adsorbents (magnetic nanoparticles).

8.3.1 Carbon-based Nanomaterials

8.3.1.1 Fullerenes

The closed-caged pentagonal and hexagonal ring structure of fullerenes, which were first identified in interstellar dust in 1985, is denoted by the formula C_{20+m}, where m is an integer (Kroto et al., 1985). In addition to the defects, decreased aggregation tendency and wide surface area make fullerenes ideal nanomaterials to be employed for the adsorption of heavy metal ions. It is thought that fullerenes adsorbed species through the penetration of adsorbates in the spaces/defects between the carbon nano-clusters (Kaneko et al., 1993; Lucena et al., 2011).

8.3.1.2 Carbon Nanotubes

In 1999, Lijima discovered the existence of carbon nanotubes. Single-wall carbon nanotubes (SWCNTs), such as layered rolled-up graphene, or multi-wall carbon nanotubes (MWCNTs), such as multi-layered rolled-up graphene, are two different types of carbon nanotubes (Lijima, 1999; Vidu et al., 2014). Due to their distinct physico-chemical characteristics, CNTs have been the focus of nanotechnology research ever since they were discovered. Adsorption mechanisms for heavy metal ions on CNTs have been the subject of extensive research. These mechanisms appear to be highly complex and involve physical adsorption, electrostatic attraction, precipitation and chemical interaction between the heavy metal ions and the surface functional groups of CNTs (such as carboxyl, hydroxyl, lactones and phenol) (Ren et al., 2011; Lu et al., 2006; Lu & Liu, 2006; Rao et al., 2007). The presence of some organic molecules may affect the adsorption of heavy metals. For example, humic acid (HA), fulvic acid (FA) and hydroxylated and carboxylated fullerenes have a considerable impact on the adsorption of copper (II) (Sheng et al., 2010; Wang et al., 2010).

8.3.1.2.1 SWCNTs

One-dimensional carbon nanomaterials called SWCNTs consist of hollow tubes with a one atom thick wall. Because of their porous structure, high surface area, simplicity of surface functionalization and nanosize, SWCNTs are also commonly employed to reduce environmental contamination. These SWCNT characteristics make them extremely useful for the removal of heavy metals. In order to determine the heavy metal ion adsorption capabilities of unfunctionalized SWCNTs and their functionalized counterparts, such as SWCNTs-OH, SWCNTs-NH2 and SWCNTs-COOH from aqueous environments, Anitha et al. (2015) performed a molecular dynamics simulation. According to the findings, SWCNTs-COOH has substantially better adsorption capabilities than bare SWCNTs, ranging from 150 to 230%. The SWCNTs-OH and SWCNTs-NH were discovered to be weak adsorbers since they only displayed 10–47%

greater adsorption than SWCNTs. Additionally, Pb^{2+}, Cu^{2+} and Cd^{2+} ion adsorption onto SWCNTs-COOH has been reported with respective adsorption capacities of 96.02, 77.00 and 55.89 mg/g. SWCNTs-polysulfone nanocomposites-based membrane for the removal of heavy metals was designed by Gupta et al. (2015). SWCNT incorporation led to a smaller membrane pore size and a smoother surface. The proposed membrane was shown to have a high capacity for rejecting metal ions, and it rejected 96.2% of Cr^{+6}, 87.5% of As^{+3}, and 94.2% of Pb^{+2} ions.

8.3.1.2.2 MWCNTs

Multi-wall carbon nanotubes are carbon nanotubes with many rolled layers of graphene (MWCNTs). Unique characteristics of the MWCNTs include their large surface area and excellent electrical, thermal and tensile strengths (Poulsen et al., 2017). According to reports, the oxidized MWCNTs exhibit a high adsorption capacity and efficiency for the ions Cr^{6+}, Pb^{2+} and Cd^{2+} from the water (Robati, 2013). The removal of heavy metal ions from water has also been accomplished using the composite MWCNT material. The removal of the heavy ions Cr^{6+}, As^{3+}, Ni^{2+}, Pb^{2+} and Cu^{2+} ions from water have been achieved using the nanocomposites of MWCNTs-Fe_2O_3, MWCNTs-ZrO_2, MWCNTs-Fe_3O_4, MWCNTs-Al_2O_3 and MWCNTs-MnO_2-Fe_2O_3 (Yang et al., 2009; Ntim & Maitra, 2012; Tang et al., 2012; Luo et al., 2013). Additionally, oxidized MWCNTs have demonstrated an extraordinarily high capacity and efficiency for the sorption of Pb^{2+}, Cd^{2+} and Cr^{6+} from water. The ability of MWCNTs to remove lead, chromium and cadmium ions with oxygen functional groups by forming ion complexes or salt precipitates on surfaces is increased by the sorption efficiency of acid treatment.

8.3.1.3 Graphene and Graphene Oxide-based Adsorbents

The hexagonal lattice of carbon atoms in graphene is one atom thick and is known as the thinnest substance with 200 times the strength of steel. Due to their distinctive physical and chemical features, particularly their electronic properties, high surface area, thermal mobility, high mechanical strength and programmable surface chemistry, these 2-D graphene-based materials are gaining increasing attention in the field of water treatment (Mohan et al., 2018). In order to remove heavy metal ions from water as well as other pollutants, Tabish et al. (2018) created porous graphene and used it as an adsorbent. They used this porous graphene material to remove As^{3+} from water and discovered an 80% effectiveness. After regeneration and recycling, it was discovered that the material still had its water treatment capabilities (Tabish et al., 2018). Guo et al. (2018) created a partly reduced graphene oxide nanocomposite by fabricating it with Fe_3O_4 using an in-situ co-precipitation process, and they used it to remove Pb^{2+} ions from water. The developed nanocomposite's adsorption capacity of 373.14 mg/g was found to be excellent for removing the Pb^{2+} ions from an aqueous solution.

8.3.2 Non-carbon-based Nanomaterials

Nanoscale metal oxides, such as titanium dioxides, iron oxides, zinc oxides, alumina, etc. have been investigated as an inexpensive, efficient adsorbent for heavy metals, enabling a more affordable remediation method because of their size and adsorption performance (Engates & Shipley, 2011; Zhang, 2003). The primary method of controlling adsorption is by creating a complex on the surface of nanoscale metal oxides and going through a one-electron oxidation reaction while being exposed to visible light (Peng et al., 2012). Due to its possible usage and display of intriguing magnetic properties (such as superparamagnetism and strong magnetic response under low applied magnetic fields), magnetic nanoparticles among the nanoscale metal oxides have attracted a great deal of attention.

Phosphate compounds can be utilized as efficient agents for in-situ immobilization of heavy metals in contaminated soils, as proved by the immobilization of lead (Pb), where phosphate was frequently given to soil either in its liquid forms like phosphoric acid or its solid forms like synthetic apatite, natural phosphate rocks and even fishbone (with apatite being the effective composition) (Yang et al., 2001).

8.4 NANOBIOREMEDIATION: A NEWER ASPECT OF MOLLIFYING HEAVY METAL STRESS

As one of the main environmental contaminants, heavy metals are naturally occurring substances which are a serious threat to food security worldwide. Heavy metals have a non-degradative character that makes them last longer in the environment and makes them potentially harmful to both human health and the environment. To address heavy metal contamination, a variety of physical, chemical and biological detoxification techniques, including soil stabilization, soil excavation, soil washing, thermal desorption, chemical extraction and phytoremediation have been used in lab and field conditions (Saleem et al., 2022). Due to the high expense of traditional physicochemical methods of removing metal, nanobioremediation has emerged as an appealing and feasible alternative. Over the past few years, nanotechnology has been extensively used in the bioremediation of heavy metals because it significantly improves the bioremediation process, is cost-effective and is environmentally friendly.

The method of using environmentally friendly nanomaterials and nanoparticles to remove toxins is known as nanobioremediation (Mathur et al., 2022). Although the impact of bioremediation has been studied and demonstrated numerous times, nanotechnology introduced a new level to this procedure. Nanobioremediation includes a wide range of techniques and methods, including nanobiosensors, nanosorbents, nanobiosurfactants and nanophytoremediation. Nanoparticles degraded, altered or absorbed the heavy metal pollutants due to their distinctive physiochemical properties which serve as a catalyst and enhance the activation energy for degrading the pollutants. The use of plants in pollutant degradation is known as phytoremediation. Hyperaccumulating plants may be utilized in phytoremediation to clean up metal-contaminated areas. A more effective way to remove heavy metals is by combining phytoremediation and nanotechnology. Plants use phytoextraction, phytodegradation, phytovolatilization, phytostabilization and rhizo-degradation as methods of bioremediation. But, the main flaw in phytoremediation is that it takes a lot of time, frequently years or even decades, to see the effects. To assess the effectiveness of nanoparticles in promoting phytoremediation and compare their impact on the process, Jesitha and Harikumar (2018) conducted a study on *Tradescantia spathacea* and *Alternanthera dentata*. In *Tradescantia spathacea*, phytoremediation accumulated about 47% Pb and 45.3% Cd, and nanophytoremediation treatment accumulated 84.4% Pb and 64.8% Cd. In *Alternanthera dentata*, phytoremediation accumulated about 60.5% Pb and 52.5% Cd, and nanophytoremediation treatment accumulated 73.7% Pb and 71.3% Cd. Ma and Wang (2010) reported that in phytoremediation, the addition of fullerene nanoparticles drastically altered TCE (Tri-chloro ethylene) fate and uptake. Inevitably, heavy metals harm plants in ways that are obvious or invisible, resulting in enormous monetary losses. Therefore, it is essential to create a mechanism for promptly observing the harm that heavy metal stress causes to plants. Nanobiosensors are modified biosensors that include biological components connected to a transducer and are often manufactured at the nanoscale (Ghaffar et al., 2020). The amount of chlorophyll, the health of the plant's water supply, and other factors can be measured using nanobiosensors to identify heavy metal phytotoxicity (Mathur et al., 2022). A living plant cell-based biosensor is constructed for the detection of the invisible damage of plants under cadmium (Cd-II) and lead (Pb-II) stress by establishing a quantitative link between the biomarker content, the electron-transfer resistance and the chlorophyll content in plant cells (Wang et al., 2019). The plant *Pteris cretica* has a built-in capacity to accumulate arsenic, which was discovered using nanobionic sensors placed in the plant tissue (Lew et al., 2021). Researchers studying the synthesis and assembly of nanoparticles have recently focussed on the utilization of microbes in the biosynthesis of nanomaterials. The employment of micro-organisms in toxic heavy metal bioremediation becomes a crucial field in biometric and bioengineering applications (Rajendran & Gunasekaran, 2007). The use of micro-organisms such as fungi, bacteria, yeast and actinomycetes is an important source for the production of the nanoparticle. For the bioremediation of the polluted soil, three separate kinds of consortia of fungi – all isolated fungi, *Ascomycota* and *Basidiomycota* – were used. The efficiency of Ni and Pb removal in soil augmented with all the isolated fungi was 52% and 44%, respectively (Hassan et al., 2019). Increases in Cr (VI) accumulation of 13.45%, 7.56% and 8.88%, respectively, were seen in the plant's roots, stems and leaves of *Pongamia pinnata* (*L.*) as a result of its association with the rhizobacteria *Paenibacilus konsidensis* SK3 (Das et al., 2022). Large, contaminated areas may be cleaned up using nanoremediation, which might cut down on clean-up time, do deal with the need to remove pollutants, and reduce contaminant concentration to a minimum. Table 8.1 depicts the types of nanobioremediation for heavy metals.

8.5 CONCLUSION

Maintenance of soil health for agricultural land which is degraded owing to multifarious problems including heavy metal stress is one of the biggest challenges, and nanotechnology offers a great opportunity to improve this situation. The well-known features of nanotechnology and nanocomposites can help to enhance micro-nutrient use-efficiency by increasing the uptake and partitioning. Scientists throughout the

TABLE 8.1
Types of nanobioremediation for heavy metals

Types of nanobioremediation	Heavy metal	Remark	References
Phytoremediation			
Isatis cappadocica	Arsenic (As)	Glutathione (GSH) modified superparamagnetic iron oxide nanoparticles (nFe_3O_4 and nFe_3O_4@GSH) reduced As content in plant shoots.	Souri et al. (2020)
Raphanus sativus L.	Lead (Pb)	The combination of TDZ and MgO nanoparticles can significantly influence secondary metabolite synthesis and Pb phytoaccumulation.	Hussain et al. (2019)
Microbial remediation by nanotechnology			
Bacillus subtilis X3	Lead (Pb)	Bacillus subtilis X3 biomass isolated from lead mine soil was used for the bio-immobilization of lead via various adsorption methods.	Qiao et al. (2019)
Bacillus thioparans strain U3	Cu and Pb	Maximum biosorption capabilities of Cu and Pb by *B. thioparans* U3 indicate that it has potential as a biosorbent for these metals.	Rodríguez-Tirado et al. (2012)
Nanobiosensor			
Vitronectin-like proteins (VN) in Arabidopsis and soybean	Cadmium [Cd(II)] or lead [Pb(II)]	The invisible damage of plants under the stress of heavy metals was detected.	Wang et al. (2019)

world are constantly trying to prepare smart fertilizer sources through nanotechnology which can improve productivity by enhancing seed germination, shoot and root growth, abiotic stress tolerance, photosynthesis and photosynthate partitioning as well as crop quality. Synthesis of nanomaterials includes various approaches like top-down approach, ball milling method, laser ablation method, sputtering method, biological synthesis or green synthesis, etc. Nanotechnology can also successfully mitigate the heavy metal stress from the soil. Large amounts of polluted soil can be treated in-situ (on-site) or ex-situ (taken and treated off-site) with the help of nanotechnology. The use of environmentally friendly nanomaterials to reduce heavy metal toxicity in the soil is also another important aspect. Nanobioremediation includes a wide range of techniques and methods, including nanobiosensors, nanosorbents, nanobiosurfactants and nanophytoremediation. Overall, nanotechnology is a boon to the agricultural system to improve soil fertility and reduce the heavy metal stresses from the soil to have higher productivity and profitability from agricultural land.

REFERENCES

Anitha, K., Namsani, S., & Singh, J. K. (2015). Removal of heavy metal ions using a functionalized single-walled carbon nanotube: A molecular dynamics study. *The Journal of Physical Chemistry A*, *119*(30), 8349–8358.

Bruchez, M., Jr., Moronne, M., Gin, P., Weiss, S., & Alivisatos, A. P. (1998). Semiconductor nanocrystals as fluorescent biological labels. *Science*, *281*(5385), 2013–2016.

Das, P. K., Das, B. P., & Dash, P. (2022). A super-tolerant bacteria strain improves phytoremediation of Cr (VI) contaminated soil with Pongamia pinnata. *Rhizosphere*, *22*(1), 100543.

Dey, P., Mahapatra, B. S., Juyal, V. K., Pramanick, B., Negi, M. S., Paul, J., & Singh, S. P. (2021). Flax processing waste: A low-cost, potential biosorbent for treatment of heavy metal, dye and organic matter contaminated industrial wastewater. *Industrial Crops and Products*, *174*, 114195. https://doi.org/10.1016/j.indcrop.2021.114195

Dubey, S., Banerjee, S., Upadhyay, S. N., & Sharma, Y. C. (2017). Application of common nano-materials for removal of selected metallic species from water and wastewaters: A critical review. *Journal of Molecular Liquids*, *240*, 656–677.

Durgude, S. A., Ram, S., Kumar, R., Singh, S. V., Singh, V., Durgude, A. G., Pramanick, B., Maitra, S., Gaber, A., & Hossain, A. (2022). Synthesis of mesoporous silica and graphene-based FeO and ZnO nanocomposites for nutritional biofortification and sustained the productivity of rice (*Oryza sativa* L.). *Journal of Nanomaterials*, *2022*, 5120307. https://doi.org/10.1155/2022/5120307

Edelstein, R. L., Tamanaha, C. R., Sheehan, P. E., Miller, M. M., Baselt, D. R., Whitman, L., & Colton, R. J. (2000). The BARC biosensor applied to the detection of biological warfare agents. *Biosensors and Bioelectronics*, *14*(10–11), 805–813.

El-Saadony, M. T., ALmoshadak, A. S., Shafi, M. E., Albaqami, N. M., Saad, A. M., El-Tahan, A. M., et al. (2021). Vital roles of sustainable nano-fertilizers in improving plant quality and quantity-an updated review. *Saudi Journal of Biological Sciences*, *28*(12), 7349–7359.

Engates, K. E., & Shipley, H. J. (2011). Adsorption of Pb, Cd, Cu, Zn, and Ni to titanium dioxide nanoparticles: Effect of particle size, solid concentration, and exhaustion. *Environmental Science and Pollution Research*, *18*(3), 386–395.

Gehrke, I., Geiser, A., & Somborn-Schulz, A. (2015). Innovations in nanotechnology for water treatment. *Nanotechnology, Science and Applications*, *8*, 1–17.

Ghaffar, N., Farrukh, M.A., & Naz, S. (2020). Applications of Nanobiosensors in Agriculture. In S. Javad (eds) *Nanoagronomy*. Springer. https://doi.org/10.1007/978-3-030-41275-3_10

Ghosh, S. K., & Bera, T. (2021). Molecular mechanism of nano-fertilizer in plant growth and development: A recent account. In *Advances in nano-fertilizers and nano-pesticides in agriculture* (pp. 535–560). Woodhead Publishing.

Guo, T., Bulin, C., Li, B., Zhao, Z., Yu, H., Sun, H., Sudisha Jogaiah, Harikesh Bahadur Singh, Leonardo Fernandes Fraceto Renata de Lima (2018). Efficient removal of aqueous Pb (II) using partially reduced graphene oxide-Fe3O4. *Adsorption Science & Technology*, *36*(3–4), 1031–1048.

Gupta, S., Bhatiya, D., & Murthy, C. N. (2015). Metal removal studies by composite membrane of polysulfone and functionalized single-walled carbon nanotubes. *Separation Science and Technology*, *50*(3), 421–429.

Han, X., Li, S., Peng, Z., Al-Yuobi, A. O., Bashammakh, A. S. O., & Leblanc, R. M. (2016). Interactions between carbon nanomaterials and biomolecules. *Journal of Oleo Science*, *65*(1), 1–7.

Haruta, M. (2003). When gold is not noble: Catalysis by nanoparticles. *The Chemical Record*, *3*(2), 75–87.

Hassan, A., Pariatamby, A., Ahmed, A., Auta, H. S., & Hamid, F. S. (2019). Enhanced bioremediation of heavy metal contaminated landfill soil using filamentous fungi consortia: A demonstration of bioaugmentation potential. *Water, Air, & Soil Pollution*, *230*(9), 1–20.

Hussain, F., Hadi, F., & Akbar, F. (2019). Magnesium oxide nanoparticles and thidiazuron enhance lead phytoaccumulation and antioxidative response in *Raphanus sativus* L. *Environmental Science and Pollution Research International*, *26*(29), 30333–30347. https://doi.org/10.1007/s11356-019-06206-7

Jesitha, K., & Harikumar, P. S. (2018). Application of nano-phytoremediation technology for soil polluted with pesticide residues and heavy metals. In A. Ansari, S. Gill, R. Gill, G. R. Lanza, & L. Newman (Eds.), *Phytoremediation*. Springer. https://doi.org/10.1007/978-3-319-99651-6_18

Kaneko, K., Ishii, C., Arai, T., & Suematsu, H. (1993). Defect-associated microporous nature of fullerene C60 crystals. *The Journal of Physical Chemistry*, *97*(26), 6764–6766.

Kosmala, A., Wright, R., Zhang, Q., & Kirby, P. (2011). Synthesis of silver nano particles and fabrication of aqueous Ag inks for inkjet printing. *Materials Chemistry and Physics*, *129*(3), 1075–1080.

Kroto, H. W., Heath, J. R., O'Brien, S. C., Curl, R. F., & Smalley, R. E. (1985). C60: Buckminsterfullerene. *Nature*, *318*(6042), 162–163.

Laik, R., Kumara, B. H., Pramanick, B., Singh, S. K., Nidhi, A. M., Gaber, A., & Hossain, A. (2021). Labile soil organic matter pools are influenced by 45 years of applied farmyard manure and mineral nitrogen in the wheat—Pearl millet cropping system in the sub-tropical condition. *Agronomy*, *11*, 2190. https://doi.org/10.3390/agronomy11112190

Lakshmi, P. V., Singh, S. K., Pramanick, B., Kumar, M., Laik, R., Kumari, A., Shukla, A. K., Abdel Latef, A. A. H., Ali, O. M., & Hossain, A. (2021). Long-term Zinc fertilization in calcareous soils improves wheat (*Triticum aestivum* L.) productivity and soil zinc status in the rice–wheat cropping system. *Agronomy*, *11*, 1306. https://doi.org/10.3390/agronomy11071306

Lee, K., Mazare, A., & Schmuki, P. (2014). One-dimensional titanium dioxide nanomaterials: Nanotubes. *Chemical Reviews*, *114*(19), 9385–9454.

Lew, T. T. S., Park, M., Cui, J., & Strano, M. S. (2021). Plant nanobionic sensors for arsenic detection. *Advanced Materials*, *33*, 2005683. https://doi.org/10.1002/adma.202005683

Lijima, S., Yudasaka, M., Yamada, R., Bandow, S., Suenaga, K., Kokai, F., & Takahashi, K. (1999). Nano-aggregates of single-walled graphitic carbon nano-horns. *Chemical Physics Letters*, *309*(3–4), 165–170.

Liu, J., Liu, Y., Liu, N., Han, Y., Zhang, X., Huang, H., et al. (2015). Metal-free efficient photocatalyst for stable visible water splitting via a two-electron pathway. *Science*, *347*(6225), 970–974.

Lu, C., Chiu, H., & Liu, C. (2006). Removal of zinc (II) from aqueous solution by purified carbon nanotubes: Kinetics and equilibrium studies. *Industrial & Engineering Chemistry Research*, *45*(8), 2850–2855.

Lu, C., & Liu, C. (2006). Removal of nickel (II) from aqueous solution by carbon nanotubes. *Journal of Chemical Technology & Biotechnology: International Research in Process, Environmental & Clean Technology*, *81*(12), 1932–1940.

Lucena, R., Simonet, B. M., Cárdenas, S., & Valcárcel, M. (2011). Potential of nanoparticles in sample preparation. *Journal of Chromatography A*, *1218*(4), 620–637.

Luo, C., Tian, Z., Yang, B., Zhang, L., & Yan, S. (2013). Manganese dioxide/iron oxide/acid oxidized multi-walled carbon nanotube magnetic nanocomposite for enhanced hexavalent chromium removal. *Chemical Engineering Journal*, *234*, 256–265.

Ma, Q. Y., Traina, S. J., Logan, T. J., & Ryan, J. A. (1993). In situ lead immobilization by apatite. *Environmental Science & Technology*, *27*(9), 1803–1810.

Ma, X., & Wang, C. (2010). Fullerene nanoparticles affect the fate and uptake of trichloroethylene in phytoremediation systems. *Environmental Engineering Science*, *27*(11), 989–992.

Mah, C., Zolotukhin, I., Fraites, T. J., Dobson, J., Batich, C., & Byrne, B. J. (2000). Microsphere-mediated delivery of recombinant AAV vectors in vitro and in vivo. *Molecular Theraphy*, *1*(5), S293.

Mahtab, R., Rogers, J. P., & Murphy, C. J. (1995). Protein-sized quantum dot luminescence can distinguish between "straight", "bent", and "kinked" oligonucleotides. *Journal of the American Chemical Society*, *117*(35), 9099–9100.

Mathur, S., Singh, D., & Ranjan, R. (2022). Remediation of heavy metal (loid) contaminated soil through green nanotechnology. *Frontiers in Sustainable Food Systems*, *6*, 932424.

Matsuoka, F., Shinkai, M., Honda, H., Kubo, T., Sugita, T., & Kobayashi, T. (2004). Hyperthermia using magnetite cationic liposomes for hamster osteosarcoma. *BioMagnetic Research and Technology*, *2*(1), 1–6.

Mohan, V. B., Lau, K. T., Hui, D., & Bhattacharyya, D. (2018). Graphene-based materials and their composites: A review on production, applications and product limitations. *Composites Part B: Engineering*, *142*, 200–220.

Molday, R. S., & Mackenzie, D. (1982). Immunospecific ferromagnetic iron-dextran reagents for the labeling and magnetic separation of cells. *Journal of Immunological Methods*, *52*(3), 353–367.

Ning, F., Shao, M., Xu, S., Fu, Y., Zhang, R., Wei, M., et al. (2016). TiO_2/graphene/NiFe-layered double hydroxide nanorod array photoanodes for efficient photoelectrochemical water splitting. *Energy & Environmental Science*, *9*(8), 2633–2643.

Ntim, S. A., & Mitra, S. (2012). Adsorption of arsenic on multiwall carbon nanotube–zirconia nanohybrid for potential drinking water purification. *Journal of Colloid and Interface Science*, *375*(1), 154–159.

Pantarotto, D., Partidos, C. D., Hoebeke, J., Brown, F., Kramer, E. D., Briand, J. P., et al. (2003). Immunization with peptide-functionalized carbon nanotubes enhances virus-specific neutralizing antibody responses. *Chemistry & Biology*, *10*(10), 961–966.

Peng, H., Feng, S., Zhang, X., Li, Y., & Zhang, X. (2012). Adsorption of norfloxacin onto titanium oxide: Effect of drug carrier and dissolved humic acid. *Science of the Total Environment*, *438*, 66–71.

Poulsen, S. S., Knudsen, K. B., Jackson, P., Weydahl, I. E., Saber, A. T., Wallin, H., & Vogel, U. (2017). Multi-walled carbon nanotube-physicochemical properties predict the systemic acute phase response following pulmonary exposure in mice. *PLoS One*, *12*(4), e0174167.

Pramanick, B., Brahmachari, K., Ghosh, A., & Zodape, S. T. (2014). Foliar nutrient management through *Kappaphycus* and *Gracilaria* saps in rice-potato-greengram crop sequence. *Journal of Scientific and Industrial Research, 73*, 613–617.

Pramanick, B., Kumar, M., Naik, B. M., Kumar, M., Singh, S. K., Maitra, S., Naik, B. S. S. S., Rajput, V. D., & Minkina, T. (2022). Long-term conservation tillage and precision nutrient management in maize–wheat cropping system: Effect on soil properties, crop production, and economics. *Agronomy, 12*, 2766. https://doi.org/10.3390/agronomy12112766

Qiao, W., Zhang, Y., Xia, H., Luo, Y., Liu, S., Wang, S., & Wang, W. (2019). Bioimmobilization of lead by *Bacillus subtilis* X3 biomass isolated from lead mine soil under promotion of multiple adsorption mechanisms. *Royal Society of Open Science, 6*(2), 181701. https://doi.org/10.1098/rsos.181701

Rajendran, P., & Gunasekaran, P. (2007). Nanotechnology for bioremediation of heavy metals. In S. N. Singh & R. D. Tripathi (Eds.), *Environmental bioremediation technologies*. Springer. https://doi.org/10.1007/978-3-540-34793-4_9

Rao, G. P., Lu, C., & Su, F. (2007). Sorption of divalent metal ions from aqueous solution by carbon nanotubes: A review. *Separation and Purification Technology, 58*(1), 224–231.

Ren, X., Chen, C., Nagatsu, M., & Wang, X. (2011). Carbon nanotubes as adsorbents in environmental pollution management: A review. *Chemical Engineering Journal, 170*(2–3), 395–410.

Robati, D. (2013). Pseudo-second-order kinetic equations for modeling adsorption systems for removal of lead ions using multi-walled carbon nanotube. *Journal of Nanostructure in Chemistry, 3*(1), 1–6.

Roco, M. C., Williams, R. S., & Alivisatos, P. (2000). *Nanotechnology research directions: IWGN workshop report: Vision for nanotechnology in the next decade*. Springer Science & Business Media.

Rodríguez-Tirado, V., Green-Ruiz, C., & Gómez-Gil, B. (2012). Cu and Pb biosorption on Bacillus thioparans strain U3 in aqueous solution: Kinetic and equilibrium studies. *Chemical Engineering Journal, 181–182*, 352–359.

Saleem, S., Rizvi, A., & Khan, M. S. (2023). Microbiome-mediated nano-bioremediation of heavy metals: A prospective approach of soil metal detoxification. *International Journal of Environmental Science and Technology, 20*, 12823–12846. https://doi.org/10.1007/s13762-022-04684-w

Saliba, N., Parker, D. H., & Koel, B. E. (1998). Adsorption of oxygen on Au (111) by exposure to ozone. *Surface Science, 410*(2–3), 270–282.

Sheng, G., Li, J., Shao, D., Hu, J., Chen, C., Chen, Y., & Wang, X. (2010). Adsorption of copper (II) on multiwalled carbon nanotubes in the absence and presence of humic or fulvic acids. *Journal of Hazardous Materials, 178*(1–3), 333–340.

Singh, R. P. (2011). Prospects of nanobiomaterials for biosensing. *International Journal of Electrochemistry, 2011*, 30.

Singh, S. P., Ma, L. Q., & Harris, W. G. (2001). Heavy metal interactions with phosphatic clay: Sorption and desorption behavior. *Journal of Environmental Quality, 30*(6), 1961–1968.

Souri, Z., Karimi, N., Norouzi, L., & Ma, X. (2020). Elucidating the physiological mechanisms underlying enhanced arsenic hyperaccumulation by glutathione modified superparamagnetic iron oxide nanoparticles in Isatiscappadocica. *Ecotoxicology and Environmental Safety, 206*, 111336.

Tabish, T. A., Memon, F. A., Gomez, D. E., Horsell, D. W., & Zhang, S. (2018). A facile synthesis of porous graphene for efficient water and wastewater treatment. *Scientific Reports, 8*(1), 1–14.

Tang, W. W., Zeng, G. M., Gong, J. L., Liu, Y., Wang, X. Y., Liu, Y. Y., et al. (2012). Simultaneous adsorption of atrazine and Cu (II) from wastewater by magnetic multi-walled carbon nanotube. *Chemical Engineering Journal, 211*, 470–478.

Vidu, R., Rahman, M., Mahmoudi, M., Enachescu, M., Poteca, T. D., & Opris, I. (2014). Nanostructures: A platform for brain repair and augmentation. *Frontiers in Systems Neuroscience, 8*, 91.

Wang, D. W., & Su, D. (2014). Heterogeneous nanocarbon materials for oxygen reduction reaction. *Energy & Environmental Science, 7*(2), 576–591.

Wang, L., Ma, W., Xu, L., Chen, W., Zhu, Y., Xu, C., & Kotov, N. A. (2010). Nanoparticle-based environmental sensors. *Materials Science and Engineering, 70*(3–6), 265–274.

Wang, X., Cheng, M., Yang, Q., Wei, H., Xia, A., Wang, L., Ben, Y., Zhou, Q., Yang, Z., & Huang, X. (2019). A living plant cell-based biosensor for real-time monitoring invisible damage of plant cells under heavy metal stress. *Science of the Total Environment, 697*, 134097.

Yang, J., Mosby, D. E., Casteel, S. W., & Blanchar, R. W. (2001). Lead immobilization using phosphoric acid in a smelter-contaminated urban soil. *Environmental Science & Technology, 35*(17), 3553–3559.

Yang, S., Li, J., Shao, D., Hu, J., & Wang, X. (2009). Adsorption of Ni (II) on oxidized multi-walled carbon nanotubes: Effect of contact time, pH, foreign ions and PAA. *Journal of Hazardous Materials, 166*(1), 109–116.

Zhang, W. X. (2003). Nanoscale iron particles for environmental remediation: An overview. *Journal of Nanoparticle Research, 5*(3), 323–332.

9 Alleviation of Heavy Metal Stress Through the Use of Nanotechnology

Subhadip Saha, Madhurima Dey, Swapnadip Bose, Ratneswar Poddar and Prasanta Kumar Bandyopadhyay*

9.1 INTRODUCTION

In this modern era, the continuous application of fertilizer and different agrichemicals on the crop field in connection with rapid and unreasonable mining and improper waste management are the main culprits of heavy metal contamination (Zhao et al., 2015; Zhai et al., 2018). In recent years, people are more conscious of heavy metal (HM) contamination because of its presence in a substantive manner in consumable food, vegetables and in the blood (Zhou et al., 2021). Thus, heavy metal pollution is one of the serious threats to mankind at present which not only hampers crop production and food security but also keeps harming human health dangerously (Javaid et al., 2020; Irshad et al., 2020). The application of nanomaterials to short out this heavy metal pollution is a new concept which has a huge opportunity to grow in the field of agriculture. The most studied materials of the century are nanomaterials, also known as "magic bullets" by Paul Ehrlich (Kreuter, 2007), which have given rise to a new field of study termed nanotechnology. This topic is a vibrant multidisciplinary scientific discipline that is fast growing and has a wide range of possible applications. To assess how nanomaterials interact with biological systems and their effects on plant ecosystems and trophic transport, it is essential to gain valuable insights into these interactions. These nanomaterials have unique properties that make their chemical actions completely different from that of the corresponding bulk materials, including small size, shape, reactivity and a higher surface area-to-volume ratio. This has helped them to garner significant attention and allowed for their wide-ranging and extensive use (Brunner et al., 2006). The nanoparticles' smaller size makes it easier for them to pass through various plant system barriers and get to specific cellular areas, while their increased surface area promotes higher adsorption and targeted administration of the required chemicals (Kashyap et al., 2015).

Among these, heavy metals, which are harmful to human health, are one notable contaminant that is exponentially increasing around the world. Food security and

* Corresponding author.

DOI: 10.1201/9781003376446-9

crop yield, especially in nations with large populations, are thus the new global problem (Rai et al., 2019). Heavy metals are defined as elements with an atomic mass larger than 20 and a density greater than 5 g/cm^3. They are the organic components of soil that come from the weathering of soil's parent material, a process known as "pedogenesis". Although they are present in low concentrations in the soil naturally, their concentration steadily rises (Bakshi & Abhilash, 2020). This growth is related to several anthropogenic activities, including the release of industrial effluents into water bodies and the landfilling of waste from pulp and paper, textile, tannery, metal mining and other industries, which then leaches and builds up in soil (Rai et al., 2019). Aerosols are produced as a result of the combustion of fossil fuels, and they play a significant part in HM accumulation (Bhuiyan et al., 2021). Pb (lead), Cr (chromium), As (arsenic), Zn (zinc), Cd (cadmium), Cu (copper), Hg (mercury) and Ni (nickel) are among the HMs that are regarded as being highly toxic, while Ba (barium), Sb (antimony), Mo (molybdenum), Tl (thallium) and Sn (tin) are among the HMs that are regarded as being less toxic (Bakshi & Abhilash, 2020). Both the plant and its associated microbiota are negatively impacted by the presence of HMs in the soil. As a result of the stress they cause, ROS (reactive oxygen species) are produced, which in turn harm the body's important macro-molecules and have an impact on both the nutrition and ecology of plants and crops. Nanotechnology has been employed in a variety of ways to help plants cope with HM stress, including reducing the bioavailability of HM concentrations in soil, enhancing physiological defence, creating protective agents and controlling the expression of heavy metal transport genes (Cao et al., 2020). In plant cell walls, nanoparticles combine with heavy metals to produce complexes that are inaccessible to organisms. Complexes' reduced biological activity and difficulty moving throughout the plant are caused by adsorption on their surface in plant cells (Wang et al., 2021).

9.2 HEAVY METAL TOXICITY IN PLANTS

Exposure to heavy metal toxicity in plants causes a wide range of physiological and metabolic alternations which ultimately are reflected through some visual toxicity symptoms. Although these toxicity symptoms differ from plant to plant due to variable sites of action, the most common possible mechanism is the production of reactive oxygen species and oxidative damage either directly through redox-active metals (Fe, Mn, Cr, Cu) or indirectly by the non-redox-active metals (Cd, Hg, Ni, Zn, Al). ROS production rate in plants reaches up to a toxicological level due to heavy metal stress via Haber-Weiss and Fenton reaction, and its concentration far exceeds the persistent production rate through main metabolic processes (photosynthesis and respiration) and disrupts the cell's homeostasis. The oxidative stress caused by ROS in plants can damage cell membranes, distort electrolytes, cause protein oxidation, lipid peroxidation and eventually the death of the cell. A high level of reactive oxygen species (ROS) production can alter different amino acids at specific sites, fragment peptide chains, an aggregate of cross-linked reaction products, alter redox and make proteins more susceptible to proteolysis (Sharma et al., 2012). A few symptomatic appearances of HM stress on plants are presented in (Table 9.1).

TABLE 9.1
Different heavy metal toxicity and their impact on plant

Heavy metals	Plant response to heavy metal stress	Reference
Chromium (Cr)	Shortening of root length and cell cycle extension in case of Sunflower (*Helianthus annuus L.*).	Fozia et al. (2008)
	Inhibits cell division and causes Ca^{2+} imbalance in cells, preventing calcium cations from entering the cytoplasm through the plasma membrane. (Root growth reduction in *Amaranthus viridis L.*).	Zou et al. (2006)
	The decline in seed germination of Cowpea (*Vigna sinensis L.*). uring seed germination, metabolic disorders occur by disrupting the conversion of food reserves into energy that is necessary for seedling emergence.	Nagajyoti et al. (2010)
Aluminium (Al)	Inhibition of root growth.	Ma et al. (2001)
	Make lateral roots thicker and browner.	Mossor-Pietraszewska (2001)
	Disruptions in enzymatic regulation of sugar phosphorylation and reduction of root respiration.	Rout et al. (2001)
	Molecular events implicated in Al-induced root depression, including the attachment of Al to the carboxylic groups of pectins within root cells, or the restriction of cell division in roots by Al ions binding to DNA.	Lidon & Barreiro (2002)
	A similar symptomatic effect like phosphorus deficiency occurs through Al-induced stress including, stunted leaves, discolouration of leaves, stems followed by yellowing and death of the leaf tips and Al toxicity can lead to rolling or curling of young leaves as well as the death of growing points like Ca deficiency.	Silva (2012); Bian et al. (2013)
	Chlorosis in the margins and centres of older leaves, with necrotic spots on the borders of young leaves.	Steiner et al. (2012)
Manganese (Mn)	Affected leaves may be crinkled, have darkened leaf veins, have chlorosis and brown spots on their surfaces and have black marks on their stems.	Reddy (2006); Maksimovi´c et al. (2012)
Nickel (Ni)	Increasing Ni concentrations will prevent plants from absorbing Fe^{2+} and Zn^{2+}, resulting in Fe or Zn deficiency and chlorosis expression.	Khan & Khan, (2010)
	Plants are adversely affected by excessive nickel in the germination process as it inhibits enzyme activities such as amylase and protease and disrupts the process of hydrolyzing the stored food.	Aydinalp and Marinova (2009); Sethy and Ghosh (2013)
	As a result of Ni agglomeration at the root apex, mitotic cell division at the root apex is greatly hindered, thereby resulting in reduced growth.	L'Huillier et al. (1996)

(Continued)

TABLE 9.1 (CONTINUED)
Different heavy metal toxicity and their impact on plant

Heavy metals	Plant response to heavy metal stress	Reference
Copper (Cu)	Initially, Cu toxicity results in hindrance of root elongation and growth reduction and subsequently chlorosis, necrosis and discolouration of leaves.	Yruela (2009)
	Excess Cu can adhere to cell membrane sulfhydryl groups and trigger lipid peroxidation, causing membrane damage and free radical generation in plant organelles and components.	Chen et al. (2000)
	Toxic levels of Cu through redox cycling between Cu^+ and Cu^{2+}, reactive oxygen species such as singlet oxygen (O^{2-}) and hydroxyl radical (HO) are formed, damaging DNA, RNA, lipids, carbohydrates and proteins.	Lombardi and Sebastiani (2005); Yurekli and Porgali (2006)
Zinc (Zn)	Chlorosis in young leaves caused by iron or manganese deficiency and the emergence of a purplish-red colour on the leaf surface caused by P deficiency are two visual indications of Zn poisoning in plants. In the active sites of enzymes or transporters, this shows that Zn^{2+} in excess can quickly replace other metals, especially those with similar ionic radii.	Yadav (2010); Baran (2013); Fukao et al. (2011)
	Extra Zn^{2+} in cells can result in the production of ROS and have a negative impact on membrane integration and permeability.	Mishra and Prakash (2010); Hosseini and Poorakbar (2013)
	Elevated Zn^{2+} reduces initial and maximal Chl fluorescence, repressing PSII activity.	Tsonev and Lidon (2012)

9.3 MECHANISMS OF NANOPARTICLES TO ALLEVIATE HEAVY METAL STRESS

A variety of approaches have been used to overcome HM stress in plants, including lowering the bioavailability of HM concentrations in soil (Moharem et al., 2019), improving plant's physiological defence, developing protective agents and regulating the expression of heavy metal transport genes (Cao et al., 2020). Upon binding with heavy metals, nanoparticles form complexes in plant cell walls that render them inaccessible to organisms. Adsorption of the complexes on the surface of plant cells hinders their movement throughout the plant and minimizes their biological activity (Wang et al., 2021).

The principal antioxidant enzymes include ascorbate peroxidase (APX), catalase (CAT), peroxidase (POD), superoxide dismutase (SOD), superoxide peroxidase (SPX), glutathione peroxidase (GPX) and glutathione reductase (GR), all of which work to keep cellular ROS levels stable. Because of this, these enzymes scavenge reactive species when there is an abundance of ROS. Nonenzymatic metabolites of low molecular weight (polyphenols, vitamins E and C) also have the potential to neutralize ROS (Ullah et al., 2020; Zhao et al., 2020).

TABLE 9.2
Mitigation of heavy metal stress through metal and non-metal-based nanoparticles

Name	Mode of action	Example
Metal-based nanoparticles on the HM stress in plants		
Cerium dioxide	As a result of the dangling bonds on the Ce^{3+} and Ce^{4+} surface of CeO_2 NPs, the oxidative stress caused by HM is reduced by scavenging ROS. (Wu et al., 2017).	• Rice grown in hydroponic conditions ($CdCl_2$ 50mM) was effectively stimulated by foliar application of CeO_2 NPs (200 mg/L) to enhance antioxidant defence and inhibit Cd accumulation (Wang et al., 2019).
Titanium dioxides	Adsorption of HM and activation of oxidative defence system	• TiO2 NPs effectively limited Cd toxicity to soybean plants by increasing physiological parameters and photosynthetic rate (Singh & Lee, 2016). • Foliar application of TiO_2 NPs reduced Cd toxicity to maize grown on soil contaminated with Cd through increased SOD and GST activities, as well as a significant increase in galactose, alanine, aspartame acid, and other metabolic pathways over soil application (Lian et al., 2020).
Iron	Because of their unique structure and electronic properties, nano-Fe has a lot of mechanisms for alleviating symptoms, including HM adsorption, promoting the formation of root surface iron films, activating oxidation defence systems and scavenging ROS (Konate et al., 2017)].	• In sunflower leaves, nanoscale zero-valent iron has been shown to reduce HM accumulation, increase SOD and POD activity and encourage plant growth. (Michálková et al., 2017). • Under Cd stress (available Cd 0.93 mg/kg), foliar and soil application of Fe_2O_3 NPs (5, 10, 15 and 20 ppm) suppressed leaf electrolyte leaks, reduction of grain Cd content and increases in antioxidant enzyme activity and dry weight (Hussain et al., 2019).

(Continued)

TABLE 9.2 (CONTINUED)
Mitigation of heavy metal stress through metal and non-metal-based nanoparticles

Name	Mode of action	Example
Zinc oxide	Its relief techniques include nutritional value, activating oxidative stress system and enhancing plant physiological and biochemical processes.	• ZnO NPs (50 mg/L) seed soaking increased glycolytic metabolism and cell wall biosynthesis, boosting corn seed germination and the development of roots and embryos (Luying et al., 2020). • ZnO NPs and Zn^{+2} considerably reduced the arsenic level in rice roots and shoots (Ma et al., 2020).
Non-metal-based nanoparticles impact on the HM stress in plants		
Selenium	Se, a key component of glutathione peroxidase, regulates antioxidant enzyme activity, removes ROS from plants and boosts their stress tolerance.	• Se treatment has been demonstrated to lessen Cd uptake and toxicity in rice (Hu et al., 2014; Lin et al., 2012). • Chinese cabbage's biomass, height, chlorophyll content, super oxide dismutase and glutathione peroxidase increased after foliar spraying of Se NPs (1.25 and 12.5 mg/L) under Cd stress (10 and 20 mg/kg).
Silicon	It is generally acknowledged that Si reduces HM stress through the following four mechanisms: triggering the body's natural defence mechanisms to increase ROS scavenging (Emamverdian et al., 2018), immobilizing and complexing HM to lessen its biological activity (Liang et al., 2007; Savvas & Ntatsi, 2015), serving as a nutrient source and encouraging the production of structural protective agents which can chelate with HM and controlling the expression of HM transport genes, such as vacuolar H+-pyrophosphatase1 (OVP1) over expression (Cao et al., 2020).	• By triggering the antioxidant defence mechanisms, de Sousa et al. (2019) discovered that Si NPs (4 mg/kg) reduced the toxicity of Al to maize. • Hussain et al. (2020) found that foliar Si NPs (10 and 20 mg/L) reduced Cd in grains and enhanced yield.
Hydroxyapatite	After being added to the soil, hydroxyapatite (HAP) can release phosphate to create complexes with HM (Mignardi et al., 2012). Ca ions on HAP could exchange with HM ions in the soil to prevent plant damage (Mingxin et al., 2019).	• Sun et al. (2018) found that HAP NPs reduced Cu and Zn in ryegrass. • After applying HAP NPs to Cd-contaminated soil, the pH increased and catalase and urease activity improved (Qingqing et al., 2017).

From a physiological point of view, apoplastic barriers in the roots of plants protect the plant and control the flow of water, ions and oxygen (Chao-Dong et al., 2013). NPs can influence the establishment of apoplastic barriers, which may reduce the quantity of HM in the root (Rossi et al., 2017). However, apoplastic barriers solely may not be an efficient technique for mitigating the damage produced by HM stress because root plasma membranes include many ions as well as protein transporters that can simultaneously transport HMs. In addition, certain NPs have the ability to regulate plant metal transport genes, which, in turn, improves the ability of the plant's extracellular barrier to prevent the entry of HMs. Listed in (Table 9.2) are some possible mechanisms of reducing heavy metal stress through metal and non-metal-based nanoparticles.

9.4 NANOBIOREMEDIATION: AN EFFECTIVE WAY TO ALLEVIATE HMS STRESS

The term "nanobioremediation" refers to the practise of cleaning polluted areas using nanomaterials and nanoparticles that have been created using environmentally friendly methods (Amin et al., 2021). It is a low-cost way of breaking down hazardous substances using plants and bacteria, with the goals of lowering pollution and increasing the condition of the soil. Through the breakdown of contaminants, this process has the potential to get rid of, drastically reduce or retain the same level of pollutants in the soil (Ashraf et al., 2019). The efficiency of bioremediation has been studied and enhanced by utilizing biotechnology (Gerhardt et al., 2017), but nanotechnology has introduced a new facet to the process (Gong et al., 2018). Nanotechnology-based phytoremediation increases phytoremediation's effectiveness by using nanoparticles to treat contaminated soil, water and organic matter (Awasthi et al., 2022). NPs have different physico-chemical properties that operate as a catalyst to minimize the activation energy needed to break down molecules, causing targeted contaminants to be destroyed, absorbed or transformed. HMs, on the other hand, present an unusual obstacle due to the fact that they do not biodegrade and are very prone to contaminating food chains and biological systems, both of which still need to be fully comprehended in order to ensure that they are effectively removed (Briffa et al., 2020). Nanobioremediation encompasses a variety of approaches and devices, including nanobiosensors, nanosorbents, nanophytoremediation and nanobiosurfactants which are described in greater depth below.

9.4.1 NANOBIOSENSORS

Nanobiosensors are modified biosensors composed of biological materials that are connected to a transducer, which is typically manufactured on a nanoscale to acquire, process and analyse data at the atomic level (Ghaffar et al., 2020). A physico-chemical transducer that is more compact and coupled to sensitive components is used to detect the analytes at very low concentrations. Nanobiosensors offer a number of fascinating characteristics, such as fictionalization, immobilization and miniaturization, which boost analytical performance by integrating biocomponents into complex structures (Usman et al., 2020). Nanobiosensors have exhibited remarkable advantages over conventional biosensors due to their enhanced specificity of detection and their ability to be applied in a wide array of fields such as bioprocessing, soil nutrient monitoring, agriculture and environment control. The phytotoxicity of heavy metals can be identified using nanobiosensors by monitoring changes in plant water status and chlorophyll content, etc.

9.4.2 Nanobiosorbents

Biosorption is the process of removing organic and inorganic substances with the assistance of materials that have been derived from biological sources. Nanobiosorbents, on the other hand, are used to get rid of pollutants by combining biosorption and nanotechnology. It is one of the most modern methods that places a significant emphasis on the removal of HMs from soil and water. Nanobiosorbents are renewable materials that are heavily utilized due to their superior pollutant removal capabilities. The presence of -COOH (carboxyl) and -OH (hydroxyl) groups is a prerequisite for a biosorbent to absorb pollutants, heavy metals, radionuclides and dyes (Rizwan et al., 2022).

9.4.3 Nanobiosurfactants

Molecules that are classified as biosurfactants are those that are either produced in living environments or are secreted extracellularly by microbes. These molecules have both lipophilic and hydrophilic activities, and they lower the surface tension. Biosurfactants have emerged as an environmentally friendly and potentially fruitful source for the synthesis of nanoparticles in the field of sustainable chemistry (Debnath et al., 2021). Biosurfactants derived from microbes are frequently used for the remediation of heavy metals such as copper, nickel and zinc. Biswas and Raichur (2008) looked into how rhamnolipid (a glycolipid biosurfactant) interacted with nanozirconia. According to the findings of the study, there is an electronegativity interaction on the surface of nanozirconia with rhamnolipid. Soil that was contaminated with HMs and hydrocarbons was cleaned up well with the help of a lipopeptide biosurfactant made from *Bacillus subtilis*.

9.4.4 Nanophytoremidiation

Phytoremediation is more advantageous than conventional technologies because of its ease of use, long-term viability and environmental compatibility. The rhizospheric microbiome is just as important as plants when it comes to purifying these polluted resources. When it comes to removing organic and inorganic pollutants, plants and microbes use the same principles of phytoremediation that can be found in nature (Bakshi & Abhilash, 2020). Through a process called phytoremediation, nanomaterials (NMs) can significantly reduce the negative effects of heavy metal contamination. It has been suggested that flax (*Linum usitatissimum*), which may be grown as flax seed, might be a good choice for phytoremediation of Cu because of its ability to remove substantial amounts of Cu from soils.

Scientists have documented how TiO2 (nanotitanium dioxide) affects the Cd accumulation in glycine max plant from soil (Usman et al., 2020). TiO_2 was able to reduce the toxicity of Cd and increase the amount of Cd that soybean plants were able to absorb by acting as a barrier against oxidative damage and as a scavenger for the free radicals that were created by Cd's toxicity. As an improved method of remediation, Liang et al. (2017) conducted a field experiment in which they studied the combined influence of NHAP (nanohydroxyapatite) and NCB (nanocarbon black) on the phytoextraction of lead from the soil by *Lolium temulentum L.* (ryegrass). The phytotoxic effects of lead on ryegrass were significantly mitigated by either NCB or NHAP, which also increased the plant's ability to phytoextract lead. For the in-situ treatment of lead-tainted soils, NHAP was found to be superior to NCB in terms of suitability (Liang et al., 2017). Nanomaterials can be used to modulate gene expressions relevant to water homeostasis, metal stress, cell wall synthesis, oxidative stress, cell division and photosynthetic processes.

So, rapid industrialization and modern lifestyle are facing the heavy metal pollution at a speedy rate which needs to be remediated as it is moving from the contaminated soil to the plant system and ultimately to the human or animal body. Different techniques applied in agriculture to counter heavy metal stress in crop fields, mainly by converting the toxic form to non-toxic form or available form to unavailable form in soil, reduce the uptake of heavy metal by the crop. Nanotechnology in alleviation of HM stress has a wide application in agriculture and now more focus can be given to synthesizing nanoparticles efficiently by using green and environmentally safe methods, which are economically viable. Thus, this is just the beginning and application of nanoscience, which still needs more indepth research to understand how NPs can improve plant resistance or keep the plant away from accumulation of heavy metals in contaminated soil and environments.

REFERENCES

Amin, I., Nazir, R., & Rather, M. A. (2021). Nano-bioremediation: An innovative approach for remedying heavy metals using fungi. *Journal of Bioremediation and Biodegradation*, *12*(6), 187.

Ashraf, S., Siddiqa, A., Shahida, S., & Qaisar, S. (2019). Titanium-based nanocomposite materials for arsenic removal from water: A review. *Heliyon*, *5*, e01577.

Awasthi, G., Nagar, V., Mandzhieva, S., Minkina, T., Sankhla, M. S., Pandit, P. P., Aseri, V., Awasthi, K. K., Rajput, V. D., Bauer, T., & Srivastava, S. (2022). Sustainable amelioration of heavy metals in soil ecosystem: Existing developments to emerging trends. *Minerals*, *12*, 85.

Aydinalp, C., & Marinova, S. (2009). The effects of heavy metals on seed germination and plant growth on alfalfa plant (*Medicago sativa*). *Bulgarian Journal of Agricultural Science*, *15*(4), 347–350.

Bakshi, M., & Abhilash, P. C. (2020). Nanotechnology for soil remediation: Revitalizing the tarnished resource. In P. Singh, A. Borthakur, P. K. Mishra, & D. Tiwary (Eds.), *Nano-materials as Photocatalysts for degradation of environmental pollutants* (pp. 345–370). Elsevier.

Baran, A. (2013). Assessment of *Zea mays* sensitivity to toxic content of zinc in soil. *Polish Journal of Environmental Studies*, *22*(1), 77–83.

Bhuiyan, M. A. H., Karmaker, S. C., Bodrud-Doza, M., Rakib, M. A., & Saha, B. B. (2021). Enrichment, sources and ecological risk mapping of heavy metals in agricultural soils of Dhaka district employing SOM, PMF and GIS methods. *Chemosphere*, *263*, 128339.

Bian, M., Zhou, M., Sun, D., & Li, C. (2013). Molecular approaches unravel the mechanism of acid soil tolerance in plants. *The Crop Journal*, *1*(2), 91–104.

Biswas, M., & Raichur, A. M. (2008). Electrokinetic and rheological properties of nano zirconia in the presence of rhamnolipid biosurfactant. *Journal of the American Ceramic Society*, *91*, 3197–3201.

Briffa, J., Sinagra, E., & Blundell, R. (2020). Heavy metal pollution in the environment and their toxicological effects on humans. *Heliyon*, *6*, e04691.

Brunner, T. J., Wick, P., Manser, P., Spohn, P., Grass, R. N., Limbach, L. K., Bruinink, A., & Stark, W. J. (2006). In vitro cytotoxicity of oxide nanoparticles: Comparison to asbestos, silica, and the effect of particle solubility. *Environmental Science and Technology*, *40*(14), 4374–4381.

Cao, F., Dai, H., Hao, P. F., & Wu, F. (2020). Silicon regulates the expression of vacuolar H+-pyrophosphatase 1 and decreases cadmium accumulation in rice (*Oryza sativa* L.). *Chemosphere*, *240*, 124907.

Chao-Dong, Y., Xia, Z., Guo-Feng, L., Jun-Wei, Z., Man-Zhu, B., & Zhi-Xiang, Z. (2013). Progress on the structure and physiological functions of apoplastic barriers in root. *Bulletin of Botanical Research*, *33*(1), 114–119.

Chen, L. M., Lin, C. C., & Kao, C. H. (2000). Copper toxicity in rice seedlings: Changes in antioxidative enzyme activities, H_2O_2 level, and cell wall peroxidase activity in roots. *Botanical Bulletin of Academia Sinica*, *41*(2), 99–103.

de Sousa, A., Saleh, A. M., Habeeb, T. H., Hassan, Y. M., Zrieq, R., Wadaan, M. A. M., Hozzein, W. N., Selim, S., Matos, M., & AbdElgawad, H. (2019). Silicon dioxide nanoparticles ameliorate the phytotoxic hazards of aluminium in maize grown on acidic soil. *Science of the Total Environment*, *693*, 133636. https://doi.org/10.1016/j.scitotenv.2019.133636

Debnath, M., Chauhan, N., Sharma, P., & Tomar, I. (2021). Potential of nano biosurfactants as an eco-friendly green technology for bioremediation. In B. Bhanvase, S. Sonawane, V. Pawade, & A. Pandit (Eds.), *Handbook of nanomaterials for wastewater treatment* (pp. 1039–1055). Elsevier.

Emamverdian, A., Ding, Y., Xie, Y., & Sangari, S. (2018). Silicon mechanisms to ameliorate heavy metal stress in plants. *BioMed Research International*, 8492898. https://doi.org/10.1155/2018/8492898

Fozia, A., Muhammad, A. Z., Muhammad, A., & Zafar, M. K. (2008). Effect of chromium on growth attributes in sunflower (*Helianthus annuus* L.). *Journal of Environmental Sciences*, *20*(12), 1475–1480.

Fukao, Y., Ferjani, A., Tomioka, R., Nagasaki, N., Kurata, R., Nishimori, Y., Fujiwara, M., & Maeshima, M. (2011). iTRAQ analysis reveals mechanisms of growth defects due to excess zinc in Arabidopsis. *Plant Physiology*, *155*(4), 1893–1907. https://doi.org/10.1104/pp.110.169730

Gerhardt, K. E., Gerwing, P. D., & Greenberg, B. M. (2017). Opinion: Taking phytoremediation from proven technology to accepted practice. *Plant Science*, *256*, 170–185.

Ghaffar, N., Farrukh, M. A., & Naz, S. (2020). Applications of nanobiosensors in agriculture. In S. Javad (Ed.), *Nanoagronomy* (pp. 179–196). Springer.

Gong, X., Huang, D., Liu, Y., Peng, Z., Zeng, G., Xu, P., Cheng, M., Wang, R., & Wan, J. (2018). Remediation of contaminated soils by biotechnology with nanomaterials: Biobehavior, applications, and perspectives. *Critical Reviews in Biotechnology*, *38*(3), 455–468.

Hosseini, Z., & Poorakbar, L. (2013). Zinc toxicity on antioxidative response in (*Zea mays* L.) at two different pH. *Journal of Stress Physiology and Biochemistry, 9*, 66–73.

Hu, Y., Norton, G. J., Duan, G., Huang, Y., & Liu, Y. (2014). Effect of selenium fertilization on the accumulation of cadmium and lead in rice plants. *Plant and Soil, 384*(1–2), 131–140.

Hussain, A., Ali, S., Rizwan, M., Rehman, M. Z. U., Qayyum, M. F., Wang, H., & Rinklebe, J. (2019). Responses of wheat (*Triticum aestivum*) plants grown in a Cd contaminated soil to the application of iron oxide nanoparticles. *Ecotoxicology and Environmental Safety, 173*, 156–164.

Hussain, B., Lin, Q., Hamid, Y., Sanaullah, M., Di, L., Hashmi, M. L. U. R., Khan, M. B., He, Z., & Yang, X. (2020). Foliage application of selenium and silicon nanoparticles alleviates Cd and Pb toxicity in rice (*Oryza sativa* L.). *The Science of Total Environment, 712*, 136497.

Irshad, H. M. K., Noman, A., Alhaithloul, H., Adeel, M., Yukui, R., Shah, T., Zhu, S., & Shang, J. (2020). Goethite-modified biochar ameliorates the growth of rice (*Oryza sativa* L.) plants by suppressing Cd and As-induced oxidative stress in Cd and As co-contaminated paddy soil. *Science of the Total Environment, 717*, 137086. https://doi.org/10.1016/j.scitotenv.2020.137086

Javaid, S., uz-Zaman, Q., Sultan, K., Riaz, U., Aslam, A., Sharif, S., Ehsan, N., Aslam, S., Jamil, A., & Ibraheem, S. 2020. Heavy metals stress, mechanism and remediation techniques in rice (*Oryza sativa* L.): A review. *Pure and Applied Biology, 9*(1), 403–426. https://doi.org/10.19045/bspab.2020.90045

Kashyap, P. L., Xiang, X., & Heiden, P. (2015). Chitosan nanoparticle based delivery systems for sustainable agriculture. *International Journal of Biological Macromolecules, 77*, 36–51.

Khan, M. R., & Khan, M. M. (2010). Effect of varying concentration of nickel and cobalt on the plant growth and yield of chickpea. *Australian Journal of Basic and Applied Sciences, 4*(6), 1036–1046.

Konate, A., He, X., Zhang, Z., Ma, Y., Zhang, P., Alugongo, G. M., & Rui, Y. (2017). Magnetic (Fe_3O_4) nanoparticles reduce heavy metals uptake and mitigate their toxicity in wheat seedling. *Sustainability, 9*(5), 790.

Kreuter, J. (2007). Nanoparticles: A historical perspective. *International Journal of Pharmaceutics, 331*(1), 1–10.

L'Huillier, L., D'Auzac, J., Durand, M., & Michaud-Ferriere, N. (1996). Nickel effects on two maize (*Zea mays*) cultivars: Growth, structure, Ni concentration, and localization. *Canadian Journal of Botany, 74*(10), 1547–1554.

Lian, J., Zhao, L., Wu, J., Xiong, H., Bao, Y., Zeb, A., Tang, J., & Liu, W. (2020). Foliar spray of TiO_2 nanoparticles prevails over root application in reducing Cd accumulation and mitigating Cd-induced phytotoxicity in maize (*Zea mays* L.). *Chemosphere, 239*, 124794.

Liang, S. X., Jin, Y., Liu, W., Li, X., Shen, S. G., & Ding, L. (2017). Feasibility of Pb phytoextraction using nano-materials assisted ryegrass: Results of a one-year field-scale experiment. *Journal of Environmental Management, 190*, 170–175.

Liang, Y., Sun, W., Zhu, Y., & Christie, P. (2007). Mechanisms of silicon-mediated alleviation of abiotic stresses in higher plants: A review. *Environmental Pollution, 147*(2), 422–428.

Lidon, F. C., & Barreiro, M. G. (2002). An overview into aluminium toxicity in maize. *Bulgarian Journal of Plant Physiology, 28*(3–4), 96–112.

Lin, L., Zhou, W., Dai, H., Cao, F., Zhang, G., & Wu, F. (2012). Selenium reduces cadmium uptake and mitigates cadmium toxicity in rice. *Journal of Hazardous Materials, 235–236*, 343–351.

Lombardi, L., & Sebastiani, L. (2005). Copper toxicity in Prunuscerasifera: Growth and antioxidant enzymes responses of in vitro grown plants. *Plant Science*, *168*(3), 797–802.

Luying, S., Fengbin, S., Xiangnan, L., Xiancan, Z., Shengqun, L., Yang, W., & Xiaoning, Q. (2020). Effects of ZnO nanoparticles on seed germination and root carbon metabolism in maize (*Zea mays* L.). *Soils Crop*, *9*, 40–49.

Ma, J. F., Ryan, P. R., & Delhaize, E. (2001). Aluminium tolerance in plants and the complexing role of organic acids. *Trends in Plant Science*, *6*(6), 273–278.

Ma, X., Sharifan, H., Dou, F., & Sun, W. (2020). Simultaneous reduction of arsenic (As) and cadmium (Cd) accumulation in rice by zinc oxide nanoparticles. *Chemical Engineering Journal*, *384*, 123802.

Maksimović, J. D., Mojović, M., Maksimović, V., Römheld, V., & Nikolic, M. (2012). Silicon ameliorates manganese toxicity in cucumber by decreasing hydroxyl radical accumulation in the leaf apoplast. *Journal of Experimental Botany*, *63*(7), 2411–2420.

Michálková, Z., Fernández, D. M., & Komárek, M. (2017). Interactions of two novel stabilizing amendments with sunflower plants grown in a contaminated soil. *Chemosphere*, *186*, 374–380. https://doi.org/10.1016/j.chemosphere.2017.08.009

Mignardi, S., Corami, A., & Ferrini, V. (2012). Evaluation of the effectiveness of phosphate treatment for the remediation of mine waste soils contaminated with Cd, Cu, Pb, and Zn. *Chemosphere*, *86*(4), 354–360.

Mingxin, W., Caicai, W., Jinyong, Z., Yang, X., & Shize, W. (2019). Joint remediation of heavy metal contaminated soil by EDTA and nano-hydroxyapatite. *Chinese Journal of Environmental Engineering*, *13*(2), 396–405.

Mishra, P. K., & Prakash, V. (2010). Response of non-enzymatic antioxidants to zinc induced stress at different pH in *Glycine max* L. cv. Merrill. *Academic Journal of Plant Sciences*, *3*(1), 1–10.

Moharem, M., Elkhatib, E., & Mesalem, M. 2019. Remediation of chromium and mercury polluted calcareous soils using nanoparticles: Sorption–desorption kinetics, speciation and fractionation. *Environmental Research*, *170*, 366–373.

Mossor-Pietraszewska, T. (2001). Effect of aluminium on plant growth and metabolism. *Acta Biochimica Polonica*, *48*(3), 673–686.

Nagajyoti, P. C., Lee, K. D., & Sreekanth, T. V. M. (2010). Heavy metals, occurrence and toxicity for plants: A review. *Environmental Chemistry Letters*, *8*(3), 199–216.

Qingqing, Z., Shuokang, W., Chenchen, Z., Qing, W., Ying, Z., Lin, N., Shu Xuan, L., & Wei, L. (2017). Adsorption and desorption of Cd on nHAP and remediation test on Cd contaminated soil. *Environmental Engineering*, *35*(3), 179–184.

Rai, P. K., Lee, S. S., Zhang, M., Tsang, Y. F., & Kim, K. H. (2019). Heavy metals in food crops: Health risks, fate, mechanisms, and management. *Environment International*, *125*, 365–385.

Reddy, K. J. (2006). Nutrient stress. In K. V. Madhava Rao, A. S. Raghavendra, & R. K. Janardhan (Eds.), *Physiology and molecular biology of stress tolerance in plants* (pp. 187–217). Springer.

Rizwan, K., Rasheed, T., & Bilal, M. (2022). Alginate-based nanobiosorbents for bioremediation of environmental pollutants. In A. Denizli, N. Ali, M. Bilal, A. Khan, & T. A. Nguyen (Eds.), *Nano-Biosorbents for decontamination of water, air, and soil pollution* (pp. 479–502). Elsevier.

Rossi, L., Zhang, W., Schwab, A. P., & Ma, X. (2017). Uptake, accumulation, and in planta distribution of coexisting cerium oxide nanoparticles and cadmium in *Glycine max* (L.) Merr. *Environmental Science and Technology*, *51*(21), 12815–12824.

Rout, G. R., Samantaray, S., & Das, P. (2001). Aluminium toxicity in plants: A review. *Agronomie*, *21*(1), 3–21.

Savvas, D., & Ntatsi, G. (2015). Biostimulant activity of silicon in horticulture. Scientia Horticulturae, *196*, 66–81. https://doi.org/10.1016/j.scienta.2015.09.010

Sethy, S. K., & Ghosh, S. (2013). Effect of heavy metals on germination of seeds. *Journal of Natural Science, Biology and Medicine*, *4*(2), 272–275.

Sharma, P., Bhatt, D., Zaidi, M. G. H., Saradhi, P. P., Khanna, P. K., & Arora, S. (2012). Silver nanoparticle-mediated enhancement in growth and antioxidant status of *Brassica juncea*. *Applied Biochemistry and Biotechnology*, *167*(8), 2225–2233.

Silva, S. (2012). Aluminium toxicity targets in plants. *Journal of Botany*, 219462. https://doi.org/10.1155/2012/219462

Singh, J., & Lee, B. (2016). Influence of nano-TiO_2 particles on the bioaccumulation of Cd in soybean plants (*Glycine max*): A possible mechanism for the removal of Cd from the contaminated soil. *Journal of Environmental Management*, *170*, 88–96.

Steiner, F., Zoz, T., Junior, A. S. P., Castagnara, D. D., & Dranski, J. A. L. (2012). Effects of aluminium on plant growth and nutrient uptake in young physic nut plants. *Semina: CiênciasAgrárias*, *33*(5), 1779–1788.

Sun, R., Chen, J., Fan, T., Zhou, D., & Wang, Y. (2018). Effect of nanoparticle hydroxyapatite on the immobilization of Cu and Zn in polluted soil. *Environmental Science and Pollution Research*, *25*(1), 73–80.

Tsonev, T., & Lidon, F. J. C. (2012). Zinc in plants: An overview. *Emirates Journal of Food and Agriculture*, *24*(4), 322–333.

Ullah, S., Adeel, M., Zain, M., Rizwan, M., Irshad, M. K., Jilani, G., Hameed, A., Khan, A., Arshad, M., Raza, A., Baluch, M. A., & Rui, Y. (2020). Physiological and biochemical response of wheat (*Triticum aestivum*) to TiO_2 nanoparticles in phosphorous amended soil: A full life cycle study. *Journal of Environmental Management*, *263*, 110365. https://doi.org/10.1016/j.jenvman.2020.110365

Usman, M., Farooq, M., Wakeel, A., Nawaz, A., Cheema, S. A., Rehman, H., Ashraf, I., & Sanaullah, M. (2020). Nanotechnology in agriculture: Current status, challenges and future opportunities. *Science of Total Environment*, *721*, 137778. https://doi.org/10.1016/j.scitotenv.2020.137778

Wang, K., Wang, Y., Wan, Y., Mi, Z., Wang, Q., Wang, Q., & Li, H. (2021). The fate of arsenic in rice plants (*Oryza sativa* L.): Influence of different forms of selenium. *Chemosphere*, *264*, 128417. https://doi.org/10.1016/j.chemosphere.2020.128417

Wang, Y., Wang, L., Ma, C., Wang, K., Hao, Y., Chen, Q., Mo, Y., & Rui, Y. (2019). Effects of cerium oxide on rice seedlings as affected by co-exposure of cadmium and salt. *Environmental Pollution*, *252*, 1087–1096. https://doi.org/10.1016/j.envpol.2019.06.007

Wu, H., Tito, N., & Giraldo, J. P. (2017). Anionic cerium oxide nanoparticles protect plant photosynthesis from abiotic stress by scavenging reactive oxygen species. *ACS Nano*, *11*(11), 11283–11297.

Yadav, S. K. (2010). Heavy metals toxicity in plants: An overview on the role of glutathione and phytochelatins in heavy metal stress tolerance of plants. *South African Journal of Botany*, *76*(2), 167–179.

Yruela, I. (2009). Copper in plants: Acquisition, transport and interactions. *Functional Plant Biology*, *36*(5), 409–430.

Yurekli, F., & Porgali, Z. B. (2006). The effects of excessive exposure to copper in bean plants. *Acta Biologica Cracoviensia Series Botanica*, *48*(2), 7–13.

Zhai, X., Li, Z., Huang, B., Luo, N., Huang, M., Zhang, Q., & Zeng, G. (2018). Remediation of multiple heavy metal-contaminated soil through the combination of soil washing and in situ immobilization. *Science of the Total Environment*, *635*, 92–99.

Zhao, F., Ma, Y., Zhu, Y., Tang, Z., & McGrath, S. P. (2015). Soil contamination in China: Current status and mitigation strategies. *Environmental Science and Technology*, *49*(2), 750–759.

Zhao, L., Lu, L., Wang, A., Zhang, H., Huang, M., Wu, H., Xing, B., Wang, Z., & Ji, R. (2020). Nanobiotechnology in agriculture: Use of nanomaterials to promote plant growth and stress tolerance. *Journal of Agricultural and Food Chemistry, 68*(7), 1935–1947. https://doi.org/10.1021/acs.jafc.9b06615

Zhou, P., Adeel, M., Shakoor, N., Guo, M., Hao, Y., Azeem, I., Li, M., Liu, M., & Rui, Y. (2021). Application of nanoparticles alleviates heavy metals stress and promotes plant growth: An overview. *Nanomaterials, 11*(1), 26. https://doi.org/10.3390/nano11010026

Zou, J., Wang, M., Jiang, W., & Liu, D. (2006). Chromium accumulation and its effects on other mineral elements in *Amaranthus viridis* L. *Acta Biologica Cracoviensia Series Botanica, 48*(1), 7–12.

10 Green Synthesis of Nanoparticles

Applications and Feasibility Over Chemical Synthesis Procedures

Anu Kalia and Rajeev K. Gupta*

10.1 INTRODUCTION

Nanomaterials represent a unique interstitial state of matter possessing any geometric shape with at least one among the three size dimensions in a nanoscale regime (1 to 100 nm) (Roduner, 2006). These materials possess properties radically and exponentially vary from their bulk counterparts due to the strict size-dependence (Roduner, 2009). The particulate forms of metal, non-metals and their oxides, sulphides and other crystalline or non-crystalline structures are considered under a sub-category of nanomaterials termed as "nanoparticles (NPs)" with all the three size dimensions, i.e. x, y and z, within the range of 1 to 100 nm (Andrievskii, 2009). These have been identified to exhibit vivid morphologies and possess novel characteristics including high reactivity owing to surface-interface (high surface area to volume ratios) effects (Palmqvist, 2017). The NPs can be synthesized through two basic synthesis protocols, i.e. top-down and bottom-up, depending on the nature of the initial substrate which can be bulk and atoms/molecules/ions respectively (Ying et al., 2022). However, based on the nature of the method, the NP synthesis techniques can be broadly categorized as physical, chemical and biological methods, though the hybrid physicochemical and biochemical approaches have also been developed and identified (Ijaz et al., 2020). The physical NP synthesis methods are quite energy and cost-intensive, and require peculiar conditions (high temperature > 350°C and pressure) and sometimes a very sophisticated assemblage of multiple components/equipment such as the plasma generating electric arc discharge, lasers, high voltage consuming sputter machines, etc. (Imran et al., 2021). Therefore, these are being predominantly replaced with chemical and biological methods (Horikoshi & Serpone, 2013). The chemical methods refer to a variety of solvo- and hydrothermal besides solgel and aero-/cryo-chemical synthesis of NPs (Imran et al., 2021). The predominance and popularity of chemical methods are due to the low

* Corresponding author.

DOI: 10.1201/9781003376446-10

cost and ease of synthesis of a voluminous quantity of NPs (Bokov et al., 2021). The steps of synthesis are also amenable to automation and adaptable for industrial-scale synthesis. However, in spite of the cost-effective and mass-scaled-up synthesis advantages associated with the chemical synthesis of NPs, the inherent drawbacks including the requirement of concentrated acid/bases, toxic reagents, organic solvents, stabilizers and surfactant compounds and the generation of environmentally corrosive chemical wastes post-synthesis of the NPs, have led to the identification of a potential alternative method (Modan & Plaiasu, 2020). Further, the eco-toxicity of these engineered NPs has escalated public concerns particularly related to the collateral interactive effects of the NPs with the abiotic and biotic components of the contaminating eco-niches (Pérez-Hernández et al., 2021), the requirement for understanding and assessing the NP life cycle (Yan et al., 2022) and safe disposal (Gupta & Bharti, 2022). The subsequent sections of the chapter delineate the possible alternative methods of NP synthesis.

10.2 GREEN SYNTHESIS TECHNIQUES

Green NP synthesis techniques span from the green chemical to unified green chemistry approaches (Miu & Dinischiotu, 2022; Duan et al., 2015). The green chemistry of nanoparticles will involve the utilization of non-toxic solvents, capping and reducing agents with an improved yield of nanoparticles. While the unified green chemical approaches include employing a variety of living cells, their extracts and isolated biomacro-molecules to obtain strict control over the morphologies and hence the properties of the generated nanomaterials (Miu & Dinischiotu, 2022). Small biomolecules such as phenolic derivatives (flavonoids, flavanols, polyphenols and similar compounds) can reduce the ionic forms of metal to lower or zero valency states (Duan et al., 2015). Medium to large molecular weight biomolecules including the catalytic proteins (a variety of enzymes) with their co-enzyme/co-factor components can carry out the metal reduction reactions through the proton/electron exchange reactions (Thapa et al., 2017). Polysaccharides, the other category of biomacro-molecules can function both as reducing and capping agents to accomplish NP synthesis. The optimum conditions for the NP biosynthesis such as pH, temperature, the concentration of the metal salt, volume of the cell/cell extracts, shaking/static incubation conditions and reaction time can have a direct impact on the homogeneity or polydispersity of nanosol (Ying et al., 2022). The general similarities and differences between the green chemistry and green synthesis techniques have been depicted in Table 10.1.

10.2.1 Green Chemical Nanoparticle Synthesis

The greener chemistry of nanoparticle synthesis is based on the utilization of a few crucial components during the NP synthesis (Kharissova et al., 2019). These include the use of a green solvent medium and non-toxic reducing and stabilizing agents (Duan et al., 2015) contrary to the conventional chemical synthesis approaches which require organic solvents, highly reactive reducing agents, and linear and dendritic surfactant polymers to ensure improved stability and shelf life of generated nanosols (Kharissova et al., 2019).

TABLE 10.1
Green chemistry and biosynthetic approaches for nanoparticle generation

Attribute	Green chemical approach	Biosynthetic approach
Utilization of green solvent (most probably water)	Other less corrosive solvents (supercritical fluids-CO_2 and water, ionic liquids, poly glycols) have also been utilized.	Intra/extracellular environment of the cell is strictly water-based.
Green reducing agents	Safe/less toxic organic compounds and biomolecules includes simple reducing sugars-Glucose, free amino acids, polyphenols.	Small, medium and high molecular biomolecules including sugars, phenols, proteins, enzymes and polysaccharides.
Green stabilizing agents	Biodegradable and safe biomolecules-starch, carboxymethyl cellulose.	Dual action reducing and stabilizing agents-starch, xanthan gum, agar.

The green reducing and capping agents can be derived from natural sources as small molecular weight biomolecules. Monosaccharides such as glucose or dextrose (Chunfa et al., 2016), fructose (Castro-Guerrero et al., 2018), disaccharide sucrose (Meshram et al., 2013) and a mixture of two or more monosaccharides i.e. honey (El-Desouky & Ammar, 2016) are the the prime examples of natural reducing agent. Furthermore, other small biomolecules such as amino acids like tyrosine and tryptophan have also been utilized for the synthesis of noble metal nanoparticles (Shankar & Rhim, 2015). There are several reports regarding the synthesis of silver (Lee et al., 2021), gold (Sun et al., 2022), zinc oxide and titanium dioxide (Samb-Joshi et al., 2019) nanoparticles by use of polyphenol compounds, particularly lignin as a reducing and capping agent. Another report documented curcumin and melanin-mediated synthesis of gold/silver (Patra & El Kurdi, 2021) and copper oxide nanoparticles (Roy & Rhim, 2019). These biomolecules can also be utilized as a capping agent to improve the stability of the prepared nanoparticles by optical (laser, UV), microwave or acoustic wave exposure (Yu et al., 2021).

10.2.2 Biosynthesis of Nanoparticles

Biosynthesis of NPs by a variety of eukaryotic and prokaryotic cells has been exponentially reported in published literature over the last three decades (Yusof et al., 2019; Priyam et al., 2019; Vardhini, 2015; Patil & Taranath, 2016). The early reports have been obtained for the microbial prokaryotic genera followed by abundant literature on plant cell-derived NP synthesis (Jeevanandam et al., 2022; Šebesta et al., 2022). The fundamental mechanism of NP biosynthesis by any cell predominantly involves a variety of redox reactions occurring through diverse biomolecules in the cell that result in the reduction of metal salts to form metal^0 or metal oxide NPs

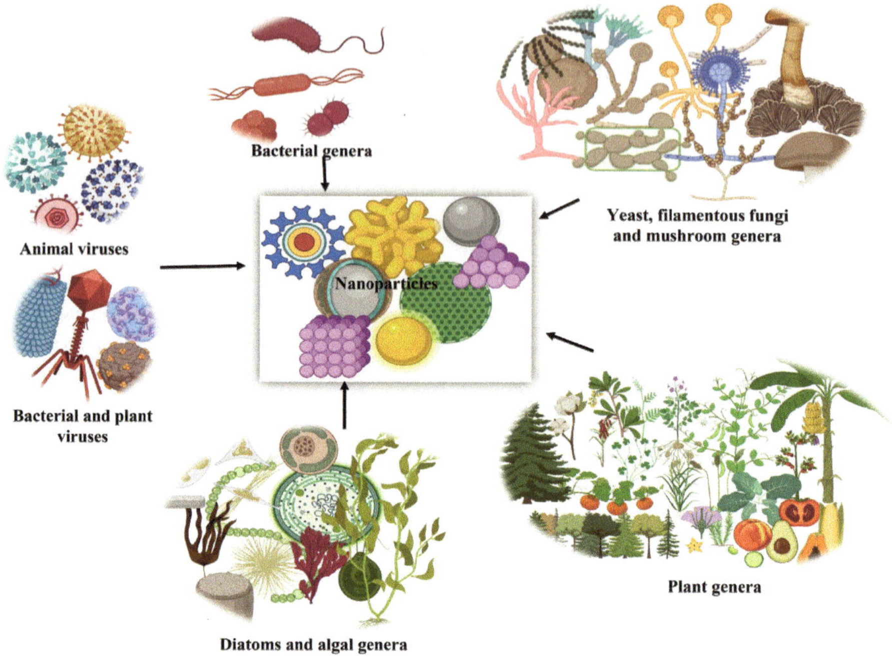

FIGURE 10.1 Biosynthesis of metal and metal oxide nanoparticles by different living cells

(Šebesta et al., 2022). The biosynthesized NPs exhibit superior and peculiar properties many times as compared to the chemically or physically synthesized nanoparticles, such as improved biocompatibility, morphological homogeneity, stability and shelf life of the NP sols. Furthermore, for NP biosynthesis, micro-organisms and plants offer a faster rate of reduction and significantly higher biocompatibility. Thus, biological approaches of nanoparticle synthesis entice higher interest in the scientific network (BalaKumaran et al., 2020). A layout of various agents involved in the biosynthesis of NP has been provided in Figure 10.1.

The use of nanomaterials is expanding over domains such as agriculture, food and medicine. It has been realized that biocompatible nanoparticles must be generated which will bear low to negligible ecological footprints and associated health hazards (Reddy et al., 2016; Choudhary et al., 2018). However, several fundamental challenges must be overcome to generate complex nanostructures through green synthesis techniques such as the need for an even distribution of size, shape and composition, reproducibility, scalability and amenability to automation for industrial or commercial applications. A variety of facile methods have been developed for rapid synthesis, allowing for control over particle size, shape, composition and other specific properties of nanoparticles. The cost of the nanomaterials synthesized through sophisticated techniques is fairly high (Pandey & Jain, 2020). As a viable alternative to physical or chemical methods, biological nanoparticle synthesis has emerged over the past two decades (Attarad et al., 2016).

10.2.2.1 Eukaryotic Cell Nanofactories

10.2.2.1.1 Unicellular Eukaryotes: Yeast, Lower Algae and Protists

Yeast genera including the *Saccharomyces cerevisiae* (Rasmussen et al., 2022; Eugenio et al., 2016; Elnagar et al., 2021b), *Pichia pastoris* (Eze & Nwabor, 2020), *Monascus purpureus* (El-Sayed et al., 2020; Sun et al., 2021) and *Candida albicans* (Romero-Núñez et al., 2019) can carry out the synthesis of silver, zinc oxide, selenium, titanium dioxide and even bimetallic cobalt ferrite nanoparticles. In these reports, the researchers have identified that the biomolecules responsible for the biosynthesis of the NPs were beta-glucans (Elnagar et al., 2021a), NADPH-dependent nitrate reductase enzymes (Dhabalia et al., 2020) and other small molecular weight biomolecules. The NP synthesis potential of the microbial cells including the NP yield per unit biomass and the rate of NP synthesis can be improved by developing recombinant strains. A recombinant *Pichia pastoris* strain Cyb5R overexpressing metal-resistant NADH-Cyt b5 reductase enzyme did not show any drastic variation in the growth rate during and after the extra/intracellular synthesis of Au and Pd NPs (Elahian et al., 2020).

Unicellular micro-algae and their cell extracts can lead to the generation of phyconanoparticles (Chokshi et al., 2016; Al-Enazi, 2022). Diverse genera such as *Chlamydomonas reinhardtii* (Barwal et al., 2011), *Ulvophyte* sp. MBIC10591 (Hamida et al., 2022), *Coelastrella aeroterrestrica* (Hamida et al., 2022) and *Acutodesmus dimorphus* (Chokshi et al., 2016) have been identified to synthesize silver nanoparticles. Another genera *Phaeodactylum tricornutum* has been reported to exhibit synthesis of TiO_2 NPs (Caliskan et al., 2022). The mechanism of synthesis of the NPs by the algal cells has been elucidated through various electron microscopy and associated spectroscopy techniques. The enzymes associated with the redox reactions (oxido-reductive proteins) including ATPase, Fd-NADP dependent reductase, superoxide dismutase and carbonic anhydrase play a predominant role in the synthesis of the NPs (Barwal et al., 2011). Likewise, cell extracts of *Bacillariophyceae* diatom protistan genera were used for the synthesis of Ag, Au and bimetallic Au/Ag NPs for environmental clean-up application (Al-Enazi, 2022). Two species of another known protist *Euglena, E. gracilis* and *E. intermedia* were identified to form AgNPs and both exhibited the formation of spherical AgNPs (Li et al., 2015).

10.2.2.1.2 Multicellular Eukaryotes

10.2.2.1.2.1 Myconanoparticle Synthesis Through Filamentous and Macro-fungi

Higher fungi including the filamentous molds and mushrooms can accomplish the reduction of the metal salt solution to generate lower valency state metal/metal oxide nanoparticles – the myconanoparticles (Prabhu et al., 2019; Bhardwaj et al., 2020). The predominant fungal genera which have been evaluated for NP synthesis include *Aspergillus sydowii* (Wang et al., 2021), *A. flavus* (Fouda et al., 2022), *A. terreus* (Lotfy et al., 2021; Priyadarshini et al., 2014), *A. versicolor* (Elgorban, 2016), *A. clavatus* (Verma et al., 2010)], *Fusarium* (Venkatesh et al., 2016; Bansal et al., 2006) and *Trichoderma* (Murillo-Rábago et al., 2022; Kalia et al., 2020; Kaur et al., 2016). Likewise, different edible mushroom genera such as *Agaricus bisporus* (Owaid &

Ibraheem, 2017), *Coprinus comatus*, *Pleurotus* (Kaur et al., 2020; Chaturvedi et al., 2020), *Laetiporus versisporus* (Fathima et al., 2022) and medicinal mushroom cultures like *Cordyceps* (Liu et al., 2022; Ji et al., 2019), *Ganoderma* (Murillo-Rábago et al., 2022), *Grifola frondosa* (Vetchinkina et al., 2018) and *Lentinus* (Owaid et al., 2019) have been used for facile and low-cost one-pot synthesis of a variety of noble metal and metal oxide nanoparticles. Several reports appeared in the recent decade for the application of rare fungi such as marine, thermophilic and acidophilic fungi (Ahmed et al., 2023).

The fungal cell/cell extract-mediated NPs synthesis mechanisms are based on the occurrence of many universal redox proteins and biocatalytic enzymes involved in oxidation-reduction reactions as discussed in the previous section 9.2.2.1.1 (Kalia & Kaur, 2018). Filamentous fungi, being multicellular in nature, exhibit a larger cell surface area to interact with the heavy metal or mineral ions in the vicinity through the processes of biosorption, biomineralization and bioaccumulation. However, certain unique fungal cell components including the beta glycans (Guo et al., 2015), polychelatins (Chaturvedi et al., 2020), glycoproteins (Evans, 2019), specific pigment molecules (Nuanaon et al., 2022; El-Sayed et al., 2020), metallothioneins (Choi & Lee, 2020) and a few enzymes such as triose phosphate isomerase (Liu et al., 2021) besides the free amino acids or tyrosine and cysteine-rich short peptides (Liu et al., 2019) are potential biological molecules involved in myconanoparticle synthesis.

10.2.2.1.2.2 Phyto Nanoparticle Synthesis Involving Plants Cells

The phytosynthesis of NPs emerged as the quickest, facile, cost-effective and reproducible green synthesis technique for generation of high volumes/quantities. Different types of plant organs/tissues such as leaf, flower, fruit, peel, seeds, root, stem bark and barkless wood can be utilized for the NP synthesis. Herbaceous, perennial woody shrubs and trees including those having the regular and medicinal utility have been evaluated for the NP phytosynthesis aspects (Bhardwaj et al., 2020). The reaction conditions affect the morphology and stability of the phytosynthesized nanosols (Tiwari et al., 2021). The ratio of the addition of metal salt solution to the plant extract affects the NP size (Ahmad et al., 2021). Most of the phytosynthesis studies report the use of 1:1 to 1:10 ratio of mixing of salt solution-plant extract (Khan et al., 2022). During interaction, if the temperature of the reaction mixture is kept high, small-sized NPs can be obtained (Rai et al., 2006; Khan et al., 2022). Dielectric heating through microwave irradiation has been reported to be more effective to obtain small-sized NPs (Kaur et al., 2020). The nanoparticle size grows at alkaline pH with generally the monodispersity and uniform morphologies improving at alkaline pH conditions (Armendariz et al., 2004). This effect though may vary with the type of the plant and even with the type of plant extract (Velázquez-Hernández et al., 2022; Khan et al., 2022). This can be identified through UV-Vis spectroscopy analysis of the lower full-width half maxima (FWHM) due to the formation of sharper UV absorbance peaks (Khan et al., 2022). The other factors affecting the size dimensions and shape of the biosynthesized NPs are the reaction time of incubation of the plant extract with salt solution and the concentration of the salt solution used (Khan et al., 2022).

10.2.2.2 Prokaryotic Cell Nanofactories: Bacteria-mediated Nanoparticle Synthesis

Micro-organisms encompass enormous taxonomic, morphological, biochemical and physiological diversity (Onen et al., 2020; Vitorino & Bessa, 2018). The existing diversity among the prokaryotic microbes occurs at each hierarchical state (Cohan & Koeppel, 2008). The omnipresence of microbes on earth can be identified through their ability to inhabit air, land and water. Microbial isolates have been obtained from extreme environments such as hot springs, sulphur vents, dry/cold deserts, saline estuaries and other similar extreme habitats (Cohan & Koeppel, 2008). The inhabitation of these diverse microbial ecological niches showcase the physiological versatility and metabolic variability of the microbial genera residing in specialized niches (Zinder & Dworkin, 2013). The occurrence of molecular/biochemical heterogeneity is the most critical factor responsible for the reduction of soluble metal salts to metal or metal oxide nanoparticles. The physiological versatility of microbes related to the natural mechanism for detoxification of metal ions through reduction can be harnessed both extracellularly or intracellularly by the discrete phenomena of bioaccumulation, precipitation, biomineralization and biosorption (Kaur et al., 2022; Vigneshwaran & Jain, 2011). Therefore, the microbial cells behave as electrochemically active entities involved in the effective reduction of the metal under variable incubation conditions (Šebesta et al., 2022; Jeevanandam et al., 2022).

Bacterial cells can be utilized for the fabrication of nanoparticles (Iravani, 2014). A variety of bacterial cells have been evaluated for the synthesis of noble and other metal/metal oxide NPs synthesis (Ali et al., 2019; Kaur & Kalia, 2016; Javaid et al., 2018; Huq & Akter, 2021). Bacteria exhibit the occurrence of specific biomachinery including the metal ion channels, enzymes and other types of biomolecules for adsorption, uptake, accumulation and bioconversion of the metal ions to metal/metal oxide NPs, which can exist as the inclusion bodies on further aggregation and agglomeration (Ali et al., 2019; Kaur et al., 2022; Javaid et al., 2018).

The NPs synthesis can be accomplished through two broad mechanisms *viz.*, extracellular and intracellular synthesis. The extracellular synthesis of NPs by bacterial cells occurs due to the production of specific biocatalytic reductase proteins which can either be secreted outside the cell or occur on the surface of the cell membrane (Moreno-Vivián et al., 1999). The extracellular NP synthesis can be accomplished by obtaining the cell biomass followed by incubation of the metal salt solution in various ratio proportions (generally two to five times higher ratio of the biomass is kept concerning the volume of the metal salt) under shaking or static growth conditions (Francis et al., 2022). Apart from the cell biomass-based incubation, another alternative technique includes the generation of the cell-free extracts by lysis of cellular biomass followed by centrifugation to obtain the cell-lysate supernatant which can be incubated with the salt solution (Mukherjee et al., 2018). Another possible alternative can be the utilization of the growth media/culture broth filtrates (Huq & Akter, 2021). However, these cell-free filtrates may also contain the culture media components as the other compounds or components that may affect the rate and morphology of the NPs synthesis (Iravani & Varma, 2020).

The primary biomolecules involved in the extracellular synthesis of NPs include the NADPH/ NADH itself (Hietzschold et al., 2019), NADPH/ NADH-dependent nitrate reductase (Javaid et al., 2018), nitrite reductase enzyme (Bharti et al., 2020), and extracellular polysaccharide compounds including the xanthan, levan, curdlan, dextran, sphingan or gellan, succino-glycan and heteropolysaccharides like FucoPol and GalactoPol (Budamagunta et al., 2023; Escárcega-González et al., 2018).

The intracellular NP synthesis requires a viable respiring cell. The cell pellet of the freshly grown test bacterial culture can be diluted in an appropriate buffer solution and incubated with filter-sterilized aqueous metal salt solution under shaking conditions at the optimum growth temperature of the test bacteria. The process begins with the adsorption of the metal ions on the surface of the bacterial cell due to the occurrence of a net negative charge of the cell wall (Lin et al., 2019). The adsorbed metal ions may form ligand complexes with the cell wall components, particularly the small molecular weight biomolecules and proteins, and can lead to the formation of bacterial cell wall metal nanocomposite (Iravani & Varma, 2020). Alternatively, the metal ions are internalized through dedicated cation uptake channels or proteins, and the metal ions are transformed and accumulated as inclusion bodies or nanostructured particles (e.g. magnetite nanoparticles existing as magnetosomes in magnetotactic bacteria) (Hansda et al., 2016). The intracellular synthesis of NPs is therefore, a way more complex process involving the interaction of the metal ions with a variety of biomolecules including the biocatalytic proteins (Shah & Rather, 2021) and small molecular weight compounds, which can act both as biotemplating and reducing agents (Alfryyan et al., 2022).

10.2.2.3 Virus-Mediated Nanoparticle Synthesis

Viruses exhibit nanoscale dimensions and refer to naturally occurring nanomaterials that can be utilized as biotemplates or scaffolds for the synthesis of a variety of nanoparticles of metal oxides with single and multiple chemical origin/nature (Zhang et al., 2018). The most unique aspect of the virus-mediated NP synthesis process is the fabrication of tailor-made advanced nanoporous structures of diverse morphologies through the use of virus capsid as a scaffolding matrix besides as functional entities (Koudelka et al., 2015). This is a relatively clean and environment-friendly synthesis process which also out-plays the predominant problems of NP biosynthesis, i.e. heterogeneity (variable size, shape and 3-D asymmetry of synthesized NPs) due to symmetrical architecture (Pokorski & Steinmetz, 2011). The viral nanoparticles may involve the virus capsid metal nanohybrid materials involving deposition or adsorption of metal atoms at specific regions such as the vertices, faces, throughout the surface or inside the cavity of the capsid (Zhang et al., 2018). The synthesis of viral nanoparticles was initiated with a variety of human pathogenic viruses (icosahedral: adenovirus, human papillomavirus, sindbis virus) (Koudelka et al., 2015) and bacteriophages (icosahedral: MS2, Qβ, P22; filamentous: M13) (Steinmetz, 2010). However, later plant pathogenic viruses (icosahedral: cowpea chlorotic mosaic virus, cowpea mosaic virus, red clover necrotic mottle virus; filamentous: tobacco mosaic virus, potato virus X) were also evaluated for the formation of viral nanoparticles/ virus metal nanohybrids (Lee et al., 2021; Zhang et al., 2018; Steinmetz, 2010).

The virus nanoparticle/virus metal nanohybrids can be generated in a precisely controlled manner through the processes of biotemplating or the use of viral capsids as a scaffold, viral self-assembly and bioconjugation/pre or post-functionalization followed by biomineralization (Koudelka et al., 2015; Zhang et al., 2018). Viral NPs can be generated on incubation of salt precursor solution with the virion particles that lead to either assembly of the virus around the nucleating metal particle (encapsulation or infusion) or adsorption and impregnation of the metal ions around the virus particle, followed by metal reduction resulting in the formation of a metal coat (Koudelka et al., 2015; Wen & Steinmetz, 2016). The filamentous viruses can lead to the synthesis of inorganic metal nanotube composites (Röder et al., 2019) while the icosahedral or spherical viruses help fabricate 2-D and 3-D patterns for sensor and energy applications (Zhang et al., 2018; Pokorski & Steinmetz, 2011). A variety of metal elements as salt precursors such as gold, silver, platinum, nickel, cobalt, iron, copper and many more have been utilized for the synthesis of hierarchically high aspect nanomaterials (nanotubes, nanowires, nanocables) besides nanoparticles (Lee et al., 2021).

10.2.2.4 Cell-Derived Macro-molecules for the Synthesis of Nanoparticles

The cell macro-molecules are linear to branched-chain polymers of a variety of subunit biomolecules such as saccharides, peptides and nucleotides. The biomacro-molecules can function both as reducing as well as capping agents. Also, the unique critical morphological features (surface topography and pores or internal cavities) of these cellular components have been aptly utilized for the generation of designer NPs with strict control over the morphology and hence the monodispersed characteristics of the nanosols.

Much published literature supports the paradigm of NP synthesis through polysaccharides such as agar (Lomelí-Rosales et al., 2019), carrageenan (Chen et al., 2018), chitosan (Yao et al., 2022), starch (Lomelí-Marroquín et al., 2019; Tarmizi et al., 2022), sodium alginate (Chunfa et al., 2016) and xanthan gum (Xu et al., 2014). Small molecular weight biomolecules produced by fungi also have been utilized for the synthesis of nanoparticles such as beta-glucans (Elnagar et al., 2021a), specialized glycans derived from medicinal mushroom-Lentinan (Guo et al., 2015), quinones (naptho-, hydro- and anthra-quinones) (Ghosh et al., 2021) and other similar compounds. Also, lipid molecules like mannosyl erythritol moiety (Bakur et al., 2019) and phosphatidyl choline membranes (Miesen et al., 2020) have been identified to help in shape and size-controlled synthesis of inorganic NPs.

The isolated proteins such as the bovine serum albumin and collagen (Mishra & Nayar, 2014), casein (Ashraf et al., 2013), ovalbumin, haemoglobin (Chakraborty et al., 2018) and phycobiliprotein (El-Naggar et al., 2018) have been reported to mediate the synthesis of silver NPs. Biocatalytic proteins including several types of enzymes have also been evaluated for their ability to exhibit a one-pot synthesis of metal nanoparticles (Thapa et al., 2017; Ovais et al., 2018; Durán et al., 2015). Metal nanoparticles have been identified to be synthesized by incubation with amylase (Durán et al., 2015), catalase (Losada-Garcia et al., 2022), horseradish peroxidase (Parashar et al., 2017), laccase (Chaurasia et al., 2022), lignin peroxidase (Singh et al., 2017), lysozyme (Dudhagara et al., 2022; Kumar et al., 2012), lipase (Naapuri et al., 2022), superoxide dismutase (Arib et al., 2021) and sulfate reductase (Ahmad et al., 2002).

Nucleic acid molecules, including the DNA and RNA sequences, also exhibit the ability to interact with the metal ions through electrostatic bonds formed with the phosphate backbone and metal conjugation with the nucleobases leading to the formation of nanoparticles (Berti & Burley, 2008). Alterations in the nucleobase type in the DNA sequence can also affect the morphology of the generated NPs. For example, in a study, this effect was identified through the use of a 30 nucleotide poly A/C and poly-T sequence as flower-shaped and spherical nano-Au NPs were generated on incubation with the former and latter sequences, respectively (Wang et al., 2010). Similar to DNA molecules, RNA sequences also exhibit metal binding abilities which can be harnessed for the fabrication of inorganic metal NPs (Carter et al., 2013). A recent review of the nucleic acid-mediated NP nucleation and synthesis further enhances the insights into the synthesis mechanisms (Graczyk et al., 2020).

10.3 POTENTIAL AGRI-APPLICATIONS AND FUTURE PROSPECTS

The biosynthesized NPs are finding diverse agri-applications, including their use as nanofertilizer nutrients (Smirnov et al., 2022). The nanonutrient fertilizers have been designed for slow/controlled and even targeted release of the nutrient to decrease their non-target losses and hence improve the nutrient use efficiency (Kalia et al., 2020). Further, nanoparticles may improve the interactions in the soil-microbe-plant continuum (Kaur et al., 2022). Also, the green synthesized metal/metal oxide nanoparticles can function as nanopesticides to curb the attack/infestation of pests and pathogens (Kalia et al., 2020). The nanoscale dimensions of the nanoparticles render these materials the ability to trespass the cell wall and the membrane and their interactions with the cellular biomolecules elicit the reactive oxygen species (ROS) cascades at sub-lethal levels that can function to positively alter various physiological and biochemical events such as seed germination and vegetative growth of the seedling (Kalia et al., 2019). This aspect may be attributed to the nanostimulant action leading to enhancement in the seed germination, vegetative growth and development of different crop plants (Tondey et al., 2021, 2022). The nanoparticle-enabled manoeuvring of the plant physiology can be useful to develop tolerance to a variety of abiotic stresses in the wake of the drastic climate change aspects through the nanoplant abiotic stress mitigation phenomena (Kaur et al., 2022; Thakur et al., 2022).

Another agri-innovation relies on the use of green nanoparticles as active components of the nanobiosensors for rapid and cost-effective plant disease diagnostics (Ahmed et al., 2023) during the pre-production and active growth and development stages of the crop plants. Preventive early and on-field detection of the pathogen will be very useful to avoid the spread of the pathogens to uninfected plants through the judicious application of pesticides. Furthermore, the nanosensors can also be utilized for pesticide and other agrochemical pollution monitoring and risk assessment. The green nanoparticles have not only been utilized for improving crop productivity, but rather the application of nanoparticles as nanofiller agents has been instrumental in the development of innovative, intelligent and antimicrobial packaging systems to depreciate food losses due to microbial contamination and spoilage (Kaur et al., 2017; Kalia et al., 2021). The sensor applications of nanoparticles as components of nanosensor modules have been also beneficial for the early and rapid identification of food quality and safety (Kaur et al., 2023).

The utilization of the synthesized nanomaterials for nanobased remediation strategies for contaminant removal and reclamation of the land for agricultural purposes (Ahuja et al., 2022) is another valuable innovative eco-clean-up application of biosynthesized nanoparticles. The conjugate nanobioremediation technologies based on teaming up biosorption, bioaccumulation and biomineralization processes of the plant/microbial cells in combination with the unique high porosity and surface area attributes of the nanoparticles can help improve the speed of the decontamination of the pollutants from diverse eco-niches (Singh et al., 2020).

Another novel innovative application of the biosynthesized NPs is as nanocatalysts for various organic chemistry (homo- and hetero- chemistry synthesis techniques) and biochemical reactions for the rapid synthesis of certain organic compounds (Roy et al., 2021; Veisi et al., 2021) and biomolecules (Roy et al., 2021) to obtain high yield rates.

As discussed in different sections, the biosynthesis of NPs from plant and microbial cell/cell extracts is a unique and cost-effective technology. However, the use of fruit, flower and other parts of the plant which can be alternatively consumed by humans and other animals as food and feed for NP synthesis will escalate the load for enhanced production. Likewise, the microbial cultures must be grown in defined rich media which have ingredients derived from plant or animal origins. Therefore, a plausible recluse is the utilization of agricultural waste material for the green synthesis of nanoparticles (Ahmad et al., 2021). Figure 10.2 depicts the layout for the probable use of agri-waste for NPs generation. Furthermore, the NPs yield

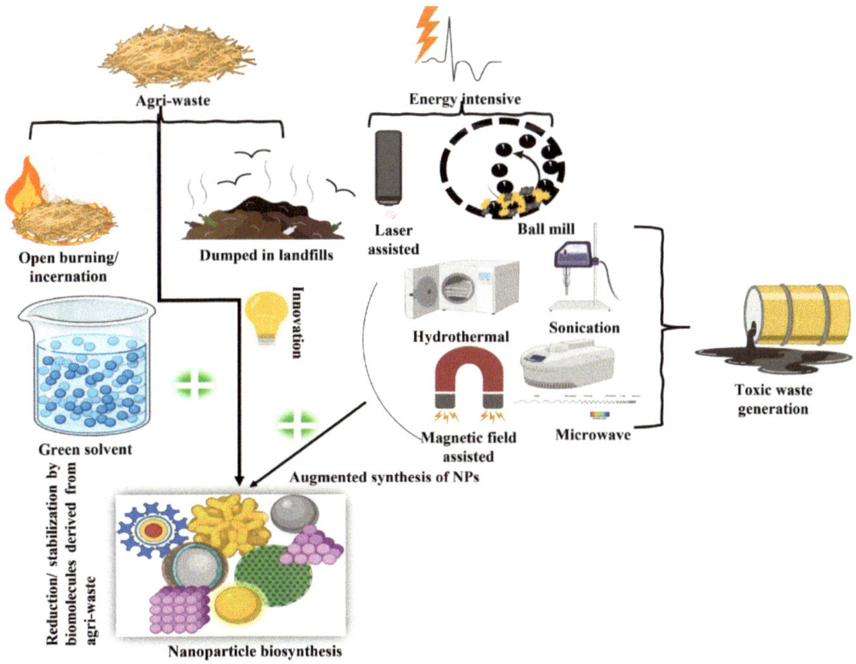

FIGURE 10.2 Innovative agri-waste valorization technique for generation of green nanoparticles

rates and thus the economics of the reaction can be improved by the use of hybrid, augmented technologies such as the sonochemical-assisted, microwave-assisted and hydrothermal-assisted biosynthesis of NPs (Duan et al., 2015).

10.4 CONCLUSION

Green synthesis of nanomaterials has evolved over the decade with initial reports which are just targeted for the NM synthesis through various types of living cells followed by validation of the nanoscale dimensions. However, as the advanced hybrid green synthesis approaches are now available, the potential for scale up synthesis of uniform-sized nanomaterials can be achieved beside the lucrative cost-effectiveness and environment-friendly attributes of this technology. Furthermore, as the comprehensive information regarding the specific mechanisms and key biomolecules involved in nanomaterial synthesis has been obtained, the dependence on living cells will soon be over-ridden and synthetically generated or biomimicked. Key molecules can be used for improved mondispersity of the generated nanomaterials. The utilization of agri-waste or other waste for nanomaterial generation can be a potential valorization circular economy approach providing high quality material from waste. This approach has high environmental significance due to reduction of pollution caused by burning or dumping of the agri-waste.

REFERENCES

Ahmad, A., Mukherjee, P., Mandal, D., Senapati, S., Khan, M. I., Kumar, R., & Sastry, M. (2002). Enzyme mediated extracellular synthesis of CDS nanoparticles by the fungus, *Fusarium oxysporum. Journal of the American Chemical Society, 124*(41), 12108–12109.

Ahmad, T., Iqbal, J., Bustam, M. A., Irfan, M., & Asghar, H. M. A. (2021). A critical review on phytosynthesis of gold nanoparticles: Issues, challenges and future perspectives. *Journal of Cleaner Production, 309*, 127460.

Ahmed, F. K., Kalia, A., Ahmad, A., Alghuthaymi, M. A., & Abd-Elsalam, K. A. (2023). Marine fungi and yeast: A green approach for production of bionanoparticles. In K. A. Abd-Elsalam (Ed.), *Fungal cell factories for sustainable Nano materials productions and agricultural applications* (pp. 337–360). Elsevier.

Ahuja, R., Kalia, A., Sikka, R., & Chaitra, P. (2022). Nano modifications of biochar to enhance heavy metal adsorption from wastewaters: A review. *ACS Omega, 7*(50), 45825–45836.

Al-Enazi, N. M. (2022). Optimized synthesis of mono and bimetallic nanoparticles mediated by unicellular algal (diatom) and its efficiency to degrade AZO dyes for wastewater treatment. *Chemosphere, 303*(1), 135068.

Alfryyan, N., Kordy, M. G. M., Abdel-Gabbar, M., Soliman, H. A., & Shaban, M. (2022). Characterization of the biosynthesized intracellular and extracellular plasmonic silver nanoparticles using Bacillus cereus and their catalytic reduction of methylene blue. *Scientific Reports, 12*(1), 12495.

Ali, J., Ali, N., Wang, L., Waseem, H., & Pan, G. (2019). Revisiting the mechanistic pathways for bacterial mediated synthesis of noble metal nanoparticles. *Journal of Microbiological Methods, 159*, 18–25.

Andrievskii, R. A. (2009). Size-dependent effects in properties of nanostructured materials. *Reviews on Advanced Materials Science, 21*, 107–133.

Arib, C., Spadavecchia, J., & de la Chapelle, M. C. (2021). Enzyme mediated synthesis of hybrid polyedric gold nanoparticles. *Scientific Reports*, *11*(1), 3208.

Armendariz, V., Herrera, I., Peralta-Videa, J. R., Jose-Yacaman, M., Troiani, H., Santiago, P., & Gardea-Torresdey, J. L. (2004). Size controlled gold nanoparticle formation by Avena sativa biomass: Use of plants in Nano biotechnology. *Journal of Nanoparticle Research*, *6*(4), 377–382.

Ashraf, S., Abbasi, A. Z., Pfeiffer, C., Hussain, S. Z., Khalid, Z. M., Gil, P. R., Parak, W. J., Hussain, I. (2013). Protein-mediated synthesis, PH-induced reversible agglomeration, toxicity and cellular interaction of silver nanoparticles. *Colloids and Surfaces, Part B: Biointerfaces*, *102*, 511–518.

Attarad, A., Zafar, H., Zia, M., Ul-Haq, I., Phull, A. R., Ali, J. S., & Hussain, A. (2016). Synthesis, characterization, applications, and challenges of iron oxide nanoparticles. *Nanotechnology, Science and Application*, *9*(9), 49–67.

Bakur, A., Niu, Y., Kuang, H., & Chen, Q. (2019). Synthesis of gold nanoparticles derived from Mannosylerythritol lipid and evaluation of their bioactivities. *AMB Express*, *9*(1), 62.

BalaKumaran, M. D., Ramachandran, R., Balashanmugam, P., Jagadeeswari, S., & Kalaichelvan, P. T. (2020). Comparative analysis of antifungal, antioxidant and cytotoxic activities of Mycosynthesized silver nanoparticles and gold nanoparticles. *Materials Technology*, *37*(6), 411–421.

Bansal, V., Ahmad, A., & Sastry, M. (2006). Fungus-mediated biotransformation of amorphous silica in rice husk to nanocrystalline silica. *Journal of the American Chemical Society*, *128*(43), 14059–14066.

Barwal, I., Ranjan, P., Kateriya, S., & Yadav, S. C. (2011). Cellular oxido-reductive proteins of *Chlamydomonas reinhardtii* control the biosynthesis of silver nanoparticles. *Journal of Nanobiotechnology*, *9*(1), 56.

Berti, L., & Burley, G. A. (2008). Nucleic acid and nucleotide-mediated synthesis of inorganic nanoparticles. *Nature Nanotechnology*, *3*(2), 81–87.

Bhardwaj, K., Dhanjal, D. S., Sharma, A., Nepovimova, E., Kalia, A., Thakur, S., Bhardwaj, S., Chopra, C., Singh, R., Verma, R., Kumar, D., Bhardwaj, P., & Kuča, K. (2020). Conifer-derived metallic nanoparticles: Green synthesis and biological applications. *International Journal of Molecular Sciences*, *21*(23), 1–22.

Bhardwaj, K., Sharma, A., Tejwan, N., Bhardwaj, S., Bhardwaj, P., Nepovimova, E., Shami, A., Kalia, A., Kumar, A., Abd-Elsalam, K. A., & Kuča, K. (2020). Pleurotus macrofungi-assisted nanoparticle synthesis and its potential applications: A review. *Journal of Fungi*, *6*(4), 1–21.

Bharti, S., Mukherji, S., & Mukherji, S. (2020). Extracellular synthesis of silver nanoparticles by *Thiosphaera pantotropha* and evaluation of their antibacterial and cytotoxic effects. *3 Biotech*, *10*(6), 237.

Bokov, D., Jalil, A. T., Chupradit, S., Suksatan, W., Ansari, M. J., Shewael, I. H., Valiev, G. H., & Kianfar, E. (2021). Nanomaterial by sol-gel method: Synthesis and application. *Advances in Materials Science and Engineering*, *2021*, 1–21.

Budamagunta, V., Shameem, N., Irusappan, S., Parray, J. A., Thomas, M., Marimuthu, S., Kirubakaran, R., Jothi, K. N. A., Sayyed, R. Z., & Show, P. L. (2023). Nanovesicle and extracellular polymeric substance synthesis from the remediation of heavy metal ions from soil. *Environmental Research*, *219*, 114997.

Caliskan, G., Mutaf, T., Agba, H. C., & Elibol, M. (2022). Green synthesis and characterization of titanium nanoparticles using microalga, *Phaeodactylum tricornutum*. *Geomicrobiology Journal*, *39*(1), 83–96.

Carter, C. J., Owczarek, A., & Feldheim, D. L. (2013). Methods for isolating RNA sequences capable of binding to or mediating the formation of inorganic materials. *Methods in Molecular Biology*, *1026*, 121–135.

Castro-Guerrero, C. F., Morales-Cepeda, A. B., Hernández-Vega, L. K., & Díaz-Guillén, M. R. (2018). Fructose-mediated gold nanoparticles synthesis. *Cogent Chemistry*, *4*(1), 1447262.

Chakraborty, I., Feliu, N., Roy, S., Dawson, K., & Parak, W. J. (2018). Protein-mediated shape control of silver nanoparticles. *Bioconjugate Chemistry*, *29*(4), 1261–1265.

Chaturvedi, V. K., Yadav, N., Rai, N. K., Abd Ellah, N. H., Bohara, R. A., Rehan, I. F., Marraiki, N., Batiha, G. E. S., Hetta, H. F., & Singh, M. P. (2020). *Pleurotus sajor-caju*-mediated synthesis of silver and gold nanoparticles active against colon cancer cell lines: A New Era of Herbonanoceutics. *Molecules*, *25*(13), 3091.

Chaurasia, P. K., Bharati, S. L., & Yadava, S. (2022). Nano-reduction of gold and silver ions: A perspective on the fate of microbial laccases as potential biocatalysts in the synthesis of metals (gold and silver) nano-particles. *Current Research in Microbial Sciences*, *3*, 100098.

Chen, X., Zhao, X., Gao, Y., Yin, J., Bai, M., & Wang, F. (2018). Green synthesis of gold nanoparticles using carrageenan oligosaccharide and their *in vitro* antitumor activity. *Marine Drugs*, *16*(8), 277.

Choi, Y., & Lee, S. Y. (2020). Biosynthesis of inorganic nanomaterials using microbial cells and bacteriophages. *Nature Reviews Chemistry*, *4*(12), 638–656.

Chokshi, K., Pancha, I., Ghosh, T., Paliwal, C., Maurya, R. K., Ghosh, A., & Mishra, S. (2016). Green synthesis, characterization and antioxidant potential of silver nanoparticles biosynthesized from de-oiled biomass of thermotolerant oleaginous microalgae *Acutodesmus dimorphus*. *RSC Advances*, *6*(76), 72269–72274.

Choudhary, M. A., Manan, R., Mirza, M. A., Khan, H. A., Qayyum, S., & Ahmed, Z. (2018). Biogenic synthesis of copper oxide and zinc oxide nanoparticles and their application as antifungal agents. *International Journal of Material Science and Engineering*, *4*(1), 1–6.

Chunfa, D., Xianglin, Z., Hao, C., & Chuanliang, C. (2016). Sodium alginate mediated route for the synthesis of monodisperse silver nanoparticles using glucose as reducing agents. *Rare Metal Materials and Engineering*, *45*(2), 261–266.

Cohan, F. M., & Koeppel, A. F. (2008). The origins of ecological diversity in prokaryotes. *Current Biology*, *18*(21), R1024–R1034.

Dhabalia, D., Ukkund, S. J., Syed, U. T., Uddin, W., & Kabir, M. A. (2020). Antifungal activity of biosynthesized silver nanoparticles from *Candida albicans* on the strain lacking the CNP41 gene. *Materials Research Express*, *7*(12), 125401.

Duan, H., Wang, D., & Li, Y. (2015). Green chemistry for nanoparticle synthesis. *Chemical Society Reviews*, *44*(16), 5778–5792.

Dudhagara, P., Alagiya, J., Bhagat, C., Dudhagara, D., Ghelani, A., Desai, J., Patel, R., Vansia, A., Nhiem, D. N., Chen, Y.-Y., & Shyu, D. J. H. (2022). Biogenic synthesis of antibacterial, hemocompatible, and antiplatelets lysozyme functionalized silver nanoparticles through the one-step process for therapeutic applications. *Processes*, *10*(4), 623.

Durán, M., Silveira, C. P., & Durán, N. (2015). Catalytic role of traditional enzymes for biosynthesis of biogenic metallic nanoparticles: A mini-review. *IET Nanobiotechnology*, *9*(5), 314–323.

Elahian, F., Heidari, R., Charghan, V. R., Asadbeik, E., & Mirzaei, S. R. (2020). Genetically modified *Pichia pastoris*, a powerful resistant factory for gold and palladium bioleaching and nanostructure heavy metal biosynthesis. *Artificial Cells, Nanomedicine, and Biotechnology*, *48*(1), 259–265.

El-Desouky, T., & Ammar, H. (2016). Honey mediated silver nanoparticles and their inhibitory effect on aflatoxins and ochratoxin A. *Journal of Applied Pharmaceutical Science*, *6*, 83–90.

Elgorban, A. M. (2016). Extracellular synthesis of silver nanoparticles using *Aspergillus versicolor* and evaluation of their activity on plant pathogenic fungi. *Mycosphere*, *7*(6), 844–852.

Elnagar, S. E., Tayel, A. A., Elguindy, N. M., Al-Saggaf, M. S., & Moussa, S. H. (2021a). Innovative biosynthesis of silver nanoparticles using yeast glucan nanopolymer and their potentiality as antibacterial composite. *Journal of Basic Microbiology*, *61*(8), 677–685.

El-Naggar, N. A., Hussein, M. H., & El-Sawah, A. A. (2018). Phycobiliprotein-mediated synthesis of biogenic silver nanoparticles, characterization, *in vitro* and *in vivo* assessment of anticancer activities. *Scientific Reports*, *8*(1), 8925.

El-Sayed, R., Abdelhakim, H. K., & Zakaria, Z. (2020). Extracellular biosynthesis of cobalt ferrite nanoparticles by *Monascus purpureus* and their antioxidant, anticancer and antimicrobial activities: Yield enhancement by gamma irradiation. *Materials Science and Engineering: Part C*, *107*, 110318.

Escárcega-González, C. E., Garza-Cervantes, J. A., Vázquez-Rodríguez, A., & Morones-Ramírez, J. R. (2018). Bacterial exopolysaccharides as reducing and/or stabilizing agents during synthesis of metal nanoparticles with biomedical applications. *International Journal of Polymer Science*, *2018*, 1–15.

Eugenio, M., Müller, N., Frasés, S., Almeida-Paes, R., Lima, L. M. T. R., Lemgruber, L., Farina, M., de Souza, W., & Sant'Anna, C. (2016). Yeast-derived biosynthesis of silver/silver chloride nanoparticles and their antiproliferative activity against bacteria. *RSC Advances*, *6*(12), 9893–9904.

Evans, J. (2019). Composite materials design: Biomineralization proteins and the guided assembly and organization of biomineral nanoparticles. *Materials*, *12*(4), 581.

Eze, F. N., & Nwabor, O. F. (2020). Valorization of pichia spent medium via one-pot synthesis of biocompatible silver nanoparticles with potent antioxidant, antimicrobial, tyrosinase inhibitory and reusable catalytic activities. *Materials Science and Engineering: Part C*, *115*, 111104.

Fathima, M. R. F., Nanthini, A. U. R., Al-Khattaf, F. S., Hatamleh, A. A., & Kabir, S. B. (2022). Mycosynthesis of noble metal nanoparticle using Laetiporus Versisporus mushroom and analysis of antioxidant activity. *Journal of Nanomaterials*, *2022*, 1–11.

Fouda, A., Awad, M. A., AL-Faifi, Z. E., Gad, M. E., Al-Khalaf, A. A., Yahya, R., & Hamza, M. F. (2022). *Aspergillus flavus*-mediated green synthesis of silver nanoparticles and evaluation of their antibacterial, anti-candida, acaricides, and photocatalytic activities. *Catalysts*, *12*(5), 462.

Francis, D. V., Aiswarya, T., & Gokhale, T. (2022). Optimization of the incubation parameters for biogenic synthesis of WO_3 nanoparticles using Taguchi method. *Heliyon*, *8*(9), e10640.

Ghosh, S., Ahmad, R., Banerjee, K., AlAjmi, M. F., & Rahman, S. (2021). Mechanistic aspects of microbe-mediated nanoparticle synthesis. *Frontiers in Microbiology*, *12*, 638068.

Graczyk, A., Pawlowska, R., Jedrzejczyk, D., & Chworos, A. (2020). Gold nanoparticles in conjunction with nucleic acids as a modern molecular system for cellular delivery. *Molecules*, *25*(1), 204.

Guo, D., Liu, G., Li, X., Tang, X., Zhang, J., Zhu, X., & Jiang, S. (2015). A facile one-pot green synthesis and antibacterial activity of silver nanoparticles using lentinan from *Lentinus edodes*. *Journal of Bionanoscience*, *9*(5), 325–329.

Gupta, S., & Bharti, M. K. (2022). Nanowaste disposal and recycling. In M. Rai & T. A. Nguyen (Eds.), *Nanomaterials recycling* (pp. 109–123). Elsevier.

Hamida, R. S., Ali, M. A., Almohawes, Z. N., Alahdal, H., Momenah, M. A., & Bin-Meferij, M. M. (2022). Green synthesis of hexagonal silver nanoparticles using a novel microalgae *Coelastrella aeroterrestrica* Strain BA_Chlo4 and resulting anticancer, antibacterial, and antioxidant activities. *Pharmaceutics*, *14*(10), 2002.

Hamida, R. S., Ali, M. A., Alkhateeb, M. A., Alfassam, H. E., Momenah, M. A., & Bin-Meferij, M. M. (2022). Algal-derived synthesis of silver nanoparticles using the unicellular *Ulvophyte* sp. MBIC10591: Optimisation, characterisation, and biological activities. *Molecules*, *28*(1), 279.

Hansda, A., Kumar, V., & Anshumali. (2016). A comparative review towards potential of microbial cells for heavy metal removal with emphasis on biosorption and bioaccumulation. *World Journal of Microbiology and Biotechnology*, *32*(10), 170.

Hietzschold, S., Walter, A., Davis, C., Taylor, A. A., & Sepunaru, L. (2019). Does nitrate reductase play a role in silver nanoparticle synthesis? Evidence for NADPH as the sole reducing agent. *ACS Sustainable Chemistry and Engineering*, *7*(9), 8070–8076.

Horikoshi, S., & Serpone, N. (2013). Introduction to nanoparticles. In S. Horikoshi & N. Serpone (Eds.), *Microwaves in nanoparticle synthesis* (pp. 1–24). Wiley-VCH Verlag GmbH & Co. KGaA.

Huq, M. A., & Akter, S. (2021). Bacterial mediated rapid and facile synthesis of silver nanoparticles and their antimicrobial efficacy against pathogenic microorganisms. *Materials*, *14*(10), 2615.

Ijaz, I., Gilani, E., Nazir, A., & Bukhari, A. (2020). Detail review on chemical, physical and green synthesis, classification, characterizations and applications of nanoparticles. *Green Chemistry Letters and Reviews*, *1*(3), 223–245.

Imran, H. J., Hubeatir, K. A., Aadim, K. A., & Abd, D. S. (2021). Preparation methods and classification study of nanomaterial: A review. *Journal of Physics: Conference Series*, *1818*(1), 012127.

Iravani, S. (2014). Bacteria in nanoparticle synthesis: Current status and future prospects. *International Scholarly Research Notices*, *2014*, 1–18.

Iravani, S., & Varma, R. S. (2020). Bacteria in heavy metal remediation and nanoparticle biosynthesis. *ACS Sustainable Chemistry and Engineering*, *8*(14), 5395–5409.

Javaid, A., Oloketuyi, S. F., Khan, M. M., & Khan, F. (2018). Diversity of bacterial synthesis of silver nanoparticles. *BioNanoScience*, *8*(1), 43–59.

Jeevanandam, J., Kiew, S. F., Boakye-Ansah, S., Lau, S. Y., Barhoum, A., Danquah, M. K., & Rodrigues, J. (2022). Green approaches for the synthesis of metal and metal oxide nanoparticles using microbial and plant extracts. *Nanoscale*, *14*(7), 2534–2571.

Ji, Y., Cao, Y., & Song, Y. (2019). Green synthesis of gold nanoparticles using a *Cordyceps militaris* extract and their antiproliferative effect in liver cancer cells (HepG2). *Artificial Cells, Nanomedicine, and Biotechnology*, *47*(1), 2737–2745.

Kalia, A., Abd-Elsalam, K. A., & Kuca, K. (2020). Zinc-based nanomaterials for diagnosis and management of plant diseases: Ecological safety and future prospects. *Journal of Fungi*, *6*(4), 222.

Kalia, A., & Kaur, G. (2018). Biosynthesis of nanoparticles using mushrooms. In B. Singh & P. A. Lallawmsanga (Eds.), *Biology of macrofungi, fungal biology* (pp. 351–360). Springer.

Kalia, A., Kaur, J., Kaur, A., & Singh, N. (2020). Antimycotic activity of biogenically synthesised metal and metal oxide nanoparticles against plant pathogenic fungus *Fusarium moniliforme* (*F. fujikuroi*). *Indian Journal of Experimental Biology*, *58*, 263–270.

Kalia, A., Kaur, M., Shami, A., Jawandha, S. K., Alghuthaymi, M. A., Thakur, A., & Abd-Elsalam, K. A. (2021). Nettle-leaf extract derived ZnO/CuO nanoparticle-biopolymer-based antioxidant and antimicrobial nanocomposite packaging films and their impact on extending the post-harvest shelf life of guava fruit. *Biomolecules*, *11*(2), 224.

Kalia, A., Manchanda, P., Bhardwaj, S., & Singh, G. (2020). Biosynthesized silver nanoparticles from aqueous extracts of sweet lime fruit and callus tissues possess variable antioxidant and antimicrobial potentials. *Inorganic and Nano-Metal Chemistry*, *50*(11), 1053–1062.

Kalia, A., Sharma, S. P., & Kaur, H. (2019). Nanoscale fertilizers: Harnessing boons for enhanced nutrient use efficiency and crop productivity. In K. A. Abd-Elsalam, R. Prasad (Eds.), *Nanobiotechnology applications in plant protection - Nanotechnology in the life sciences* (pp. 191–208). Springer.

Kalia, A., Sharma, S. P., Kaur, H., & Kaur, H. (2020). Novel nanocomposite-based controlled-release fertilizer and pesticide formulations: Prospects and challenges. In K. A. Abd-Elsalam (Ed.), *Multifunctional hybrid nanomaterials for sustainable agri-food and ecosystem* (pp. 99–134). Elsevier Inc.

Kalia, A., & Singh, S. (2020). Myco-Decontamination of Azo Dyes: Nano-Augmentation Technologies. *3 Biotech*, *10*(9), 384.

Kaur, A., Kaur, J., Kalia, A., & Singh, N. (2016). Effect of media composition on extent of antimycotic activity of silver nanoparticles against plant pathogenic fungus *Fusarium moniliforme*. *Plant Disease Research*, *31*(1), 1–5.

Kaur, G., Kalia, A., & Sodhi, H. S. (2020). Size controlled, time-efficient biosynthesis of silver nanoparticles from *Pleurotus florida* using ultra-violet, visible range, and microwave radiations. *Inorganic and Nano-Metal Chemistry*, *50*(1), 35–41.

Kaur, H., Kalia, A., Sandhu, J. S., Dheri, G. S., Kaur, G., & Pathania, S. (2022). Interaction of TiO_2 nanoparticles with soil: Effect on microbiological and chemical traits. *Chemosphere*, *301*, 134629.

Kaur, H., Kaur, J., Kalia, A., & Kuca, K. (2022). The Janus face of nanomaterials: Physiological responses as inducers of stress or promoters of plant growth? In J.-T. Chen (Ed.), *Plant and nanoparticles* (pp. 395–426). Springer Nature Singapore.

Kaur, J., Dhiman, D., & Kalia, A. (2023). Nanobiotechnological strategies for detection of mycotoxins in food products. In K. A. Abd-Elsalam (Ed.), *Fungal cell factories for sustainable nanomaterials productions and agricultural applications* (pp. 511–541). Elsevier.

Kaur, M., & Kalia, A. (2016). Role of salt precursors for the synthesis of zinc oxide nanoparticles and in imparting variable antimicrobial activity. *Journal of Applied and Natural Science*, *8*(2), 1039–1048.

Kaur, M., Kalia, A., & Thakur, A. (2017). Effect of biodegradable chitosan–Rice-Starch nanocomposite films on post-harvest quality of stored peach fruit. *Starch/Staerke*, *69*(1–2), 1–12.

Kaur, P., Sharma, S., Albarakaty, F. M., Kalia, A., Hassan, M. M., & Abd-Elsalam, K. A. (2022). Biosorption and bioleaching of heavy metals from electronic waste varied with microbial genera. *Sustainability*, *14*(2), 935.

Khan, S., Almarhoon, Z. M., Bakht, J., Mabkhot, Y. N., Rauf, A., & Shad, A. A. (2022). Single-step acer Pentapomicum-mediated green synthesis of silver nanoparticles and their potential antimicrobial and antioxidant activities. *Journal of Nanomaterials*, *2022*, 1–10.

Kharissova, O. V., Kharisov, B. I., González, C. M. O., Méndez, Y. P., & López, I. (2019). Greener synthesis of chemical compounds and materials. *Royal Society Open Science*, *6*(11), 191378.

Koudelka, K. J., Pitek, A. S., Manchester, M., & Steinmetz, N. F. (2015). Virus-based nanoparticles as versatile nanomachines. *Annual Review of Virology*, *2*(1), 379–401.

Kumar, U., Ranjan, A. K., Sharan, C., Hardikar, A. A., Pundle, A., & Poddar, P. (2012). Green approach towards size controlled synthesis of biocompatible antibacterial metal nanoparticles in aqueous phase using lysozyme. *Current Nanoscience*, *8*(1), 130–140.

Lee, K. Z., Pussepitiyalage, V. B., Lee, Y. H., Loesch-Fries, L. S., Harris, M. T., Hemmati, S., & Solomon, K. V. (2021). Engineering tobacco mosaic virus and its virus-like-particles for synthesis of Biotemplated nanomaterials. *Biotechnology Journal*, *16*(4), 2000311.

Lee, S. J., Begildayeva, T., Yeon, S., Naik, S. S., Ryu, H., Kim, T. H., & Choi, M. Y. (2021). Eco-friendly synthesis of lignin mediated silver nanoparticles as a selective sensor and their catalytic removal of aromatic toxic nitro compounds. *Environmental Pollution*, *269*, 116174.

Li, Y., Tang, X., Song, W., Zhu, L., Liu, X., Yan, X., Jin, C., & Ren, Q. (2015). Biosynthesis of silver nanoparticles using Euglena gracilis, Euglena intermedia and their extract. *IET Nanobiotechnology*, *9*(1), 19–26.

Lin, D., Ji, R., Wang, D., Xiao, M., Zhao, J., Zou, J., Li, Y., Qin, T., Xing, B., Chen, Y., Liu, P., Wu, Z., Wang, L., Zhang, Q., Chen, H., Qin, W., Wu, D., Liu, Y., Liu, Y., & Li, S. (2019). The research progress in mechanism and influence of biosorption between lactic acid bacteria and Pb(II): A review. *Critical Reviews in Food Science and Nutrition*, *59*(3), 395–410.

Liu, F., Csetenyi, L., & Gadd, G. M. (2019). Amino acid secretion influences the size and composition of copper carbonate nanoparticles synthesized by ureolytic fungi. *Applied Microbiology and Biotechnology*, *103*(17), 7217–7230.

Liu, F., Shah, D. S., & Gadd, G. M. (2021). Role of protein in fungal biomineralization of copper carbonate nanoparticles. *Current Biology*, *31*(2), 358–368.

Liu, L., Jing, Y., Guo, A., Li, X., Li, Q., Liu, W., & Zhang, X. (2022). Biosynthesis of platinum nanoparticles with cordyceps flower extract: Characterization, antioxidant activity and antibacterial activity. *Nanomaterials*, *12*(11), 1904.

Lomelí-Marroquín, D., Cruz, D. M., Nieto-Argüello, A., Crua, A. V., Chen, J., Torres-Castro, A., Webster, T. J., & Cholula-Díaz, J. L. (2019). Starch-mediated synthesis of mono- and bimetallic silver/gold nanoparticles as antimicrobial and anticancer agents. *International Journal of Nanomedicine*, *14*, 2171–2190.

Lomelí-Rosales, D. A., Zamudio-Ojeda, A., Cortes-Llamas, S. A., & Velázquez-Juárez, G. (2019). One-step synthesis of gold and silver non-spherical nanoparticles mediated by eosin methylene blue agar. *Scientific Reports*, *9*(1), 19327.

Losada-Garcia, N., Rodriguez-Otero, A., Ortega-Nieto, C., Azarmi, A., & Palomo, J. M. (2022). Catalase like-activity of metal NPs–enzyme biohybrids. *Applied Nano*, *3*(3), 149–159.

Lotfy, W. A., Alkersh, B. M., Sabry, S. A., & Ghozlan, H. A. (2021). Biosynthesis of silver nanoparticles by Aspergillus terreus: Characterization, optimization, and biological activities. *Frontiers in Bioengineering and Biotechnology*, *9*, 633468.

Meshram, S. M., Bonde, S. R., Gupta, I. R., Gade, A. K., & Rai, M. K. (2013). Green synthesis of silver nanoparticles using White sugar. *IET Nanobiotechnology*, *7*(1), 28–32.

Miesen, T. J., Engstrom, A. M., Frost, D. C., Ajjarapu, R., Ajjarapu, R., Lira, C. N., & Mackiewicz, M. R. (2020). A hybrid lipid membrane coating 'shape-locks' silver nanoparticles to prevent surface oxidation and silver ion dissolution. *RSC Advances*, *10*(27), 15677–15693.

Mishra, S., & Nayar, S. (2014). Protein-polymer matrix mediated synthesis of silver nanoparticles. *Nanobiomedicine*, *1*, 3.

Miu, B. A., & Dinischiotu, A. (2022). New green approaches in nanoparticles synthesis: An overview. *Molecules*, *27*(19), 6472.

Modan, E. M., & Plaiasu, A. G. (2020). Advantages and disadvantages of chemical methods in the elaboration of nanomaterials. *The Annals of "Dunarea de Jos" University of Galati. Fascicle IX, Metallurgy and Materials Science*, *43*(1), 53–60.

Moreno-Vivián, C., Cabello, P., Martínez-Luque, M., Blasco, R., & Castillo, F. (1999). Prokaryotic nitrate reduction: Molecular properties and functional distinction among bacterial nitrate reductases. *Journal of Bacteriology*, *181*(21), 6573–6584.

Mukherjee, K., Gupta, R., Kumar, G., Kumari, S., Biswas, S., & Padmanabhan, P. (2018). Synthesis of silver nanoparticles by Bacillus clausii and computational profiling of nitrate reductase enzyme involved [in press]. *Journal of Genetic Engineering and Biotechnology*, *16*(2), 527–536.

Murillo-Rábago, I., E., Vilchis-Nestor, A. R., Juarez-Moreno, K., Garcia-Marin, L. E., Quester, K., & Castro-Longoria, E. (2022). Optimized synthesis of small and stable silver nanoparticles using intracellular and extracellular components of fungi: An alternative for bacterial inhibition. *Antibiotics*, *11*(6), 800.

Naapuri, J. M., Losada-Garcia, N., Deska, J., & Palomo, J. M. (2022). Synthesis of silver and gold nanoparticles–enzyme–polymer conjugate hybrids as dual-activity catalysts for chemoenzymatic cascade reactions. *Nanoscale*, *14*(15), 5701–5715.

Nuanaon, N., Bhatnagar, S., Motoike, T., & Aoyagi, H. (2022). Light-emitting-diode-assisted, fungal-pigment-mediated biosynthesis of silver nanoparticles and their antibacterial activity. *Polymers*, *14*(15), 3140.

Onen, O. I., Aboh, A. A., Mfam, A. N., Akor, M. O., Nweke, C. N., & Osuagwu, A. N. (2020). Microbial diversity: Values and roles in ecosystems. *Asian Journal of Biology*, *9*, 10–22.

Ovais, M., Khalil, A. T., Ayaz, M., Ahmad, I., Nethi, S., & Mukherjee, S. (2018). Biosynthesis of metal nanoparticles via microbial enzymes: A mechanistic approach. *International Journal of Molecular Sciences*, *19*(12), 4100.

Owaid, M. N., & Ibraheem, I. J. (2017). Mycosynthesis of nanoparticles using edible and medicinal mushrooms. *European Journal of Nanomedicine*, *9*(1), 5–23.

Owaid, M. N., Rabeea, M. A., Aziz, A. A., Jameel, M. S., & Dheyab, M. A. (2019). Mushroom-assisted synthesis of triangle gold nanoparticles using the aqueous extract of fresh *Lentinula edodes* (Shiitake), Omphalotaceae. *Environmental Nanotechnology, Monitoring and Management*, *12*, 100270.

Palmqvist, N. G. M. (2017). *Nanoparticles: Case studies of their synthesis, properties and biological interaction* (Doctoral Thesis). Swedish University of Agricultural Sciences, Uppsala.

Pandey, G., & Jain, P. (2020). Assessing the nanotechnology on the grounds of costs, benefits, and risks. *Beni-Suef University Journal of Basic and Applied Sciences*, *9*(1), 63.

Parashar, A., Kedare, P. S., Alex, S. A., Chandrasekaran, N., & Mukherjee, A. (2017). A novel enzyme-mediated gold nanoparticle synthesis and its application for in situ detection of horseradish peroxidase inhibitor phenylhydrazine. *New Journal of Chemistry*, *41*(24), 15079–15086.

Patil, B. N., & Taranath, T. C. (2016). *Limonia acidissima* L. Leaf mediated synthesis of zinc oxide nanoparticles: A potent tool against *Mycobacterium tuberculosis*. *International Journal of Mycobacteriology*, *5*(2), 197–204.

Patra, D., & El-Kurdi, R. (2021). Curcumin as a novel reducing and stabilizing agent for the green synthesis of metallic nanoparticles. *Green Chemistry Letters and Reviews*, *14*(3), 474–487.

Pérez-Hernández, H., Pérez-Moreno, A., Sarabia-Castillo, C. R., García-Mayagoitia, S., Medina-Pérez, G., López-Valdez, F., Campos-Montiel, R. G., Jayanta-Kumar, P., & Fernández-Luqueño, F. (2021). Ecological drawbacks of nanomaterials produced on an industrial scale: Collateral effect on human and environmental health. *Water, Air, and Soil Pollution*, *232*(10), 435.

Pokorski, J. K., & Steinmetz, N. F. (2011). The art of engineering viral nanoparticles. *Molecular Pharmaceutics*, *8*(1), 29–43.

Prabhu, N., Karunakaran, S., Devi, S. K., Ravya, R., Subashree, P., & Vadivu, M. S. (2019). Biogenic synthesis of Myconanoparticles from mushroom extracts and its medical applications : A review biogenic synthesis of Myconanoparticles from mushroom extracts and its medical applications : A review. *International Journal of Pharmaceutical Sciences and Research*, *10*(5), 2108–2118.

Priyadarshini, E., Pradhan, N., Sukla, L. B., & Panda, P. K. (2014). Controlled synthesis of gold nanoparticles using *Aspergillus terreus* IF0 and its antibacterial potential against Gram negative pathogenic bacteria. *Journal of Nanotechnology*, *2014*, 1–9.

Priyam, A., Das, R. K., Schultz, A., & Singh, P. P. (2019). A new method for biological synthesis of agriculturally relevant nanohydroxyapatite with elucidated effects on soil bacteria. *Scientific Reports*, *9*(1), 1–14.

Rai, A., Singh, A., Ahmad, A., & Sastry, M. (2006). Role of halide ions and temperature on the morphology of biologically synthesized gold nanotriangles. *Langmuir*, *22*(2), 736–741.

Rasmussen, L., Shi, H., Liu, W., & Shannon, K. B. (2022). Quantification of silver nanoparticle interactions with yeast *Saccharomyces cerevisiae* studied using single-cell ICP-MS. *Analytical and Bioanalytical Chemistry, 414*(9), 3077–3786.

Reddy, P. V. L., Hernandez-Viezcas, J. A., Peralta-Videa, J. R., & Gardea-Torresdey, J. L. (2016). Lessons learned: Are engineered nanomaterials toxic to terrestrial plants? *Science of the Total Environment, 568*, 470–479.

Röder, J., Dickmeis, C., & Commandeur, U. (2019). Small, smaller, Nano: New applications for potato virus X in nanotechnology. *Frontiers in Plant Science, 10*, 1–17.

Roduner, E. (2006). Size matters: Why nanomaterials are different. *Chemical Society Reviews, 35*(7), 583–592.

Roduner, E. (2009). Physics and chemistry of nanostructures: Why Nano is different. In V. Kharkin (Ed.), *Encyclopedia of life support systems (EOLSS)* (pp. 1–35). EOLSS Publisher.

Romero-Núñez, A., González, G., Romero-Ibarra, J. E., Vega-González, A., Cruz-Jiménez, G., Hernández-Cristóbal, O., Zárraga-Núñez, R. A., Obregón-Herrera, A., López-Romero, E., Pedraza-Reyes, M., & Cuéllar-Cruz, M. (2019). Synthesis of bimetallic nanoparticles of Cd_4HgS_5 by *Candida* species. *Crystals, 9*(2), 61.

Roy, A., Elzaki, A., Tirth, V., Kajoak, S., Osman, H., Algahtani, A., Islam, S., Faizo, N. L., Khandaker, M. U., Islam, M. N., Emran, T. B., & Bilal, M. (2021). Biological synthesis of nanocatalysts and their applications. *Catalysts, 11*(12), 1494.

Roy, S., & Rhim, J.-W. (2019). Melanin-mediated synthesis of copper oxide nanoparticles and preparation of functional agar/CuO NP nanocomposite films. *Journal of Nanomaterials, 2019*, 1–10.

Samb-Joshi, K. M., Sethi, Y. A., Ambalkar, A. A., Sonawane, H. B., Rasale, S. P., Panmand, R. P., Patil, R., Kale, B. B., & Chaskar, M. G. (2019). Lignin-mediated biosynthesis of ZnO and TiO_2 nanocomposites for enhanced antimicrobial activity. *Journal of Composites Science, 3*(3), 90. https://doi.org/10.3390/jcs3030090.

Šebesta, M., Vojtková, H., Cyprichová, V., Ingle, A. P., Urík, M., & Kolenčík, M. (2022). Mycosynthesis of metal-containing nanoparticles-fungal metal resistance and mechanisms of synthesis. *International Journal of Molecular Sciences, 23*(22), 14084.

Shah, A. H., & Rather, M. A. (2021). Intracellular and extracellular microbial enzymes and their role in nanoparticle synthesis. In M. A. Ansari & S. Rehman (Eds.), *Microbial nanotechnology: Green synthesis and applications* (pp. 41–59). Springer Singapore.

Shankar, S., & Rhim, J. W. (2015). Amino acid mediated synthesis of silver nanoparticles and preparation of antimicrobial agar/silver nanoparticles composite films. *Carbohydrate Polymers, 130*, 353–363.

Singh, R., Shedbalkar, U. U., Nadhe, S. B., Wadhwani, S. W., & Chopade, B. A. (2017). Lignin peroxidase mediated silver nanoparticle synthesis in *Acinetobacter* sp. *AMB Express, 7*(1), 226.

Singh, S., Kuca, K., & Kalia, A. (2020). Alterations in Growth and Morphology of *Ganoderma lucidum* and *Volvariella volvaceae* in response to nanoparticle supplementation. *Mycobiology, 48*(5), 383–391.

Smirnov, O., Kalynovskyi, V., Yumyna, Y., Zelena, P., Levenets, T., Kovalenko, M., Dzhagan, V., & Skoryk, M. (2022). Potency of phytosynthesized silver nanoparticles from Lathraea squamaria as anticandidal agent and wheat seeds germination enhancer. *Biologia, 77*(9), 2715–2724.

Steinmetz, N. F. (2010). Viral nanoparticles as platforms for next-generation therapeutics and imaging devices. *Nanomedicine: Nanotechnology, Biology and Medicine, 6*(5), 634–641.

Sun, Y., Li, Q., Du, X., Thipe, V., Vardhanabhuti, B., Sengupta, S., Katti, K., & Wan, C. (2022). Lignin-containing nanocellulose for in situ chemical-free synthesis of AgAu-based nanoparticles with potent antibacterial activities. *ACS Omega, 7*(45), 41548–41558.

Sun, Y., Shi, Y., Jia, H., Ding, H., Yue, T., & Yuan, Y. (2021, December). Biosynthesis of selenium nanoparticles of Monascus purpureus and their inhibition to *Alicyclobacillus acidoterrestris*. *Food Control*, *130*, 108366.

Tarmizi, Z. I., Burhanuddin, F. M. B., Shahrul, M., Mokhtar, M. A. M., Zhe, J. C., Taib, S. H. M., & Sukri, S. N. A. (2022). Preparation and characterization of acetylated starch mediated silver nanoparticles: The effect of solution ratio and time-varying exposure. *Journal of Physics: Conference Series*, *2259*(1), 012005.

Thakur, S., Asthir, B., Kaur, G., Kalia, A., & Sharma, A. (2022). Zinc oxide and titanium dioxide nanoparticles influence heat stress tolerance mediated by antioxidant defense system in wheat. *Cereal Research Communications*, *50*(3), 385–396. https://doi.org/10.1007/s42976-021-00190-w

Thapa, R., Bhagat, C., Shrestha, P., Awal, S., & Dudhagara, P. (2017). Enzyme-mediated formulation of stable elliptical silver nanoparticles tested against clinical pathogens and MDR bacteria and development of antimicrobial surgical thread. *Annals of Clinical Microbiology and Antimicrobials*, *16*(1), 39.

Tiwari, S., Verma, S. K., Bhagat, P., Yadav, S., Sharma, R., Aseri, G. K., Sohal, J. S., Sharma, D., Dwivedi, U. K., Singh, R., Singh, D., & Khare, N. (2021). An overview of the phytosynthesis of various metal nanoparticles. *3 Biotech*, *11*(11), 478.

Tondey, M., Kalia, A., Singh, A., Abd-Elsalam, K. A., Hassan, M. M., & Dheri, G. S. (2022). A comparative evaluation of the effects of seed invigoration treatments with precursor zinc salt and nano-sized zinc oxide (ZnO) particles on vegetative growth, grain yield, and quality characteristics of *Zea mays*. *Journal of Analytical Science and Technology*, *13*(1), 40. https://doi.org/10.1186/s40543-022-00346-1

Tondey, M., Kalia, A., Singh, A., Dheri, G. S., Taggar, M. S., Nepovimova, E., Krejcar, O., & Kuca, K. (2021). Seed priming and coating by nano-scale zinc oxide particles improved vegetative growth, yield and quality of fodder maize (*Zea mays*). *Agronomy*, *11*(4), 729.

Vardhini, B. V. (2015). Synthesis of nanoparticles from plant extracts. *International Journal of Modern Chemistry and Applied Science*, *2*(3), 195–203.

Veisi, H., Karmakar, B., Tamoradi, T., Hemmati, S., Hekmati, M., & Hamelian, M. (2021). Biosynthesis of CuO nanoparticles using aqueous extract of herbal tea (*Stachys lavandulifolia*) flowers and evaluation of its catalytic activity. *Scientific Reports*, *11*(1), 1983.

Velázquez-Hernández, A. M., García-Rivas, J. L., Martínez-Gallegos, S., González-Juárez, J. C., Schabes-Retchkiman, P., & Albiter, V. (2022, December). Phytosynthesis of TiO_2 nanoparticles using *E. Crassipes* Leaf extracts, their photocatalytic evaluation and microbicide effect. *International Journal of Photoenergy*, 20221–7. https://doi.org/10.1155/2022/5177859

Venkatesh, K. S., Gopinath, K., Palani, N. S., Arumugam, A., Jose, S. P., Bahadur, S. A., & Ilangovan, R. (2016). Plant pathogenic fungus *F. solani* mediated biosynthesis of nanoceria: Antibacterial and antibiofilm activity. *RSC Advances*, *6*(48), 42720–42729.

Verma, V. C., Kharwar, R. N., & Gange, A. C. (2010). Biosynthesis of antimicrobial silver nanoparticles by the endophytic fungus *Aspergillus clavatus*. *Nanomedicine*, *5*(1), 33–40.

Vetchinkina, E., Loshchinina, E., Kupryashina, M., Burov, A., Pylaev, T., & Nikitina, V. (2018, July). Green synthesis of nanoparticles with extracellular and intracellular extracts of basidiomycetes. *PeerJ*, *6*, e5237.

Vigneshwaran, N., & Jain, P. (2011). Metal nanoparticles in microbiology. In: Rai, M. & Duran, N. (eds.) *Metal nanoparticles in microbiology* (pp. 135–150). Springer.

Vitorino, L., & Bessa, L. (2018). Microbial diversity: The gap between the estimated and the known. *Diversity*, *10*(2), 46.

Wang, D., Xue, B., Wang, L., Zhang, Y., Liu, L., & Zhou, Y. (2021). Fungus-mediated green synthesis of nano-silver using *Aspergillus sydowii* and its antifungal/antiproliferative activities. *Scientific Reports*, *11*(1), 10356.

Wang, Z., Zhang, J., Ekman, J. M., Kenis, P. J. A., & Lu, Y. (2010). DNA-mediated control of metal nanoparticle shape: One-pot synthesis and cellular uptake of highly stable and functional gold nanoflowers. *Nano Letters*, *10*(5), 1886–1891.

Wen, A. M., & Steinmetz, N. F. (2016). Design of virus-based nanomaterials for medicine, biotechnology, and energy. *Chemical Society Reviews*, *45*(15), 4074–4126.

Xu, W., Jin, W., Lin, L., Zhang, C., Li, Z., Li, Y., Song, R., & Li, B. (2014). Green synthesis of xanthan conformation-based silver nanoparticles: Antibacterial and catalytic application. *Carbohydrate Polymers*, *101*, 961–967.

Yan, X., Pan, Z., Chen, S., Shi, N., Bai, T., Dong, L., Zhou, D., White, J. C., Zhao, L. (2022). Rice exposure to silver nanoparticles in a life cycle study: Effect of dose responses on grain metabolomic profile, yield, and soil bacteria. *Environmental Science: Nano*, *9*(6), 2195–2206.

Yao, Y., Cao, S., Zhang, X., Yan, J., Li, W., Whittaker, A. K., & Zhang, A. (2022). Microconfinement from dendronized chitosan oligosaccharides for mild synthesis of silver nanoparticles. *ACS Applied Nano Materials*, *5*(3), 4350–4359.

Ying, S., Guan, Z., Ofoegbu, P. C., Clubb, P., Rico, C., He, F., & Hong, J. (2022). Green synthesis of nanoparticles: Current developments and limitations. *Environmental Technology and Innovation*, *26*, 102336.

Yu, Y., Naik, S. S., Oh, Y., Theerthagiri, J., Lee, S. J., & Choi, M. Y. (2021). Lignin-mediated green synthesis of functionalized gold nanoparticles via pulsed laser technique for selective colorimetric detection of lead ions in aqueous media. *Journal of Hazardous Materials*, *420*, 126585.

Yusof, H. M., Mohamad, R., Zaidan, U. H., & Rahman, N. A. A. (2019). Microbial synthesis of zinc oxide nanoparticles and their potential application as an antimicrobial agent and a feed supplement in animal industry: A review. *Journal of Animal Science and Biotechnology*, *10*(1), 1–22.

Zhang, Y., Dong, Y., Zhou, J., Li, X., & Wang, F. (2018). Application of plant viruses as a biotemplate for nanomaterial fabrication. *Molecules*, *23*(9), 2311.

Zinder, S. H., & Dworkin, Z. (2013). Morphological and physiological diversity. In: Rosenberg, E., DeLong, E.F., Lory, S., Stackebrandt, E., Thompson, F. (eds) *The prokaryotes* (pp. 89–122). Springer.

11 Application of Nanotechnology in Rice

*Sanjoy K Debsharma, H Khatun, M Khatun, AKM Shalahuddin, MA Rahman, T Halder, Popy R Roy, N Akter and MA Syed**

11.1 INTRODUCTION

In the past 100 years, nanotechnology has been ranked as the sixth most revolutionary technology (Chhipa, 2017). In the past two decades, the agricultural industry has transferred and adapted the knowledge amassed in these fields in order to use it for the creation of plant-protecting agrochemicals (Mattos et al., 2017). Worldwide, an increasing number of unexpected hazards are posing challenges to agricultural systems. Nanotechnology, or the sophisticated agronomic application of small-scale agricultural practices in plants, is crucial for preserving sustainable agriculture and food supply (Wang et al., 2016). By reducing critical losses and boosting input efficiency, nanotechnology can boost agricultural output, which is a key factor in maintaining the viability of the expansion of agriculturally based and connected industries (Pandey, 2018). By increasing grain yields and production while sustaining environmental adaptability, ecological stability and economic viability, the potential use of nanotechnology can revolutionize agrosystems (Acharya & Pal, 2020).

Nanotechnology refers to the study and use of structures, devices and systems with dimensions on the scale of a few nanometers to several hundreds of nanometers. In the context of rice, nanotechnology has been used to boost-up the different characteristics of rice. Fortification of rice is one of the latest concepts where rice is boosted with essential different micronutrients like iron and vitamin B12, making it a more nutritious food source. Nanotechnology has the potential to impact the properties of rice in various ways. For example, by modifying the surface characteristics of rice grains at the nanoscale level, it is possible to improve their water-absorption and cooking properties, leading to the development of more easily cooked and tasty rice. Furthermore, the use of nanoparticles as food packaging materials has gained attention as a means of preserving the freshness of rice and other food products. These nanopackaging materials can help to extend the shelf-life of rice by preventing oxidation, moisture loss and the growth of bacteria. Additionally, the use of nanotechnology in agriculture can also lead to the development of more efficient fertilizers and pesticides, which can help to improve crop yields and reduce environmental damage. Overall, the impact of nanotechnology on rice properties is continually

* Corresponding author

DOI: 10.1201/9781003376446-11

developing, and to fully grasp its potential, more study is required to ensure its safe and effective application in the food industry. The construction of numerous significant goals, such as nanocatalysts, nanosensors, nanodevices and nanosystems is also a phase in the process of nanomaterial synthesis (Kulkarni, 2015). Numerous methods exist for creating a wide variety of nanomaterials, including nanoparticles, nanotubes, nanofibers, nanorods and nanoporous materials. Green nanotechnology has been defined as the advancement of renewable technologies, the production and application of nanomaterials and the replacement of existing products with novel eco-friendly nanoproducts (Sarong et al., 2020).

Because of biotic and abiotic aspects like insects, weeds, pests and illnesses, one of the key difficulties is the shortage of high-quality food for humans (Sinha et al., 2017). According to recent data, pests caused losses of 8–10% in wheat harvests, 20% in sugar, 25% in rice, 35% in pulses and 50% in cotton. Pests damage over 25% of the world's crop production each year (Rajendran & Islam, 2017). Due to its transdisciplinary applications in the disciplines of pharmacology, agriculture, pest and disease management and parasitology, nanotechnology has attracted particular attention and has emerged as a new field in recent years (Sayed et al., 2017). Due to their cost-effectiveness, eco-friendly behaviour, tiny size (1–100 nm), and vast surface space, nanoparticles – the branch of nanotechnology that is developing the fastest – provide solutions to numerous environmental issues. New research indicates that nanotechnology can be used to create new nanocomposites that are effective against pests and to improve crop varieties (Chaloupka et al., 2010). Therefore, this chapter shows the role of nanotechnology in genetic engineering, biofortification, precision agriculture, pest insect management and different biotic and abiotic stress in rice for recent working activities to overcome the difficulties. Also, the current chapter elucidates the challenges or major problems as well as future prospectives for environmental eco-friendliness and the sustainability of upcoming nanotechnologies.

11.2 ROLE OF NANOTECHNOLOGY IN RICE

A "key enabling technology", as defined by the European Commission, nanotechnology is one of the most inventive and promising instruments available today (Singh et al., 2021). Studies in the field of nanotechnology and its utilization in the field of plant science can deliver as a crucial substitute to reach the objective of sustainable agriculture for future advancements in the era of weather change, contaminated environments and universal food insecurity (Rastogi et al., 2019; Fincheira et al., 2020). Agricultural output has decreased during the past few decades due to a decrease in the agricultural land area caused by population growth and a rise in environmental stressors. As a result, newer innovations in the agricultural sector are essential, including the use of new technologies (Usman et al., 2020). Agriculture also requires constant innovation. Although other technologies, including genetic engineering, plant breeding and the use of synthetic chemicals have increased agricultural output, they are not without drawbacks. Therefore, it is strongly encouraged that more inventions using nanotechnology be made in order to meet the rising international food mandate and to also have positive effects on the economy, the environment and human health (Rastogi et al., 2019; Fincheira et al., 2020).

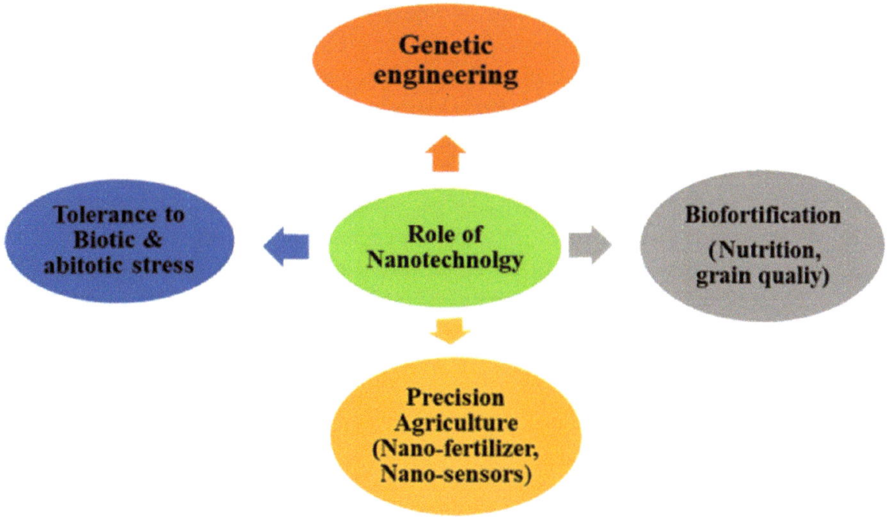

FIGURE 11.1 Diagram showing role of nanotechnology in rice

Rice is a complex plant, and breeders have traditionally relied on conventional methods of breeding to improve yields and disease resistance. However, conventional methods for crop enhancement are difficult, laborious and challenging. Consequently, it is urgently necessary to use cutting-edge plant breeding technologies to address the shortcomings of traditional plant breeding techniques. These technologies include gene knockout or knockin, epigenetic modifications and the generation of genetically targeted alterations in specific genomic sections (Chen et al., 2019; Fiaz et al., 2021). Major improvements in the field of biotechnology have occurred during the past ten years, including the emergence of third-generation genome editing (GE) methods, genome sequencing, plant-based synthetic biology and bioengineering (Wang et al., 2019; Zhang et al., 2019). In order to ensure grain output, excellence and resistance to biotic and abiotic challenges as well as climatic variation, these strategies have been effectively used to generate elite germplasm (Cunningham et al., 2018; Fiaz et al., 2021). One of the most significant staple crops in the world is rice, providing food for over half of the global population. As such, the ability to improve rice breeding through nanotechnology has the potential to have a significant impact on food security. The different aspects about the roles of nanotechnology are shown in Figure 11.1.

11.3 APPLICATION OF NANOTECHNOLOGY IN RICE

11.3.1 Nanotechnology Usage in Genome Editing

Genome editing is one kind of nanotechnology. The most practical gene-editing method, CRISPR/Cas9, is used by numerous researchers to introduce various features into rice cultivars. According to (Dong et al., 2020), a 5.2 kb carotenoid biosynthetic cassette was inserted at two genomic sites in rice using the CRISPR-Cas9 system. These findings show rice plants without markers, with a significant carotenoid concentration in the seeds and no noticeable impact on yield or morphology. Gene OsSPO11-2, but not OsSPO11-4, plays a crucial role in rice meiosis, as shown by the CRISPR/Cas9 deletion technique (Fayos et al., 2020). The functional genomic investigations in rice are sped up using CRISPR applicable functional redundancy inspector (CAFRI) based target selection, which is also easily extended to other plant species (Hong et al., 2020). In addition, the *OsFWL4* gene is a negative regulator of tiller number and grain yield, and the CRISPR/Cas9 system mediated by agrobacterium can produce transgene-free rice plants with targeted mutations in the T1 generation (Gao et al., 2020). By using CRISPR/Cas9-targeted mutagenesis, novel eIF4G alleles are created that confer resistance to the rice tungro spherical virus (Manghwar et al., 2019).

11.3.2 Nanotechnology Usage in Biofortification

The only method to lessen rice deficiency is micronutrient improvement through biofortification. In the poor world that relied on cereal-based staple foods, notably in Asian countries where more than half of the world's population utilized rice as a staple food, micro-nutrient malnutrition, often known as hidden hunger, has been well-recognized (Alam et al., 2020). India's East and North East are the main growing regions for black rice, *Oryza sativa.* Due to its exorbitant price and low yield, this cultivar was unable to become popular. It has high levels of three different anthocyanin types, each of which has a significant impact on human health. At varied temperatures ranging from 250 to 500°C, four different types of ZnO nanomaterial were created. It was discovered that seeds germinated quickly as a result of the interaction between black rice grain and nanozinc material. Rooting growth was greater. When they reacted with 350°C nanomaterial, better impacts were seen (Mahajan et al., 2019). Orange-fleshed sweet potatoes (OFSP) and quality protein maize have received widespread acceptance and can be considered intriguing examples of conventionally developed biofortified crops, according to Sharma et al. (2017). Other examples include zinc rice and iron beans. It is clear that traditional breeding cannot improve all micro-nutrients in crops. This makes the use of genetic nutrient modification in nutrient augmentation necessary. The use of chemical fertilizers can be reduced due to nanochelating technology, which also has the ability to biofortify crops with essential micro-nutrients (Fakharzadeh et al., 2020). ZnO NPs treatment considerably raised the Zn content of edible polished rice and promoted the translocation of Zn from the aleurone layer for the Zn nutrition in rice grain. This study successfully proved that ZnO NPs may be a high-performing fertilizer option for raising rice output and quality (Zhang et al., 2021).

11.3.3 Nanotechnology Usage in Precision Agriculture

Nanotechnology provides researchers with a unique approach to various scientific fields, including agriculture (Ekinci et al., 2014). Agriculture is being revolutionized around the world by precision agriculture (PA), which has a number of potential benefits for the production, grain quality, profit, resilience, environmental protection, food security, on-farm quality of life and rural economic growth (Zhang et al., 2021).

11.3.3.1 Nanopriming Used in Rice for Better Growth

Seed priming has biological impacts that provide assistance to farmers decreasing the time, fertilization and expenditure of re-seeding. Nanopriming increases α-amylase activity in rice plants and ensures greater soluble sugar concentration for supportive seedling growth. However, higher ROS production was seen in germinating seeds of rice plants treated with nanopriming compared to control plants, demonstrating that both ROS and aquaporins play important roles in promoting seed germination (Mahakham et al., 2017; do Espirito Santo Pereira et al., 2021). For risk reduction in DSR, seed priming with the mixture of ZnO NPs, Na selenite and Na selenate may be an alternative (Adhikary et al., 2022). Rice seeds were primed with various concentrations of MWCNT (70, 80 and 90 g/mL) that have been oxidized to a diameter of 14 to 30 nm and a length of 200 to 300 nm (Joshi et al., 2020). The levels of proline amino acids are increased after priming with ZnONPs, helping the plant resist the stress of water scarcity. In rice plants grown from ZnONPs primed seeds, SOD, CAT and POD activities are additionally increased by 11%, 13% and 38%, respectively. Following seed priming with ZnONPs, it was shown that the activities of antioxidant enzymes increased and the levels of oxidative stress markers reduced. The 25 ppm concentration of ZnONPs was revealed to be the most effective for reducing drought-related damages (Mazhar et al., 2022). According to Mahakham et al. (2017), priming rice seeds with biogenic silver nanoparticles made from kaffir lime leaf extract improved water uptake and α-amylase activity in the seeds, which improved germination and plant biomass.

11.3.3.2 Uses of Nanofertilizers in Rice

The plant's physiology was affected and its growth rate was accelerated by nanoparticles. Consequently, applying nanoscale fertilizers through nanotechnology represents a novel and cutting-edge method for crop management (Hussien et al., 2015; Valojai et al., 2021). Nanofertilizers can boost plant development with lower concentrations than mass fertilizers (Herrera et al., 2016). Rice crops may benefit from using nano-macro-micro-nutrients to increase growth and production, increase fertilizer use effectiveness and reduce adverse environmental effects.

According to a study on rice crops, using nano-urea of half the recommended rate of regular urea fertilizer led to better growth and yield parameters, including the highest plant height, number of tillers, filled grains per panicle and grain yield. This indicated that using nano-urea in addition to less regular nitrogen fertilization may boost growth and yield while requiring fewer fertilizer inputs (Midde et al., 2021). Yan et al. (2010) stated that the use of nanomaterial urea for optimal fertilizer management resulted in higher rice grain yield than with ordinary urea. According to Wu (2013), nanocarbon slow-release fertilizer produced positive effects on nitrogen loss from paddy soil as well as lessening the pollution of water. Nitrogen fertilizer is more efficient and cost-effective when applied with nanocarbon (Fan et al., 2012). Therefore nano-urea has the potential to increase rice production productivity and profitability. The application of nano-KH_2PO_4 increased rice plants' instant water-use efficiency. The application of 50% of the acceptable dose of phosphorus (P) from nano-KH_2PO_4 resulted in a better photosynthetic rate than 100% of the necessary quantity of P from traditional fertilizer, according to Miranda-Villagómez et al., (2019). This suggests that using nano-KH_2PO_4 can minimize the amount of fertilizer required, reducing production costs and the likelihood of pollution. Using nano-KH_2PO_4 may give a more sustainable and effective method for sustaining the health of rice plants.

Nano-K fertilizer is an excellent source of potassium (K) for rice. Nano-K fertilizer has been shown to increase the number of grains per panicle, the percentage of fill grains and overall rice grain production (Sirisena et al., 2013). Studies have shown that nano-K fertilizer can boost grain yield by up to 10% compared to regular potassium chloride (MoP) fertilizer. Furthermore, the amount of nano-K fertilizer needed is smaller than that of MoP, making it a more efficient solution for sustaining healthy rice crops. Applying nano-K fertilizer is a viable alternative for optimizing rice growth and output. Nano-sulfur (nano-S) has been demonstrated to have a significant potential for the gradual release of sulphate (SO_4^{2-}), enhancing sulphur utilization efficiency. The delayed release of sulphate via nano-S can give a prolonged supply of sulphur to plants (Thirunavukkarasu & Subramanian, 2014), minimizing the requirement for frequent reapplication and enhancing the overall efficacy of sulphur as a fertilizer. Nano-S can potentially increase sulphur use efficiency, making it a promising alternative for enhancing plant growth and production.

The foliar spray application of copper (Cu) and zinc (Zn) micro-nutrients in the form of nanofertilizer suspension has been found to increase the yield of both traditional and inbred rice varieties (Sandanayake et al., 2022). This presents a viable option for reducing the excessive use of bulk fertilizers, which can negatively impact the environment. The efficacy of the foliar spray application of Cu and Zn nanofertilizers in boosting rice yield while reducing the need for bulk fertilizers makes it a valuable alternative for sustainable rice production. The use of silicon (Si) and zinc (Zn) as nanoparticles have also been found to improve rice yield (Kheyri et al., 2019). This approach increases yield and lowers fertilizer costs and pollution while enriching the grain with Si and Zn. The spraying of Si and Zn NPs improves agronomic and physiological parameters, resulting in higher GY and increased nutrient accumulation in the grain. This highlights the possibility for Si and Zn nanoparticles to be used as a sustainable way to optimize rice production while improving grain nutritional quality.

Nanoparticles of silicon (NPs-Si) and selenium (NPs-Se) have been recommended for their positive impact on rice growth and yield (Badawy et al., 2021). Additionally, nano-Si and nano-Se play crucial roles in mitigating salt stress's adverse effects, a common challenge rice growers face. By improving the plant's tolerance to salt, NPs-Si and NPs-Se can help to ensure better crop yields and reduce crop losses due to salt stress.

Nanofertilizers are a new generation of plant nutrients designed to provide plants with essential nutrients in a highly soluble form that offers improved benefits over traditional fertilizers in rice production. The tiny nanoparticles in nanofertilizers allow for the efficient uptake of essential nutrients by the plant roots, leading to enhanced growth and higher yields. In addition, nanofertilizers are highly stable, reducing the risk of leaching or denitrification and promoting a more sustainable and efficient fertilizer option. Moreover, nanofertilizers can improve soil health by maintaining a balanced soil micro-organism population, resulting in better soil structure and fertility for rice plants. Nanofertilizers can be applied directly to the soil or mixed with irrigation water for convenient and flexible use by farmers. The small size of the nanoparticles also allows for quicker penetration and response to the application. Nanofertilizers are a promising solution for farmers looking to optimize their rice production and achieve higher yields and quality of grain.

11.3.3.3 Geographic Information System (GIS) in Precision Agriculture

Geographic information system (GIS) plays a significant role in precision agriculture. It is a farming method that utilizes advanced technologies to improve the efficiency and productivity of crop production. It helps farmers make informed decisions about crop production by providing a visual representation of various agricultural data such as soil properties, crop yields, weather patterns and other relevant information. With the help of GIS, farmers can analyze and map data, creating a comprehensive view of their land and crops. This enables them to implement precision farming practices, such as site-specific application of inputs like fertilizer and pesticides, more effectively and efficiently. Precision agriculture supported by GIS can increase crop yields, reduce waste and improve resource use, leading to more sustainable and cost-effective farming practices.

Field mapping: GIS can be used to create detailed maps of fields showing topography, soil conditions and crop growth patterns. This information can be used to optimize planting, fertilizing and harvesting operations (Udoumoh et al., 2021).

Data analysis: GIS can be used to analyze various data sources such as remote sensing imagery, soil maps and weather data to generate insights about crop growth, soil fertility and potential yield (Mani et al., 2021).

Decision support: GIS can provide farmers with decision support tools such as yield prediction models and nutrient management plans, which can be used to optimize crop production and increase efficiency (Hedley, 2015).

Environmental impact assessment: GIS can be used to assess the environmental impact of farming operations.

11.3.3.4 Remote Sensing Used in Precision Agriculture

Remote sensing plays a crucial role in precision agriculture to advance the efficiency and productivity of crop production. Remote sensing technologies, such as satellite imaging, drones and aerial photography, provide farmers with detailed information about their crops, fields and soil conditions, enabling them to make informed decisions about planting, fertilizing and harvesting. Some of the ways remote sensing contributes to precision agriculture are as follows.

Crop monitoring: Remote sensing technologies provide high-resolution images of crops, which can be used to monitor crop growth and detect potential problems, such as disease outbreaks, pest infestations and water stress (Yang, 2020).

Soil mapping: Remote sensing can be used to generate detailed maps of soil properties such as nutrient and moisture levels, which can be used to optimize fertilizer and irrigation management (Liaghat & Balasundram, 2010).

Yield prediction: Remote sensing can provide data on crop yield potential, enabling farmers to make more informed decisions about planting and harvesting. To maximize agricultural output and lower input losses, it is helpful to optimize agricultural inputs (Sishodia et al., 2020).

Water management: Remote sensing can be used to monitor soil moisture levels and predict water requirements for crops, leading to more efficient irrigation management and water conservation (Aragon et al., 2018).

Overall, remote sensing is a key tool in precision agriculture, helping farmers to optimize crop production and increase efficiency while reducing costs and environmental impact.

11.3.3.5 Use of Robotics in Precision Agriculture

Robotics plays an important role in precision agriculture, a farming method that utilizes advanced technologies to improve the efficiency and productivity of crop production. Robotics technologies, such as drones, autonomous tractors and soil sensors, can provide farmers with detailed information about their crops, fields and soil conditions.

Crop monitoring: Robotics technologies, such as drones and autonomous tractors, can provide high-resolution images of crops and fields, which can be used to monitor crop growth and detect potential problems, such as disease outbreaks, pest infestations and water stress (Satish & Sudeep, 2007).

Soil analysis: Robotics technologies, such as soil sensors, can be used to measure soil properties, such as nutrient and moisture levels, which can be used to optimize fertilizer and irrigation management (Linaza et al., 2021).

Automated planting and harvesting: Robotics technologies, such as autonomous tractors, can be used to automate planting and harvesting operations, leading to increased efficiency and reduced labour costs.

Field mapping: Robotics technologies, such as drones, can be used to generate detailed maps of fields, providing farmers with information about the topography, soil conditions and crop growth patterns (Meshram et al., 2022).

11.3.4 Application of Nanotechnology for Biotic Stress Management

Nanotechnology offers a cutting-edge strategy for pest management and is capable of reducing post-harvest storage losses caused by insect infestation (Bhattacharyya et al., 2010). Nanopesticides use nanoparticles as their primary ingredient to kill insects. Compared to conventional pesticides, it may be a more focussed and efficient method of controlling pests on rice. Several kinds of research have been undertaken to evaluate the effectiveness of nanoparticles in reducing insect pests and protecting stored grains. In the control of rice weevils (caused by *Sitophilus oryzae*), silver nanoparticles, aluminium oxide, zinc oxide and titanium dioxide were found to be 100% effective (Goswami et al., 2010). A comparison was made between silica nanoparticles (SNP) and bulk-sized silica (larger particles than 1 m) for entomotoxicity against the rice weevil. More than 90% mortality was reported from amorphous SNP treatments against this insect pest (Debnath et al., 2011). Zahir et al. (2012) investigated the effectiveness of silver nanoparticles against the rice weevil. Their study found that silver nanoparticles were more effective in controlling the weevil than silver nitrate and an aqueous extract from the medicinal plant *Euphorbia prostrata* for management. According to research, nano-aluminium oxide is extremely good in controlling the pest *S. oryzae*; however, nanozinc oxide is less effective against the rice weevil (El-Bendary et al., 2016). Younas et al. (2020) found that alumina nano-insecticides were effective against the rice weevil and the lesser grain borer (*Rhyzopertha dominica* F.). These findings indicate that nano-aluminium oxide may be a promising approach for managing a major stored grain pest, like the rice weevil. This study suggests that nanotechnology may be an essential alternative to insecticides. Study findings indicate that the optimal dose of silica nanoparticles (SNPs) to protect stored rice from the rice weevil was 0.5 g/kg of grain (Kar et al., 2021). Moreover, SNP-treated rice exhibited no significant differences in physicochemical or cooking quality. After washing, there was no residual silicon in the rice treated with SNP, so it is safe to cook and consume.

Within a suitable concentration range, SiO_2 NPs might promote plant immunity to defend rice against *M. oryzae* through foliar treatment that considerably reduced disease severity by about 70% (Du et al., 2022). Bhargava et al. (2018) reported nanoparticle mediated (aluminum, silver, titanium dioxide, silicon carbide nanoparticles) fungicides were very effective for controlling false smut disease (*Ustilaginoidea virens*) of rice. Furthermore, Cui et al. (2016) reported that SiO_2-Ag composite nanoparticles showed antibacterial activity against *Xanthomonas oryzae pv. Oryzae*. Different concentrations of silver nanoparticles have an inhibition capacity against rice blast disease (*Magnaporthe grisea*) (Elamawi & El-shafey, 2013). Hence, nanomaterials or nanopesticides have potential as a pest control strategy for rice. This novel technique to rice pest control may prove to be a valuable tool for enhancing rice's productivity and sustainability.

11.3.5 Nano-technology Used for Abiotic Stress Management in Rice

By controlling the transcription factors involved in the chilling response, ZnO nanoparticles (30 nm) reduced the effects of chilling stress (Song et al., 2021). Researchers found that ZnO nanoparticles significantly reduced metal uptake in rice (Akhtar et al., 2021). In rice, ZnO arsenic (As) increased growth and phytochelatin concentration while reducing assimilation by seedlings (Yan et al., 2021). In rice

(*O. sativa* L.), root to shoot As transport was reported to be reduced by applying Fe_3O_4 nanoparticles (Huang et al., 2018). In roots and shoots of rice, CuO nanoparticles reduced total As by 23% and 45%, respectively (Wang et al., 2019). Zinc oxide nanoparticles (30–40 nm) applied to rice increased plant growth and reduced the toxicity of Cd (Zhang et al., 2019). Cui et al. (2020) have shown that SiO_2 nanoparticles (100 nm) inhibit As uptake into rice suspension cells by improving pectin synthesis in rice suspension cells. Utilizing TiO2 nanoparticles (36–140 nm), arsenic bioaccumulation and toxicity in rice seedlings were reduced by 40–90% (Wu et al., 2020). Jiang et al. (2021) showed that Au nanoparticles inhibit Cd uptake and alleviate Cd toxicity. It was reported that ZnO nanoparticles (20–30 nm) were added to rice to alleviate As toxicity and decrease As accumulation (Yan et al., 2021). Modulation of early growth and modification of biochemical and metabolic properties of rice was achieved by applying ZnO nanoparticles (20-40 nm) (Li et al., 2021). Rice was shown to be more tolerant to arsenic when Fe NPs up-regulate antioxidant enzymes and glutathione through the accumulation of phytochelatin and glutathione. Bidi et al. (2021) found that FeNPs reduced arsenic uptake and oxidative stress.

Application of iron nanoparticles in combination with hydrogel nanoparticles improved growth, biomass composition and nutrient intake, increased antioxidant enzymes and photosynthetic properties, and decreased the absorption and transfer of Cd in rice (Ahmed et al., 2021). SiO2 and TiO2 NP foliar sprays of 0, 5, 10, 15 and 20 mg L^{-1} reduced Cd uptake and protected the antioxidant system when plants were grown in Cd-contaminated soil (Rizwan et al., 2019). As CuO NP concentrations increased, the expression of PS-I proteins was slowed and PS-II was killed (Da Costa & Sharma, 2016). Melatonin in the form of 200 M Au NPs, a 40 nm nanoparticle, was used in a hydroponic system. Chlorophyll production that was hampered by Cd (which fell by 51.6%) was recovered (85.1%). Uptaking of Cd decreased by 33.0% and 46.2%, respectively (Jiang et al., 2021). ZnO NP via seed priming at 20 mg L^{-1} (arsenic-stressed) total chlorophyll concentration in 20–30 nm nanoparticles rose by 40.1%. The higher doses (100 and 200 mg L^{-1}) had no noticeable impact on the chlorophyll content (Wu et al., 2020). The foliar application of two nano-silica solutions reduces the amount of cadmium that accumulates in rice grains (Liu et al., 2009). SiO_2 NPs applied to the roots might also encourage root development, which was preferable for dealing with drought (Du et al., 2022).

11.4 MAJOR CHALLENGES IN RICE NANOTECHNOLOGY

Although nanotechnology provides many benefits, there are also some drawbacks. The major challenges are included below.

Lack of understanding: Rice nanotechnology is a relatively new field and there is limited knowledge about its potential applications and challenges.

Safety concerns: There are concerns about the safety of consuming rice products that have been modified using nanotechnology, as the potential effects of these materials on human health and the environment are not yet fully understood.

Regulatory barriers: The regulatory framework for nanotechnology-based products, including rice, is not yet well-established, which can pose a challenge to commercialization and wider adoption.

Technical challenges: There are technical challenges associated with the production of rice-based nanomaterials, including scaling up the process for commercial production and ensuring the stability and reproducibility of the nanomaterials.

Cost: The high cost of developing and producing nanotechnology-based products, including rice, can be a barrier to wider adoption and commercialization.

Public perception: There is a need to build public trust and address concerns about the safety and ethics of using nanotechnology in food products, including rice.

11.5 FUTURE PROSPECTS OF NANOTECHNOLOGY IN RICE

The prospects of nanotechnology in rice are promising and include the following.

Improved crop yields: Rice nanotechnology has the potential to improve crop yields by enhancing nutrient uptake and pest resistance, leading to better crop performance and higher yields.

Better food quality: Rice nanotechnology can boost the quality of rice, including increasing its nutritional value, extending its shelf life and improving its taste and texture.

Increased food security: A significant section of the world's population relies on rice as a staple diet, and nanotechnology has the potential to increase food security by improving crop yields and reducing food waste.

Enhanced sustainability: Rice nanotechnology has the power to enhance the sustainability of rice production by alleviating the use of fertilizers and pesticides, improving water efficiency and reducing soil degradation.

New applications: Rice nanotechnology has the potential to create new applications beyond food, such as the development of new materials and advanced biomedical applications.

Economic benefits: The development and commercialization of rice-based nanotechnology products have the potential to create new economic opportunities, including job creation and increased exports.

The future of nanotechnology in rice holds great promise and is expected to have a positive impact on the food and agriculture industries. However, it will require continued investment in research and development and careful consideration of potential risks and benefits.

11.6 CONCLUSION

Nanotechnology has the opportunity to modernize agriculture and also increase agricultural productivity. Nanoparticles are unique in that they are inexpensive, only need to be produced in small quantities and are simple to produce from a variety of sources. Nanoparticle based micronutrient formulations, nanopesticides for insect control, nanosensors for disease detection, nanocoated fertilizers and nanosensor base management help to increase rice grain yield production. Nanotechnology can provide an environmentally friendly, effective, and sustainable agricultural environment. The tools of contemporary nanotechnology can offer a bright prospect and can also help to create a sustainable agricultural industry. However, nanotechnology has opened up a world of possibilities for rice breeders. Nanoparticles can be utilized to transport beneficial genes to plants, allowing breeders to create new varieties with improved characteristics. Nanotechnology is a rapidly developing field of science that has been used to improve the breeding of rice for many years. Overall, nanotechnology has had a significant impact on the breeding of rice. By improving the absorption of nutrients, increasing the resistance of rice plants to pests and diseases and improving the processing of rice, nanotechnology has enabled the production of higher yields and improved quality. Therefore, the role of nanotechnology in rice has the potential to have a profound impact on food security and the global economy.

REFERENCES

Acharya, A., & Kumar Pal, P. (2020). Agriculture nanotechnology: Translating research outcome to field applications by influencing environmental sustainability. *NanoImpact*, *19*, 100232.

Adhikary, S., Biswas, B., Chakraborty, D., Timsina, J., Pal, S., Tarafdar, J. C., Banerjee, S., Hossain, A., & Roy, S. (2022). Seed priming with selenium and zinc nanoparticles modifies germination, growth, and yield of direct-seeded rice (*Oryza sativa* L.). *Scientific Reports*, *12*(1), 7103.

Ahmed, T., Noman, M., Manzoor, N., Shahid, M., Abdullah, M., & Ali, L. (2021). Nanoparticle-based amelioration of drought stress and cadmium toxicity in rice via triggering the stress responsive genetic mechanisms and nutrient acquisition. *Ecotoxicology and Environmental Safety*, *209*, 111829.

Akhtar, N., Khan, S., Shafiq Ur Rehman, Z. U. R., Khatoon, A., Rha, E. S., Jamil, M., & Jamil, M. (2021). Synergistic effects of zinc oxide nanoparticles and bacteria reduce heavy metals toxicity in rice (*Oryza sativa* L.) plant. *Toxics*, *9*(5), 113.

Aragon, B., Houborg, R., Tu, K., Fisher, J. B., & McCabe, M. (2018). CubeSats enable high spatiotemporal retrievals of crop-water use for precision agriculture. *Remote Sensing*, *10*(12), 1867.

Alam, A. M., Vemuri, H., Hossain, A., Abu Syed, M., Khorshed Alam, M., & Rafiqul Islam, M. (2020). Biofortification of iron, zinc, and selenium in rice for better quality. In A. Roychoudhury (Eds.), *Rice research for quality improvement: Genomics and genetic engineering*. Springer. https://doi.org/10.1007/978-981-15-5337-0_30

Badawy, S. A., Zayed, B. A., Bassiouni, S. M. A., Mahdi, A. H. A., Majrashi, A., Ali, E. F., & Seleiman, M. F. (2021). Influence of Nano silicon and Nano selenium on root characters, growth, ion selectivity, yield, and yield components of rice (*Oryza sativa* L.) under salinity conditions. *Plants*, *10*(8), 1657.

Bhargava, P., Kumar, A., Kumar, S., & Azad, C. S. (2018). Impact of fungicides and nanoparticles on Ustilaginoidea virens causing false smut disease of rice. *Journal of Pharmacognosy and Phytochemistry*, *7*(1), 1541–1544.

Bhattacharyya, A., Bhaumik, A., Rani, P. U., Mandal, S., & Epidi, T. T. (2010). Nanoparticles: A recent approach to insect pest control. *African Journal of Biotechnology*, *9*(24), 3489–3493.

Bidi, H., Fallah, H., Niknejad, Y., & Tari, D. B. (2021). Iron oxide nanoparticles alleviate arsenic phytotoxicity in rice by improving iron uptake, oxidative stress tolerance and diminishing arsenic accumulation. *Plant Physiology and Biochemistry*, *163*, 348–357.

Chaloupka, K., Karla, Malam, Y., & Seifalian, A. M. (2010). Nanosilver as a new generation of nanoproduct in biomedical applications. *Trends in Biotechnology*, *28*(11), 580–588.

Chen, K., Wang, Y., Zhang, R., Zhang, H., & Gao, C. (2019). CRISPR/Cas genome editing and precision plant breeding in agriculture. *Annual Review of Plant Biology*, *70*(1), 667–697. https://doi.org/10.1146/annurev-arplant-050718-100049

Chhipa, H. (2017). Nanofertilizers and nanopesticides for agriculture. *Environmental Chemistry Letters*, *15*(1), 15–22.

Cui, J., Li, Y., Jin, Q., & Li, F. (2020). Silica nanoparticles inhibit arsenic uptake into rice suspension cells via improving pectin synthesis and the mechanical force of the cell wall. *Environmental Science: Nano*, *7*(1), 162–171.

Cui, J., Liang, Y., Yang, D., & Liu, Y. (2016). Facile fabrication of rice husk based silicon dioxide nanospheres loaded with silver nanoparticles as a rice antibacterial agent. *Scientific Reports*, *6*(1), 21423. https://doi.org/10.1038/srep21423

Cunningham, F. J., Goh, N. S., Demirer, G. S., Matos, J. L., & Landry, M. P. (2018). Nanoparticle-mediated delivery towards advancing plant genetic engineering. *Trends in Biotechnology*, *36*(9), 882–897.

Da Costa, M. V. J., & Sharma, P. K. (2016). Effect of copper oxide nanoparticles on growth, morphology, photosynthesis, and antioxidant response in *Oryza sativa*. *Photosynthetica*, *54*(1), 110–119.

Debnath, N., Das, S., Seth, D., Chandra, R., Bhattacharya, S. Ch., & Goswami, A. (2011). Entomotoxic effect of silica nanoparticles against *Sitophilus oryzae* (L.). *Journal of Pest Science*, *84*(1), 99–105.

do Espirito Santo Pereira, A., Caixeta Oliveira, H., Fernandes Fraceto, L., Santaella, C., (2021). Nanotechnology Potential in seed priming for sustainable agriculture. *Nanomaterials*, *11*(2), 267. https://doi.org/10.3390/nano11020267.

Dong, O. X., Shu, Y., Jain, R., Zhang, N., Duong, P. Q., & Butler, C. (2020). Marker-free carotenoid-enriched rice generated through targeted gene insertion using CRISPR-Cas9. *Nature Communications*, *11*(1), 1178.

Du, J., Liu, B., Zhao, T., Xu, X., Lin, H., Ji, Y., Li, Y., Li, Z., Lu, C., & Li, P. (2022). Silica nanoparticles protect rice against biotic and abiotic stresses. *Journal of Nanobiotechnology*, *20*(1), 1–18.

Ekinci, M., Dursun, A., Yildirim, E., & Parlakova, F. (2014). Effects of nanotechnology liquid fertilizers on the plant growth and yield of cucumber (*Cucumis sativus* L.). *Acta Scientiarum Polonorum: Hortorum Cultus*, *13*(3), 135–141.

Elamawi, R., & El-Shafey, R. A. (2013). Inhibition effects of silver nanoparticles against rice Blast Disease Caused by Magnaporthe grisea. *Egyptian Journal of Agricultural Research*, *91*(4), 1271–1283.

El-Bendary, H. M., Abdel-Wahab, A. S., & El-Helaly, A. A. (2016). Entomo-toxic assay of nano-aluminum oxide and nano-zinc oxide against Sitophilus oryzae as a promising insecticide. *Egyptian Journal of Zoology*, *174*(4083), 1–8.

Fakharzadeh, S., Hafizi, M., Baghaei, M. A., Etesami, M., Khayamzadeh, M., Kalanaky, S., Akbari, M. E., & Nazaran, M. H. (2020). Using nanochelating technology for biofortification and yield increase in rice. *Scientific Reports*, *10*(1), 4351.

Fan, L., Wang, Y., Shao, X., Geng, Y., Wang, Z., Ma, Y., & Liu, J. (2012). Effects of combined nitrogen fertilizer and nano-carbon application on yield and nitrogen use of rice grown on saline-alkali soil. *Journal of Food, Agriculture and Environment*, *10*(1, Part 1), 558–562.

Fayos, I., Meunier, A. C., Vernet, A., Navarro-Sanz, S., Portefaix, M., Lartaud, M., Bastianelli, G., Périn, C., Nicolas, A., & Guiderdoni, E. (2020). Assessment of the roles of SPO11-2 and SPO11-4 in meiosis in rice using CRISPR/Cas9 mutagenesis. *Journal of Experimental Botany*, *71*(22), 7046–7058. https://doi.org/10.1093/jxb/eraa391

Fiaz, S., Khan, S. A., Anis, G. B., Gaballah, M. M., & Riaz, A. (2021). Chapter 11 - CRISPR/Cas techniques: A new method for RNA interference in cereals. In K. A. Abd-Elsalam & Ki-T. Lim (Eds.), *CRISPR and RNAi systems Nanobiotechnology for plant protection* (pp. 233–252). Elsevier. https://doi.org/10.1016/B978-0-12-821910-2.00032-1

Fincheira, P., Tortella, G., Duran, N., Seabra, A. B., & Rubilar, O. (2020). Current applications of nanotechnology to develop plant growth inducer agents as an innovation strategy. *Critical Reviews in Biotechnology*, *40*(1), 15–30. https://doi.org/10.1080/07388551.2019.1681931

Gao, Q., Li, G., Sun, H., Xu, M., Wang, H., & Ji, J. (2020). Targeted mutagenesis of the rice FW 2.2-like gene family using the CRISPR/Cas9 system reveals OsFWL4 as a regulator of tiller number and plant yield in rice. *International Journal of Molecular Sciences*, *21*(3), 809. https://doi.org/10.3390/ijms21030809

Goswami, A., Roy, I., Sengupta, S., & Debnath, Nitai (2010). Novel applications of solid and liquid formulations of nanoparticles against insect pests and pathogens. *Thin Solid Films, Biomolecular Electronics and Organic Nanotechnology for Environmental Preservation*, *519*(3), 1252–1257. https://doi.org/10.1016/j.tsf.2010.08.079

Hedley, C. (2015). The role of precision agriculture for improved nutrient management on farms. *Journal of the Science of Food and Agriculture*, *95*(1), 12–19. https://doi.org/10.1002/jsfa.6734

Herrera, J. M., Rubio, G., Häner, L. L., Delgado, J. A., Lucho-Constantino, C. A., Islas-Valdez, S., & Pellet, D. (2016). Emerging and established technologies to increase nitrogen use efficiency of cereals. *Agronomy*, *6*(2), 25. https://doi.org/10.3390/agronomy6020025

Hong, W. J., Woo-Jong, Y. J. K., Kim, E. J., Chandran, A. K. N., Moon, S., Gho, Y.-S., Yoou, M. H., Kim, S. T., & Jung, K.-H. (2020). CAFRI-rice: CRISPR applicable functional redundancy inspector to accelerate functional genomics in rice. *Plant Journal*, *104*(2), 532–545. https://doi.org/10.1111/tpj.14926

Huang, Q., Liu, Q., Lin, L., Li, F. J., Han, Y., & Song, Z. G. (2018, September). Reduction of arsenic toxicity in two rice cultivar seedlings by different nanoparticles. *Ecotoxicology and Environmental Safety*, *159*, 261–271. https://doi.org/10.1016/j.ecoenv.2018.05.008.

Hussien, M. M., El-Ashry, S. M., Haggag, W. M., & Mubarak, D. (2015, January). Response of mineral status to nano-fertilizer and moisture stress during different growth stages of cotton plants. *International Journal of ChemTech Research*, *8*, 643–650.

Jiang, M., Dai, S., Wang, B., Xie, Z., Li, J., Wang, L., Li, S., et al. (2021). Gold nanoparticles synthesized using melatonin suppress cadmium uptake and alleviate its toxicity in rice. *Environmental Science: Nano*, *8*(4), 1042–1056. https://doi.org/10.1039/D0EN01172J

Joshi, A., Sharma, L., Kaur, S., Dharamvir, K., Nayyar, H., & Verma, G. (2020). Plant nanobionic effect of multi-walled carbon nanotubes on growth, anatomy, yield and grain composition of rice. *BioNanoScience*, *10*(2), 430–445. https://doi.org/10.1007/s12668-020-00725-1

Kar, S., Nayak, R. N., Sahoo, N. R., Bakhara, C. K., Panda, M. K., Pal, U. S., & Bal, L. M. (2021, December). Rice weevil management through application of silica Nano particle and physico-chemical and cooking characterization of the treated rice. *Journal of Stored Products Research*, *94*, 101892. https://doi.org/10.1016/j.jspr.2021.101892

Kheyri, N., Norouzi, H. A., Mobasser, H. R., & Torabi, B. (2019). Effects of silicon and zinc nanoparticles on growth, yield, and biochemical characteristics of rice. *Agronomy Journal*, *111*(6), 3084–3090. https://doi.org/10.2134/agronj2019.04.0304

Kulkarni, S. K. (2015). *Nanotechnology: Principles and practices*. Springer International Publishing. https://doi.org/10.1007/978-3-319-09171-6

Li, Y., Yuzhan, Liang, Luxin, Li, W., Ashraf, U., Ma, L., & Tian, H. (2021). ZnO nanoparticle-based seed priming modulates early growth and enhances physio-biochemical and metabolic profiles of fragrant rice against cadmium toxicity. *Journal of Nanobiotechnology*, *19*(1), 75. https://doi.org/10.1186/s12951-021-00820-9

Liaghat, S., & Balasundram, S. (2010, January). A review: The role of remote sensing in precision agriculture. *American Journal of Agricultural and Biological Science*, *5*, 50–55. https://doi.org/10.3844/ajabssp.2010.50.55

Linaza, M. T., Posada, J., Bund, J., Eisert, P., Quartulli, M., Döllner, J., Pagani, A., et al. (2021). Data-driven artificial intelligence applications for sustainable precision agriculture. *Agronomy*, *11*(6), 1227. https://doi.org/10.3390/agronomy11061227

Liu, C., Li, F., Luo, C., Liu, X., Wang, S., Liu, T., & Li, X. (2009). Foliar application of two silica Sols reduced cadmium accumulation in rice grains. *Journal of Hazardous Materials*, *161*(2), 1466–1472. https://doi.org/10.1016/j.jhazmat.2008.04.116

Mahajan, S., Barthwal, S., Attri, M. K., Bajpai, S., Dabral, S., Khanuja, M., & Varma, A. (2019). Impact of ZnO nano materials on medicinal black rice seed germination. *Journal of Minerals and Materials Characterization and Engineering*, *7*(4), 180–192. https://doi.org/10.4236/jmmce.2019.74014

Mahakham, W., Sarmah, A. K., Maensiri, S., & Theerakulpisut, P. (2017). Nanopriming technology for enhancing germination and starch metabolism of aged rice seeds using phytosynthesized silver nanoparticles. *Scientific Reports*, *7*(1), 8263. https://doi.org/10.1038/s41598-017-08669-5

Manghwar, H., Lindsey, K., Zhang, X., & Jin, S. (2019). CRISPR/Cas system: Recent advances and future prospects for genome editing. *Trends in Plant Science*, *24*(12), 1102–1125. https://doi.org/10.1016/j.tplants.2019.09.006

Mani, P. K., Mandal, A., Biswas, S., Sarkar, B., Mitran, T., & Meena, R. S. (2021). Remote sensing and geographic information system: A tool for precision farming. In T. Mitran, R. S. Meena, & A. Chakraborty (Eds.), *Geospatial technologies for crops and soils* (pp. 49–111). Springer. https://doi.org/10.1007/978-981-15-6864-0_2

Mattos, B. D., Tardy, B. L., Magalhães, W. L. E., & Rojas, O. J. (2017, September). Controlled release for crop and wood protection: Recent progress toward sustainable and safe nanostructured biocidal systems. *Journal of Controlled Release*, *262*, 139–150. https://doi.org/10.1016/j.jconrel.2017.07.025

Mazhar, M. W., Ishtiaq, M., Hussain, I., Parveen, A., Bhatti, K. H., Azeem, M., Thind, S., et al. (2022). Seed nano-priming with zinc oxide nanoparticles in rice mitigates drought and enhances agronomic profile. *PLoS One*, *17*(3), e0264967. https://doi.org/10.1371/journal.pone.0264967

Meshram, A. T., Vanalkar, A. V., Kalambe, K. B., & Badar, A. M. (2022). Pesticide spraying robot for precision agriculture: A categorical literature review and future trends. *Journal of Field Robotics*, *39*(2), 153–171. https://doi.org/10.1002/rob.22043

Midde, S. K., Perumal, M. S., Murugan, G., Sudhagar, R., Mattepally, V. S., & Bada, M. R. (2021). Evaluation of Nano urea on growth and yield attributes of rice (*Oryza sativa* L.). *Chemical Science Review and Letters*, *11*(42), 211–214.

Miranda-Villagómez, E., Trejo-Téllez, L. I., Gómez-Merino, C., Sandoval-Villa, M., Sánchez-García, P., & Aguilar-Méndez, M. A. (2019, November). Nanophosphorus fertilizer stimulates growth and photosynthetic activity and improves p status in rice. *Journal of Nanomaterials*, e5368027. https://doi.org/10.1155/2019/5368027

Pandey, G. (2018, August). Challenges and future prospects of Agri-nanotechnology for sustainable agriculture in India. *Environmental Technology and Innovation*, *11*, 299–307. https://doi.org/10.1016/j.eti.2018.06.012

Rajendran, G., & Islam, R. (2017). *'Plant clinics in Bangladesh: Are farmers losing less and feeding more?' CABI impact case study series, No. 19.* https://www.cabdirect.org/cabdirect/abstract/20183309819

Rastogi, A., Tripathi, D. K., Yadav, S., Chauhan, D. K., Živčák, M., Ghorbanpour, M., El-Sheery, N. I., & Brestic, M. (2019). Application of silicon nanoparticles in agriculture. *3 Biotech*, *9*(3), 90. https://doi.org/10.1007/s13205-019-1626-7

Rizwan, M., Ali, S., Zia ur Rehman, M., Malik, S., Adrees, M., Qayyum, M. F., Alamri, S. A., Alyemeni, M. N., & Ahmad, P. (2019). Effect of foliar applications of silicon and titanium dioxide nanoparticles on growth, oxidative stress, and cadmium accumulation by rice (*Oryza sativa*). *Acta Physiologiae Plantarum*, *41*(3), 35. https://doi.org/10.1007/s11738-019-2828-7

Sandanayake, C. L. T., Weerakoon, S. R., Karthikeyan, N., & Somaratne, S. (2022). Yield performances of rice varieties (*Oryza sativa*) under Nano-CuO and Nano-ZnO micronutrient fertilizers. *Nusantara Bioscience*, *14*(1). https://doi.org/10.13057/nusbiosci/n140112

Sarong, M., Orge, R., Eugenio, P. J., & Monserate, J. (2020, February). Utilization of rice husks into biochar and nanosilica: For clean energy, soil fertility and green nanotechnology. *International Journal of Design and Nature and Ecodynamics*, *15*, 97–102. https://doi.org/10.18280/ijdne.150113

Satish Kumar, K. N., & Sudeep, C. S. (2007). Robots for precision agriculture. In *Electronic proceedings of 13th national conference on mechanisms and machines (NaCoMM07)* (pp. 12–13).

Sayed, A. M. M., Kim, S., & Behle, R. W. (2017). Characterisation of silver nanoparticles synthesised by *Bacillus thuringiensis* as a nanobiopesticide for insect pest control. *Biocontrol Science and Technology*, *27*(11), 1308–1326. https://doi.org/10.1080/09583157.2017.1397597

Sharma, P., Aggarwal, P., & Kaur, A. (2017). Biofortification: A new approach to eradicate hidden hunger. *Food Reviews International*, *33*(1), 1–21. https://doi.org/10.1080/87559129.2015.1137309

Singh, A., Tiwari, S., Pandey, J., Lata, Charu, & Singh, I. K. (2021, August). Role of nanoparticles in crop improvement and abiotic stress management. *Journal of Biotechnology*, *337*, 57–70. https://doi.org/10.1016/j.jbiotec.2021.06.022

Sinha, K., Ghosh, J., & Sil, P. C. (2017). 2 - New pesticides: A cutting-edge view of contributions from nanotechnology for the development of sustainable agricultural pest control. In A. Mihai Grumezescu (Ed.), *New pesticides and soil sensors* (pp. 47–79). Academic Press. https://doi.org/10.1016/B978-0-12-804299-1.00003-5

Sirisena, D. N., Dissanayake, D. M. N., Somaweera, K. A. T. N., Karunaratne, V., & Kottegoda, N. (2013). *Use of Nano-K Fertilizer as a Source of Potassium in Rice Cultivation. Annals of Sri Lanka Department of Agriculture*, *15*, 257–262.

Sishodia, R. P., Ray, R. L., & Singh, S. K. (2020). Applications of remote sensing in precision agriculture: A review. *Remote Sensing*, *12*(19), 3136. https://doi.org/10.3390/rs12193136

Song, Y., Jiang, M., Zhang, H., & Li, R. (2021). Zinc oxide nanoparticles alleviate chilling stress in rice (*Oryza sativa* L.) by regulating antioxidative system and chilling response transcription factors. *Molecules*, *26*(8), 2196. https://doi.org/10.3390/molecules26082196

Thirunavukkarasu, M., & Subramanian, K. S. (2014). Surface modified nano-zeolite used as carrier for slow release of sulphur. *Journal of Applied and Natural Science*, *6*(1), 19–26. https://doi.org/10.31018/jans.v6i1.369

Udoumoh, U. I., Ikrang, E. G., & Ehiomogue, P. O. (2021). Precision farming and fertilizer recommendation using Geographic Information System (GIS): A review. *International Journal of Agriculture and Earth Science*, *7*(2), 1–8.

Usman, M., Farooq, M., Wakeel, A., Nawaz, A., Cheema, S. A., Rehman, H. U., Ashraf, I., Sanaullah, M., & Sardar Alam Cheema, H. (2020, June). Nanotechnology in agriculture: Current status, challenges and future opportunities. In I. A. Rehman, M. Sanaullah et al. (Eds.) *Science of the total environment* (Vol. 721, p. 137778). https://doi.org/10.1016/j.scitotenv.2020.137778

Valojai, S., Taghi, S., Niknejad, Y., Amoli, H. F., & Tari, D. B. (2021). Response of rice yield and quality to nano-fertilizers in comparison with conventional fertilizers. *Journal of Plant Nutrition*, *44*(13), 1971–1981. https://doi.org/10.1080/01904167.2021.1884701

Wang, J. W., Grandio, E. G., Newkirk, G. M., Demirer, G. S., Butrus, S., Giraldo, J. P., & Landry, M. P. (2019). Nanoparticle-mediated genetic engineering of plants. *Molecular Plant*, *12*(8), 1037–1040. https://doi.org/10.1016/j.molp.2019.06.010

Wang, P., Lombi, E., Zhao, F.-J., & Kopittke, P. M. (2016). Nanotechnology: A new opportunity in plant sciences. *Trends in Plant Science*, *21*(8), 699–712. https://doi.org/10.1016/j.tplants.2016.04.005

Wu, F., Fang, Q., Yan, S., Pan, L., Tang, X., & Wenling, Y. (2020). Effects of zinc oxide nanoparticles on arsenic stress in rice (*Oryza sativa* L.): Germination, early growth, and arsenic uptake. *Environmental Science and Pollution Research*, *27*(21), 26974–26981. https://doi.org/10.1007/s11356-020-08965-0

Wu, M.-Y. (2013). *Effects of incorporation of nano-carbon into slow-released fertilizer on rice yield and nitrogen loss in surface water of paddy soil.* Third International Conference on Intelligent System Design and Engineering Applications (pp. 676–681). https://doi.org/10.1109/ISDEA.2012.161

Yan, S., Wu, F., Zhou, S., Yang, J., Tang, X., & Wenling, Y. (2021). Zinc oxide nanoparticles alleviate the arsenic toxicity and decrease the accumulation of arsenic in rice (*Oryza sativa* L.). *BMC Plant Biology*, *21*(1), 150. https://doi.org/10.1186/s12870-021-02929-3

Yan, X., Wang, M. G. H., Di Hao, T. X. M., & Yi, W. (2010). Effects of NMUrea on rice yield and agronomic nitrogen efficiency. *Plant Nutrition and Fertilizer Science*, *16*(6), 1479–1485.

Yang, C. (2020). Remote sensing and precision agriculture technologies for crop disease detection and management with a practical application example. *Engineering*, *6*(5), 528–532. https://doi.org/10.1016/j.eng.2019.10.015

Younas, A., Yousaf, Z., Riaz, N., Rashid, Madiha, Razzaq, Z., Tanveer, Maliha, & Huang, S. (2020). Role of nanotechnology for enhanced rice production. In R. S. Meena (Ed.), *Nutrient dynamics for sustainable crop production* (pp. 315–350). Springer. https://doi.org/10.1007/978-981-13-8660-2_11

Zahir, A. A., Bagavan, A., Kamaraj, C., Elango, G., & Rahuman, A. A. (2012). Efficacy of plant-mediated synthesized silver nanoparticles against *Sitophilus oryzae*. *Journal of Biopesticides*, *5*(Suppl.), 95–102.

Zhang, H., Demirer, G. S., Zhang, Honglu, Tianzheng, Y., Goh, N. S., Aditham, A. J., Cunningham, F. J., Fan, C., & Landry, M. P. (2019). DNA nanostructures coordinate gene silencing in mature plants. *Proceedings of the National Academy of Sciences*, *116*(15), 7543–7548. https://doi.org/10.1073/pnas.1818290116

Zhang, P., Guo, Z., Ullah, Sami, Melagraki, G., Afantitis, A., & Lynch, I. (2021). Nanotechnology and artificial intelligence to enable sustainable and precision agriculture. *Nature Plants*, *7*(7), 864–876. https://doi.org/10.1038/s41477-021-00946-6

Zhang, W., Long, J., Li, J., Zhang, M., Xiao, G., Xingyin, Y., Chang, W., & Zeng, H. (2019). Impact of ZnO nanoparticles on Cd toxicity and bioaccumulation in rice (*Oryza sativa* L.). *Environmental Science and Pollution Research*, *26*(22), 23119–23128. https://doi.org/10.1007/s11356-019-05551-x

12 Prospects and Challenges of Nanofilms-based Edible Food Coatings for Enhancement of Their Shelf Life

*Richa Prasad Mahato, Pallavi Singh and Suchi Srivastava**

12.1 INTRODUCTION

The Greek term "nano", which means dwarf, is the origin of the name "nanotechnology". The Nobel Prize winner Richard Feynman famously said, "There is plenty of room at the bottom", which has a significant impact on the development of nanotechnology (Feynman, 2018). Nanotechnology is the study of science, technology and engineering at the nanoscale (about 1 to 100 nm) (Satalkar et al., 2016; Servin & White, 2016; Di Sia, 2017). In fact, because of their high surface to volume ratio, nanoparticles exhibit fascinatingly unique characteristics from those of larger particles (Di Sia, 2017; Singh & Prasad, 2017; Khan et al., 2018; Mahato et al., 2023). In many fields of science, notably chemistry, physics, health, material science, aeronautics, pharmaceuticals, agriculture, food and now even horticulture, nanotechnology has been thriving as an efficient instrument (Duhan et al., 2017). Horticulture is a discipline of agricultural science and art that focusses on growing and managing beautiful plants, fruits and vegetables (Rana et al., 2021). For the development of substitutive edible film materials for the post-harvest preservation of fresh horticultural goods, it would be a ground-breaking technique by shrinking the edible film particles' size to the nanometre (nm) range (Parisi et al., 2015). This nanotechnology-based edible film is the most effective method for increasing the storage period of fresh produce as the nanoparticles utilized in this edible film exhibits antibacterial and antifungal features (Bajpai et al., 2018; Singh et al., 2020).

Wax, plastic, paper, glass, metal and composites are traditional food packaging materials that have been employed in the past. Unfortunately, there are significant downsides to these conventional preservation methods, including short shelf lives, expensive manufacture, undesired waste, etc. Wax coating is the most common technique for preserving

* Corresponding author.

 DOI: 10.1201/9781003376446-12

fruits and vegetables. Wax is combined with compounds known to cause cancer for this reason (Singh & Packirisamy, 2022). Plastics employed in food packaging like polyethylene, polypropylene and polyethylene terephthalate are not environmentally safe (Sundqvist-Andberg & Akerman, 2021). A developing field in the food industry is the creation of innovative packaging substances. Modern food packaging includes edible traits, as it is obtained through plants, animals, or aquatic life or even by naturally occurring food-grade polymers like polysaccharides, proteins or lipids (Mohamed et al., 2020; De Oliveira Júnior et al., 2021; Oloye et al., 2021; Kurt & Cekmecelioglu, 2021).

Biopolymers have many benefits over traditional coating techniques, like biodegradability, recycling and sustainability but they have a few disadvantages because of their inferior mechanical and barrier qualities (Mahato et al., 2023; Singh et al., 2021a). The mechanical, flexible and gas barrier qualities of packaging materials can all be enhanced by additives (Kumar et al., 2021). Therefore, extending the storage period of horticulture products, notably fruits and vegetables by increasing the stability and quality of these consumables is a challenge for nanotechnology (Song et al., 2016; Li et al., 2017).

Earlier this decade, research was conducted on the influence of adding nanoparticles (NPs) to edible films. These findings have shown that adding NPs to edible films has created new possibilities for enhancing their antimicrobial, antifungal and antiviral properties as well as their cost-effectiveness (Sorrentino et al., 2007). Fruits with a brief lifespan are coated with organic and synthetic coating containing NPs such chitosan, Si, TiO_2 and their derivative blends. As an example, these nanomaterials lengthen the storability of strawberries (Li et al., 2017), loquats (Nasrollahzadeh et al., 2020) and Chinese bayberries (Wang et al., 2010) by modifying their physical and antibacterial characteristics. A few noteworthy NPs and their possible benefits for extending the storage period of agricultural products was reported by Rana et al. (2021). Antimicrobial coatings can be applied to surfaces utilizing nanotechnology to prevent or stop the growth of biofilms (Bajpai et al., 2018; Nile et al., 2020). The environment of the food is perfect for the survival of many microbes, particularly dangerous bacteria. Bacteria frequently cling to surfaces and create biofilms (Singh et al., 2022). This biofilm is made up of a complicated group of microbes that stick together to produce a biofilm that is difficult to remove off surfaces, including surfaces that come into contact with food. Biofilms are typically difficult to eliminate since they are robust to the majority of disinfectants. As a result, this biofilm turns into a reservoir of tainted food products that can cause foodborne illnesses (Myszka & Czaczyk, 2011). In this context, nano-engineered surfaces with antimicrobial coatings were discovered as one of the most effective antimicrobials towards biofilms, thereby enhancing both the safety and quality of food items in addition to protection for exposed surfaces (Bajpai et al., 2018). To achieve this, the food sector uses nanocoatings like nanoscale silver, TiO_2 and ZnO as disinfectants for surfaces (Bajpai et al., 2018). Furthermore, it has been discovered that UV-C ultraviolet light stimulated TiO_2 is efficient for reducing biocontamination problems in poultry activities, food processing and transportation (Khan et al., 2015).

This chapter emphasizes the opportunities and difficulties associated with the use of edible films coated with nanoparticles to prolong post-harvest storage and maintenance of the quality of post-harvest food stuffs. Antimicrobial nanomaterial treatments have the capability to offer a solution for surface contamination problems in the food business by limiting microbial adhesion and the development of biofilms. Unfortunately, there are still issues with the efficiency of edible coatings made of nanoparticles when applied to post-harvest products.

12.2 PROSPECT AND CHALLENGES OF NANOFILM-BASED EDIBLE FOOD COATINGS

Traditional plastic-derived food coating materials have a number of drawbacks when it comes to their effect on the environment and non-renewable resources. Additionally, since the majority of food packaging is made of one-time use plastic, customers are more conscious of pollution control. Because of the negative effects of conventional food packaging, a substitute packaging that produces little wastage has been developed by using organic resources. Edible coating is a useful strategy for fulfilling all conditions of this alternative packaging. The creation of edible packaging, like films and coatings, is presently the focus of considerable research and advancement.

The quality and prolonging of the storage period of horticulture fruits, vegetables and cut flowers are promoted by nanotechnology. Their strength and techniques of dispersion along with their antibiosis properties are used in nanotechnology to increase the yield of horticulture crops. These nanomaterials are easy to make, inexpensive and environmentally safe. Bio-nanocomposite coatings using PLA and $AgNO_3$ having strong antifungal and antibacterial activities with a higher silver content demonstrated absence of *Salmonella enterica* in lettuce after storage in refrigerator (Martinez-Abad et al., 2013).

Additionally, edible film containing nanochitosan and copper administered on strawberries enhance the shelf life over a period of 20 days. This demonstrated three week postponement in respiration along with delayed rate of decline in antioxidant properties and lag in weight reduction process. Unlike uncoated fruits, these coatings dramatically reduce the rate of fungal degradation of fruits throughout preservation. The sensory qualities of coated strawberries demonstrated increased customer acceptance (Kumar et al., 2021).

According to Bodaghi et al. (2013), thin coatings of TiO_2 nanocomposite inhibited *Rhodotorula mucilaginosa* and *Pseudomonas spp.* function after storage at 5°C for 17 days of pears. There were considerably less mesophilic bacterial and yeast cells in contrast to samples preserved in the control coatings. Silver montmorillonite (Ag-MMT) NPs were produced with a calcium-alginate film to extend the storage period of freshly cut carrots over two months with reduced bacterial loads (Costa et al., 2012; Jafarzadeh & Jafari 2021,).

In packaged foods, the word "nanoparticle" describes a broad range of materials, such as carbon materials (Hermawan et al., 2019; Dwivany et al., 2020), metals (Cakmak et al., 2018; Pirozzi et al., 2020), metal oxides (Hermawan et al., 2019; Dwivany et al., 2020), mixed metal oxides (Biswas et al., 2019) and nanolayers (Dairi et al., 2019; Sobhan et al., 2020). The complex patterns of nanomaterials with food make it difficult to identify them in packaged foods (if they are discharged or mix with nutritional quality) (Jafarzadeh et al., 2017; Chong et al., 2019). Henceforth, though nanotechnology was already applied for food packaging, it hasn't been extensively adopted mostly because of worries about its impact on the environment and public healthcare, especially during the preservation phases of food.

The ultimate qualities of the nanocomposites are determined by the type of the polymeric components (organic phase) and NPs (inorganic phase) (Swaroop & Shukla, 2019). Inorganic nanoparticles including gold, silver, copper and zinc in nanocomposite innovations have attracted a lot of attention from both academics and business. Because of their distinctive physical, chemical, optical, antibacterial and biological activities, NPs, with metals and metal oxides, are frequently utilized in the packing and coating of fruits and vegetables. NPs have opened up numerous possibilities for improving bio/polymer features such as oxygen scavenging, thermal conductivity and mechanical qualities.

However, because of their nanoscale's prospect for movement into packaged foods during processing procedures and preservation period, public worries regarding its influence on human health has grown rapidly in the food industry (Berekaa, 2015; Gallocchio et al., 2015; Chaudhry et al., 2017). Mohammad et al. (2022) reported that NPs which migrate at sizes of 70 nm, 50 nm and 30 nm can cross through the blood-brain barrier as well as the lungs and cells, respectively. Because of their increased rate of absorption and the ease with which they can move throughout the organs, small-sized NPs pose a larger threat to human health over larger units. Echegoyen and Nern (2013) observed that packaging materials, containing silver nanoparticles, may contaminate human food products, although nothing is known regarding their toxicity. However, they are known to alter DNA, which can lead to genotoxicity. Moreover, titanium dioxide treatment and dumping may have an impact on people and the environment, hence raising concerns for both human health and ecosystem (Yang et al., 2010). They can reach human organs and create damage due to their microscopic size. Therefore, during treatment of nanoparticles to edible packaging, the size and quantity are highly important (Chawla et al., 2021; Kumar et al., 2021).

Technologies based on nanotechnology are chemical-dependent structures, and like any chemical-dependent systems, they may or may not have toxins or have a negative impact on human health (Eleftheriadou et al., 2017). As a result, creating a regulatory framework that aims to manage and regulate possible NP toxicity and hazards is essential for guaranteeing the security in the field of nanotechnology-based food sector. Investigation of the possible negative impacts of NPs on human health and the environment is in need of more research (Mohammad et al., 2022).

12.3 THE FUNDAMENTAL CHARACTERISTICS OF BIO-NANOCOMPOSITE FILMS AND COATINGS FOR FRUITS AND VEGETABLES

12.3.1 Barrier/Scavenging Characteristics

12.3.1.1 Barrier to Moisture

Edible coatings are recommended as a feasible substitute for maintaining fresh fruits and vegetables because of their capacity to avoid dehydration, scent disappearance, water uptake by the food component or oxygen penetration. Bio-nanocomposite coatings can further minimize weight reduction because of their tendency to prevent moisture loss. Additionally, by establishing semi-permeable barriers to gases and moisture, like CO_2 and O_2, edible films improve the texture, raise the morphology of the product and lengthen the storage period (Duran et al., 2016; Jafarzadeh et al., 2018).

Emamifar and Bavaisi (2020) created a bio-nanocomposite film using sodium alginate and nano-ZnO for strawberries. Research findings showed that nano-ZnO considerably enhanced the moisture barrier of films, which in turn decreased strawberry weight reduction. Mangoes packed in nanopackaging using PLA films infused with essential oils and nano-TiO_2/nano-Ag showed the least weight reduction after 15 days of storage at room temperature, probably due to their water barrier properties (Chi et al., 2019). According to Zhang et al. (2018), the PLA sheets coated with Ag NPs demonstrated an exceptional moisture barrier capability resulting in less fruit loss as compared to control.

12.3.1.2 Scavengers of Oxygen

Some packaged fruits and vegetables may suffer severe negative impact due to the presence of oxygen. This could result in a variety of degrading processes, including nutrient loss, fading in colour and the development of bad flavours (Lee et al., 2015). Additionally, it exerts significant influence on the rate of respiration and ethylene formation in fruits and vegetables. Several studies have focussed on excluding or removing oxygen because it can slow down the deterioration and increase the storage period of particular fruits and vegetables (Pereira de Abreu et al., 2012; Lee et al., 2015; Bodbodak & Rafiee 2016). Active packaging with edible films and coatings that incorporate O_2 scavengers as well as humidity regulated set ups are crucial for minimizing gases that cause fruits and vegetables to decay.

Furthermore, it has been found that the production of polymeric films incorporated by the nanoparticles (NPs) to boost gas barrier qualities could decrease the rate of respiration, transpiration, and other metabolic activities in fruit (Kaur et al., 2017). According to Resende et al. (2018), the chitosan/cellulose nanofibril (CNF) coating reduces O_2 transport, lowers respiration and slows the ascorbic acid reaction that causes strawberries to oxidize. Since films exhibits a higher CNF content demonstrated a greater barrier to O_2 and this property of edible films were advantageous for the post-harvest quality retention of strawberry. According to Li et al. (2017), the elevation in nano-ZnO caused a decrease in the diffusion coefficient in PLA sheets. Oxidative stress in the texture of the fruit was unavoidably lowered by the decreased oxygen distributed through its surface. It resulted in the removal of reactive oxygen and preservation of the cell membrane integrity of freshly sliced apples. In conclusion, oxygen scavengers are primarily responsible for providing proper protection to the fruits and vegetables. Utilization of oxygen scavenging products in packaging can prevent the aerobic bacterial growth, colour loss, mould growth, and oxidative hazards to fruits and vegetables.

12.3.1.3 Ethylene Scavengers

Ethylene regulation during preservation is crucial for increasing the lifespan of harvested fruits and vegetables. Since ethylene accelerates ripening, causes chlorophylls to loosen and degrade, and finally decline the quality of fruit and vegetables (Kaewklin et al., 2018). Therefore, the protection of fruits and vegetables that are susceptible to ethylene, such as bananas, apples, mangoes, tomatoes, onions and carrots, is made possible by ethylene scavengers (Prasad & Kochhar, 2014; Lee et al., 2015; Siripatrawan & Kaewklin, 2018). By coating the fruit with a semi-permeable substance, it is feasible to postpone fruit maturation by regulating the amount of ethylene in the fruit. The film may be a suitable substitute for the administration of chemical methods and disinfectants because it can minimize the quantity of oxygen in breathing and restrict the distribution of carbon dioxide from outside tissue (Meindrawan et al., 2018).

Chitosan-TiO_2 nanocomposites were found to have ethylene prevention action on tomatoes as reported by Kaewklin et al. (2018). On the basis of the findings by de Chiara et al. (2015), tomatoes coated in bio-nanocomposite packaging showed a much reduced ethylene content throughout preservation in contrast to those in chitosan and control films. They claimed that a mixture of TiO_2 and silica would photocatalytically remove ethylene, which prevented mature green tomatoes from maturing. According to new findings and studies, ethylene scavengers using NPs in active packaging could postpone ethylene breakdown and maintain the quality of produce throughout preservation. Additionally, it aids in enhancing the product's quality and longevity.

12.3.2 Antimicrobial Properties

Due to their large specific surface area, antibacterial qualities and matrix strengthening activities, NPs are used to reduce fruit and vegetable spoiling and manage microbial development. According to Emamifar and Bavaisi (2020), strawberries treated with 1.5% sodium alginate and 1.25 g/L nano-ZnO showed the least amount of microbial development. Fresh fruit can be kept for approximately 20 days longer due to the coating with the formulation. The effects of the chitosan/nano-TiO_2 composite film on mangoes were investigated by Xing et al. (2020), which stated that covering was essential for preserving the fruit's integrity at 13°C by inhibiting microbial development and retain the original nutrient content of mangoes.

12.4 EXECUTION OF BIO-NANOCOMPOSITE FILMS/COATINGS ON FRUITS AND VEGETABLES

Various nanomaterials are used in the food sector for different purposes. This section gives a brief explanation of each type and the potential application of each one in the food industry (Figure 12.1).

12.4.1 Bio-nanocomposite Coatings for Tropical Fruits

The customized environment packaging, wax coating and bio-nanocomposites have all been proposed by researchers as ways of preserving the integrity of different tropical fruits (Artés et al., 2006; Mangaraj & Goswami, 2009). Among them,

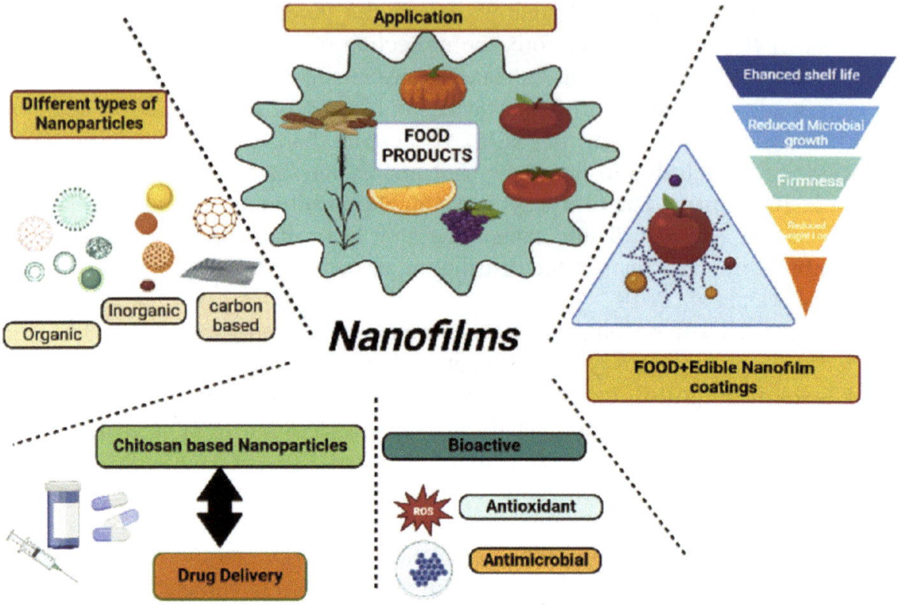

FIGURE 12.1 Current use of nanotechnology in the food preservation area

bio-nanocomposite coatings or edible films are broadly acknowledged for meeting the purpose of protecting post-harvest values. Application of edible coatings is easy, secure, inexpensive and environmentally benign, making it a viable method for protecting tropical fruits. Li et al. (2017) looked into the effects of the SPI/cinnamaldehyde/ZnONP coating on the physical, chemical and microbiological durability of bananas and demonstrated that the bio-nanocomposite sheets may significantly slow down the rate of weight loss and ripening.

Shi et al. (2013) studied nanosilica and chitosan-based semi-permeable films which significantly improved the storage period, browning index and weight retention of *Dimocarpus longan* fruit with reduced peroxidase activity.

12.4.2 Bio-nanocomposite Films/Coatings for Climacteric Fruits

Maturation of climacteric fruits, which reach their peak physiological development after being cut at a particular stage depends on the production trends of CO_2 and ethylene. These fruits *viz.* peaches, musk-melon, kiwi, banana, plum and pears keep developing flavour and get sweeter through transformation of starch into sugar.

They manufacture ethylene gas on their own to quicken the ripening phase, but they are extremely sensitive to it. Therefore, precautionary methods to control production of CO_2 and ethylene, aesthetic quality and biological safety of climacteric fruits during the preservation period (Chen et al., 2018) are crucial to employ. In order to extend the lifespan of freshly cut apples throughout cold storage, Naknaen (2014) produced biodegradable packaging with various ZnO NP levels with a storage period of 9–18 days. Zinc oxide and silver nanoparticles incorporated in chitosan-rice films were also found to increase the peaches' shelf life. Peaches with these nanocomposite coatings extended the expiration date and reduced overall surface bacteria unlike those packed with NPs (Kaur et al., 2017). For the kiwi fruit, Jamróz et al. (2019) created a biopolymer coating that incorporates Se-Ag NPs. Se-Ag NPs mixed with the FUR/GEL matrix produced a bio-nanocomposite with entirely unique characteristics that could be helpful in extending the storage period of kiwi.

12.4.3 Bio-nanocomposite Films/Coatings for Non-climacteric Fruits

Non-climacteric fruits do not exhibit a rise in respiration rate and ethylene production. Cherry, grape, orange, raspberry, cucumber, pomegranate and watermelon are examples of non-climacteric fruits that need to complete their physiological maturity on the tree. Once harvested, these fruits will stop maturing, ripening and developing sugar or flavour. Due to their nutritional value, unique sensory characteristics and bioactive components, such fruits are extremely desired to prolong their life span with edible nanocomposite coatings (Chen et al., 2018). Ortiz-Duarte et al. (2018) investigated the impact of Ag-chitosan nanocomposites on the fresh melon pieces' quality during a 13 day period of refrigeration. This nanocomposite film considerably reduced the rate of respiration of coated objects and prevented softening and received higher sensory ratings. Pomegranate coated with nano-ZnO/CMC by Saba and Amini (2017) showed reduced weight loss, total yeast and mould growth with higher vitamin C and antioxidants after 12 day storage. Additionally, Kanikireddy et al. (2020) looked at the potential of Na-alginate gum and silver-acacia nanocomposite films to protect black grapes from microbial development with increased preservation period. The grapes coated in nanocomposite films displayed less weight loss over the course of 18 days, and the nanofilms increased the ability of grapes to be stored at room temperature. Tavakoli et al. (2019) evaluated the effect of Ag NPs on the storage period of citrus limon and found that it greatly increased vitamin C content, fruit weight and reduce deterioration percentage and electrolyte leakage. Kumar et al. (2019) reported, bio-nanocomposite coatings made of agar and ZnO NPs to package the green grapes for their fresh quality up to 21 days.

12.4.4 Bio-nanocomposite Films for Vegetables

The increased urge for garden-fresh vegetables has compelled makers to maintain their freshness and avoiding damage. Recent innovations like active packaging have been introduced to the market, which interacts with food and prolong product's storage period or give it specific features (Kaewklin et al., 2018). Saeknow et al. (2019) looked at how CMC+ZnO nanoparticles improved the quality of tomato fruits. The study reported improved insistence, respiration rate, antioxidant chemicals and reduced weight reduction in tomatoes in comparison to control throughout storage. Aside from that, the bio-nanocomposite covering successfully slowed the disease's progression.

Issa (2017) investigated the physical, chemical, sensory and antibacterial qualities of small spinach leaves stored in eco-friendly packaging with nanoclay and sweet potato starch throughout refrigeration. Rezaee et al. (2018) studied the quality parameters of edible mushrooms packaged in PLA packaging stored for a long-time incorporating clay-nanoparticles. A high amount of nanoparticles in the wrapping extended the expiration date of mushrooms with increased moisture content, freshness, colour, quality and shape. Kaewklin et al. (2018) described the vigorous packaging with chitosan and TiO_2 nanoparticles on tomatoes under 20°C with a postponement in maturing process. Additionally, ZnO NPs were added to olive flounder bone gelatin (OBG) nanofilms and coated to spinach in the form of antibacterial wrapping (Beak et al., 2017). The findings revealed that microbial growth on packed spinach has decreased despite impacting overall quality (Beak et al., 2017).

12.5 TYPES OF NANOTECHNOLOGY-BASED FOOD PACKAGING

The right choice of packing techniques and technology has a big impact on the value and originality of the items during the transport and packing phases. Fruits and vegetables are frequently packaged to prevent them against mechanical pressure during distribution, O_2, moisture, water vapour, light, dust and pathogens (Jafarzadeh et al., 2017). Thus, these are the critical factors for the components that should be employed for the packing of fruits and vegetables. Different systems have been created for this reason, as detailed below.

12.5.1 Active Packaging Systems

Some of the examples of active packaging include the use of actively working polymers or the incorporation of active substances such as metal nanoparticles, antibiotics, enzymes, essential oils, antioxidants, volatile compounds, fungicides and fatty acids in food packaging. Active packaging techniques uphold and improve the quality and security of fruits and vegetables using absorbers, releasing mechanisms and other techniques.

Fresh fruits and vegetables should be properly packaged with materials that may absorb ethylene for increasing the storage quality of the foods. Duran et al. (2016) created an active packaging using nisin, chitosan, natamycin and seed extract of fruits (grapes and pomegranate) for maintaining the quality and storage period of fruits. Unlike untreated fruits, all covering materials preserved the fresh strawberry's quality and extended its life span. Cherries and strawberries were packed in an active packaging by Chiabrando et al. (2019), and the findings showed that prepared packing lowered weight loss, maintain colour, acidity and vitamin C content, and improved sensory attributes as well as delayed post-harvest senescence. Several studies have shown that active packaging is primarily alarmed for maintaining the nutrient status by addressing different chemical reactions, physiological functions and microbiological factors (Gómez-Estaca et al., 2014; Jafarzadeh et al., 2020).

12.5.2 Smart Packaging Systems

In general, smart packaging can be described as a packaging method that makes it easier to decide whether to enhance storage period, improve safety, promote quality and alerts concern about potential issues by using scientific logic (Jafarzadeh et al., 2020).

Different sensors were built inside as intelligent packaging. Enzymatic, mechanical, chemical or immunochemical processes are used to power these sensors. In addition to identifying particular foodborne pathogens, it can provide details about microbial growth, CO_2 and O_2 levels, spoiling and product freshness (Ahmad et al., 2016). Intelligent packaging tools can provide tamper evidence, good quality and more.

An intelligent package (pH sensor) made of polyvinyl alcohol and a natural colour from red cabbage extract demonstrated well that the smart packaging can be utilized as a pH biosensor (pH2 to pH12) in fresh date fruit packaging (Maftoonazad & Ramaswamy, 2019).

An intelligent packaging system created by Dirpan et al. (2018) assesses the freshness of mangoes and inform consumers about the fruit's quality. Bananas turning overripe shifted the colour of the indicator from blue to green. The pH of the mango inside the package is reflected by the colour variations of the indicators.

12.5.3 Chitosan Nanoparticle/Films in Food Packaging

Higher molecular weight polycationic heteropolysaccharide chitosan has a reputation for being biodegradable, biocompatible and capable of complexing with metals. Its widespread antibacterial activity is mostly caused by its polycationic composition (Arora et al., 2016). There are several uses for chitosan mostly in food sciences, biological and chemical disciplines and also in the pharmaceutical and food industries.

Research has been conducted on the use of chitosan nanoparticles and chitosan-based food coatings in past few years. Shiekh et al. (2013) have found many useful and advantageous uses for chitosan in these areas. Since chitosan, having positive charge, get attached to the negatively charged cell wall of bacteria, adherence to on the surface resulted in the increased permeability and disruption of the cell wall. Further Ahmed and Aljaeid (2016) emphasizes electrostatic interaction of amino groups with the polyanions to create chitosan nanoparticles by the process of ionic gelation. In accordance to that, Tripathi et al. (2009) presented an antibacterial coating made of polyvinyl alcohol and chitosan nanoparticles. De Moura et al. (2009) observed that the addition of a chitosan-derived nanocomposite improved the barrier attributes of hydroxypropyl methylcellulose. Additionally, the antibacterial properties of the nanocomposite gold-chitosan and silver-chitosan against microbes *viz; Escherichia coli, Aspergillus niger, Staphylococcus aureus* has also been described in other reports (Youssef et al., 2014).

Chitosan-derived coatings offer a superior alternative for extending the storage period of the food without degrading its sensory qualities (No et al., 2007). Superior coating with chitosan probably associated with antifungal, antibacterial and antioxidant properties that could give them a viable substitute for synthetic chemicals used in food packaging (Yuan et al., 2016). Numerous works have documented the application of chitosan-based byproducts in the storage, clarity and encapsulation of products obtained from dairy and beverages (alcoholic and non-alcoholic beverages) (Rocha et al., 2017). Additionally, few studies have experimented with combining other nanomaterials (nanocomposites) for use in food packaging and storage associated good thermal and barrier qualities.

12.5.4 Nanocomposites

Nanocomposites and nanolaminates, in addition to nanoparticles, are extensively used in the packing of food products. They have been widely implemented in the antimicrobial packaging of food products in recent times. Nanocomposites are primarily employed to prolong the expiry date of foods and prevent them from mechanical and high temperature stunning (Kim et al., 2020). For instance, the food sector has paid close attention to the zinc oxide nanocomposite because it possesses antioxidant properties and is frequently utilized in the active packaging of food materials (Kim et al., 2020). Usually, nanocomposites are made of polymers and nanoparticles, and the nanoparticles' existence enhances the polymer's characteristics (Yotova et al., 2013). By offering extremely flexible and chemical activity, nanocomposites often assist in the preservation of food products and delaying infectious decomposition for an extended duration. These are therefore largely employed in the creation of coating materials with high barrier characteristics (Pandey et al., 2013).

Nanocomposites serve as a barrier to stop carbon dioxide seepage from bottles and cans of carbonated beverages. The nanocomposite layer can be utilized in the form of bottles rather than glass to reduce the gas seepage (Pandey et al., 2013). Nanoclay nanocomposites are employed to create these gas barriers. There are many commercialized nanocomposite-derived solutions on the market today that aid in the storage, packaging and extension of the shelf life of fresh produce. Few of the commercialized goods as nanoclay derivatives are Guard IN Fresh, Top Screen DS13, and nanoceram PAC (Burdo, 2005).

12.6 APPROACHES TO PREPARE EDIBLE FILMS AND COATINGS

Films and coatings are the two approaches to pack edible food materials. Biodegradable substances like proteins, lipids or polysaccharides are used to create edible coatings (Cheng et al., 2021). Various techniques for preparation of coatings are covered in this part.

12.6.1 Edible Film Fabrication Methods

Food is packed in edible coatings, which are the main type of edible packaging. The ability of additives to dissolve in biopolymers is an essential characteristic for their function. Cohesive forces within biopolymers impact the entire mechanical characteristics. Wet and dry are the two primary procedures that can be used to create edible coatings such as casting and extrusion respectively (Cerqueira et al., 2017).

12.6.1.1 Solution Casting Method

A popular and affordable technique for making films is casting. There are three stages to this technique. Firstly, there is solubilization where the biopolymer substance is dissolved in an appropriate solvent. The solvent must be edible and non-toxic due to the fact that the film is edible by character. Water and ethyl alcohol are frequently used as solvents. The second stage includes the preparation of mould using a casting solution. The last stage involves the drying of the solution. During the drying of the cast film, the polysaccharides create a glutinous-like formation, transforming the film into a finished product. Drying is important because it has a more significant impact on the biopolymer's intramolecular bonding and physical properties (Suhag et al., 2020). In the case of the wet method, a homogeneous film will be produced with stronger biopolymer molecule bonding. There is zero chance of molecular damage because of utilization of low temperature in the production method. This technique can't be applied at commercial levels because it takes more time to dry.

12.6.1.2 Hot Melt Extrusion Method

Producing edible films by the extrusion method can boost output on a commercial scale. Since little to no solvent is used, the extrusion technique is a dry approach. There is no need of solvent evaporation during drying period. The extruder is first fed with biopolymer and additives in the feeding followed by the kneading zone in which the constituents are correctly blended with the help of a screw. The last one is a heating region, in which a tiny quantity of heat is passed using the oven. Here, the polymers combine with the additives. The form and width of the extruded film are determined by a dye that is stabilized at the extruder's end. The biopolymer's structure is transformed by high temperature, as there is the degradation of temperature-sensitive components caused by high temperature (Suhag et al., 2020; Cheng et al., 2021). Important factors include the extruder screw's speed, the heat intensity, the duration of the heating and the solvent level, if any. These have a significant impact on the mechanical and optical characteristics of the film. The multi-layered film can be created with better qualities by the use of more than one extruder, also referred as co-extrusion. The drawback of this process is that only biopolymers that can withstand heat or high temperatures can be employed as the temperature zone will enable heat-susceptible material to deteriorate (Suhag et al., 2020).

12.7 APPLICATION OF EDIBLE COATING APPROACH

The edible coating is used as the main type of edible packaging on food products *viz*; fruits, vegetables, dairy products etc. The four main coating methods are dipping, spraying, fluidized beds and panning. The external layer characteristics of the food product and the future use of the covering layer are just two of the many variables that influence the choice of coating techniques. The components must first disperse across the external layer or surface of the food prior to the adherence between the covering material and the external layer of the food (Cerqueira et al., 2017). The primary covering methods are briefly discussed in this part.

12.7.1 Dipping Method

For fruits and vegetables, the dipping technique is most frequently applied. There are three phases to this process. The first phase entails totally drenching the food product in the coating/covering-forming liquid. The covering material is then utilized on the surface of edible items. The final phase includes the removal of the solvent from the covering, producing a solution and leaving a shrill coating on the surface (Guerreiro et al., 2015). Solvent evaporation can take place at room temperature or with the aid of heat. Fruits that have just been cut are soaked in an antibacterial solution made of aqueous coating (Suhag et al., 2020). The frozen salmon fish is coated with chitosan using the dipping technique. This coating prolongs the storage period of fish by preventing the growth of harmful germs (Soares et al., 2016). On the freshly cut mangoes, malic acid dipping and alginate coatings used to be done. This coating makes coated food products firmer and prevents quality loss during preservation (Salinas-Roca et al., 2016).

12.7.2 Spraying Method

This technique involves spraying the food product with the liquid solution. Small droplets of the liquid solution are created with a larger surface area for a similar volume of liquid solution. Due to this, additional portions of the product get covered with the droplets. This technique can be separated into air spray atomization and pressure atomization depending on the droplets' origin. High-speed air is used in air spray atomization to turn liquids into droplets, whereas high pressure is employed in pressure atomization to turn liquids into droplets. The inability to spray excessively viscous biopolymers is a significant limitation of this approach. Spraying an edible coating made of citric acid, xantham gum and glycerol as a plasticizer on the lotus root has shown less colour fading and less *Bacillus subtilis* microbial development thereby extending the expiration date (Lara et al., 2020).

12.7.3 Panning Method

This technique works well for covering food and confectionery products. This technique allows a sizeable number of food items with a round or oval form to be coated in a single batch. Food items are rotated inside a large, round pan while it is rotated. With the spinning of the pan, the coating fluid is poured upon this food product's surface. The volume of solution sprayed determines the last coating thickness. The air movement helps in the evaporation of the solvent and drying of the coats (Campos et al., 2011).

12.7.4 Fluidized Bed Processing Method

This technique makes it easier to coat the exterior of dried foodstuff like wheat or nuts with a thin film of covering. With the aid of nozzles, the coating solution is sprayed, which facilitates the movement of the smaller-sized foods along with the solution. The solution on the food builds a sheet that gradually transforms into the film followed by drying process. Compared to other coating techniques, this one is more expensive (Chawla et al., 2021). This procedure lowers the likelihood of accumulation and helps in slowing the release of active chemicals (Lipin & Lipin, 2022).

12.8 CONCLUSION

Nanotechnology has proven to be a cutting-edge tool in modern agriculture by supporting sustainable agricultural production. It has widespread applications for horticulture as it supports growth, boosts yields, lowers production prices and prevents post-harvest damage. It maintains the excellent quality of fresh and processed fruits and vegetables during long-term storage. The power of nanoparticles and their techniques of distribution are utilized by nanotechnology to increase the yield of horticultural crops. Nanomaterials are easy to use, inexpensive and environmentally safe, making it possible to make them quickly with little efforts without endangering the environment. While improper utilization of nanomaterials will damage the environment and agricultural plants. As a result, beneficial nanoparticle supply and risk assessments of their field application are required to ensure food and nutritional security for the world's ever-growing population in a changing climate scenario. This chapter has so far covered the uses of nanotechnology and its antimicrobial capabilities in ensuring the quality and safety of food. We may conclude that nanotechnology, with its strong antimicrobial capabilities and ability to solve food safety problems plays a significant role in food standards.

REFERENCES

Ahmad, M., Nirmal, N. P., Danish, M., Chuprom, J., & Jafarzedeh, S. (2016). Characterisation of composite films fabricated from collagen/chitosan and collagen/soy protein isolate for food packaging applications. *RSC Advances*, *6*(85), 82191–82204.

Ahmed, T. A., & Aljaeid, B. M. (2016). Preparation, characterization, and potential application of chitosan, chitosan derivatives, and chitosan metal nanoparticles in pharmaceutical drug delivery. *Drug Design, Development and Therapy*, *10*, 483. https://doi.org/10.2147/DDDT.S99651

Arora, D., Sharma, N., Sharma, V., Abrol, V., Shankar, R., & Jaglan, S. (2016). An update on polysaccharide-based nanomaterials for antimicrobial applications. *Applied Microbiology and Biotechnology*, *100*(6), 2603–2615. https://doi.org/10.1007/s00253-16-7315-0

Artés, F., Gómez, P. A., & Artés-Hernández, F. (2006). Modified atmosphere packaging of fruits and vegetables. *Stewart Postharvest Review*, *2*(5), 1–13.

Bajpai, V. K., Kamle, M., Shukla, S., Mahato, D. K., Chandra, P., Hwang, S. K., Kumar, P., Huh, Y. S., & Han, Y. K. (2018). Prospects of using nanotechnology for food preservation, safety, and security. *Journal of Food and Drug Analysis*, *26*(4), 1201–1214. https://doi.org/10.1016/j.jfda.2018.06.011

Beak, S., Kim, H., & Song, K. B. (2017). Characterization of an olive flounder bone gelatin-zinc oxide nanocomposite film and evaluation of its potential application in spinach packaging. *Journal of Food Science*, *82*(11), 2643–2649.

Berekaa, M. M. (2015). Nanotechnology in food industry; advances in food processing, packaging and food safety. *International Journal of Current Microbiology and Applied Sciences*, *4*(5), 345–357.

Biswas, M. C., Tiimob, B. J., Abdela, W., Jeelani, S., & Rangari, V. K. (2019). Nano silica-carbon-silver ternary hybrid induced antimicrobial composite films for food packaging application. *Food Packaging and Shelf Life*, *19*, 104–113.

Bodaghi, H., Mostofi, Y., Oromiehie, A., Zamani, Z., Ghanbarzadeh, B., Costa, C., & Del Nobile, M. A. (2013). Evaluation of the photocatalytic antimicrobial effects of a TiO_2 nanocomposite food packaging film by *in vitro* and *in vivo* tests. *LWT-Food Science and Technology*, *50*(2), 702–706.

Bodbodak, S., & Rafiee, Z. (2016). Recent trends in active packaging in fruits and vegetables. In: Siddiqui, M. A. (eds.) *Eco-friendly technology for postharvest produce quality* (pp. 77–125). Academic Press.

Burdo, O. G. (2005). Nanoscale effects in food-production technologies. *Journal of Engineering Physics and Thermophysics*, *78*(1), 90–96. https://doi.org/101007/s10891-005-0033-6

Cakmak, H., Kumcuoglu, S., & Tavman, S. (2018). Production of edible coatings with twin-nozzle electrospraying equipment and the effects on shelf-life stability of fresh-cut apple slices. *Journal of Food Process Engineering*, *41*(1), e12627.

Campos, C. A., Gerschenson, L. N., & Flores, S. K. (2011). Development of edible films and coatings with antimicrobial activity. *Food and Bioprocess Technology*, *4*(6), 849–875. https://doi.org/10.1007/S11947-010-0434-1

Cerqueira, M. A. P. R., Pereira, R. N. C., da Silva Ramos, O. L., Teixeira, J. A. C., & Vicente, A. A. (Eds.) (2017). *Edible food packaging: Materials and processing technologies* (p. 469). CRC Press.

Chaudhry, Q., Watkins, R., & Castle, L. (2017). *Nanotechnologies in food: What, why and how?* Royal Society of Chemistry. https://doi.org/10.1039/9781782626879-00001

Chawla, R., Sivakumar, S., & Kaur, H. (2021). Antimicrobial edible films in food packaging: Current scenario and recent nanotechnological advancements - A review. *Carbohydrate Polymer Technologies and Applications*, *2*, 100024. https://doi.org/10.1016/j.carpta.2020.100024

Chen, Y., Grimplet, J., David, K., Castellarin, S. D., Terol, J., Wong, D. C., & Chervin, C. (2018). Ethylene receptors and related proteins in climacteric and non-climacteric fruits. *Plant Science*, *276*, 63–72.

Cheng, Y., Sun, C., Zhai, X., Zhang, R., Zhang, S., Sun, C., & Hou, H. (2021). Effect of lipids with different physical state on the physicochemical properties of starch/gelatin edible films prepared by extrusion blowing. *International Journal of Biological Macromolecules*, *185*, 1005–1014. https://doi.org/10.1016/j.ijbiomac.2021.06.203

Chi, H., Song, S., Luo, M., Zhang, C., Li, W., Li, L., & Qin, Y. (2019). Effect of PLA nanocomposite films containing bergamot essential oil, TiO_2 nanoparticles, and Ag nanoparticles on shelf life of mangoes. *Scientia Horticulturae*, *249*, 192–198.

Chiabrando, V., Garavaglia, L., & Giacalone, G. (2019). The postharvest quality of fresh sweet cherries and strawberries with an active packaging system. *Foods*, *8*(8), 335. https://doi.org/10.3390/foods8080335

Chong, E. W. N., Jafarzadeh, S., Paridah, M. T., Gopakumar, D. A., Tajarudin, H. A., Thomas, S., & Abdul Khalil, H. P. S. (2019). Enhancement in the physico-mechanical functions of seaweed biopolymer film via embedding fillers for plasticulture application—A comparison with conventional biodegradable mulch film. *Polymers*, *11*(2), 210.

Costa, C., Conte, A., Buonocore, G. G., Lavorgna, M., & Del Nobile, M. A. (2012). Calcium-alginate coating loaded with silver-montmorillonite nanoparticles to prolong the shelf-life of fresh-cut carrots. *Food Research International*, *48*(1), 164–169. https://doi.org/10.1016/j.foodres.2012.03.001

Dairi, N., Ferfera-Harrar, H., Ramos, M., & Garrigós, M. C. (2019). Cellulose acetate/AgNPs-organoclay and/or thymol nano-biocomposite films with combined antimicrobial/antioxidant properties for active food packaging use. *International Journal of Biological Macromolecules*, *121*, 508–523.

de Chiara, M. L. V., Pal, S., Licciulli, A., Amodio, M. L., & Colelli, G. (2015). Photocatalytic degradation of ethylene on mesoporous TiO_2/SiO_2 nanocomposites: Effects on the ripening of mature green tomatoes. *Biosystems Engineering*, *132*, 61–70. https://doi.org/10.1016/j.biosystemseng.2015.02.008

De Moura, M. R., Aouada, F. A., Avena-Bustillos, R. J., McHugh, T. H., Krochta, J. M., & Mattoso, L. H. (2009). Improved barrier and mechanical properties of novel hydroxypropyl methylcellulose edible films with chitosan/tripolyphosphate nanoparticles. *Journal of Food Engineering*, *92*(4), 448–453. https://doi.org/10.1016/j.jfoodeng.2008.12.015

de Oliveira Júnior, S. D., de Araújo, J. S., de Asevedo, E. A., de Medeiros, F. G. M., dos Santos, V. S., de Sousa Júnior, F. C., de Araújo, N. K., & dos Santos, E. S. (2021). Exploiting films based on pectin extracted from yellow mombin (*Spondias mombin* L.) peel for active food packaging. *Biomass Conversion and Biorefinery*, 1–15. https://doi.org/10.1007/S13399-021-01321-3

Di Sia, P. (2017). Nanotechnology among innovation, health and risks. *Procedia: Social and Behavioral Sciences*, *237*, 1076–1080.

Dirpan, A., Latief, R., Syarifuddin, A., Rahman, A. N. F., Putra, R. P., & Hidayat, S. H. (2018). The use of colour indicator as a smart packaging system for evaluating mangoes Arummanis (*Mangifera indica* L. var. Arummanisa) freshness. In *IOP conference series. Earth and environmental science* (Vol. 157, No. 1, p. 012031). IOP Publishing.

Duhan, J. S., Kumar, R., Kumar, N., Kaur, P., Nehra, K., & Duhan, S. (2017). Nanotechnology: The new perspective in precision agriculture. *Biotechnology Reports*, *15*, 11–23.

Duran, M., Aday, M. S., Zorba, N. N. D., Temizkan, R., Büyükcan, M. B., & Caner, C. (2016). Potential of antimicrobial active packaging 'containing natamycin, nisin, pomegranate and grape seed extract in chitosan coating' to extend shelf life of fresh strawberry. *Food and Bioproducts Processing*, *98*, 354–363.

Dwivany, F. M., Aprilyandi, A. N., Suendo, V., & Sukriandi, N. (2020). Carrageenan edible coating application prolongs Cavendish banana shelf life. *International Journal of Food Science*, *2020*, 11.

Echegoyen, Y., & Nerín, C. (2013). Nanoparticle release from nano-silver antimicrobial food containers. *Food and Chemical Toxicology*, *62*, 16–22.

Eleftheriadou, M., Pyrgiotakis, G., & Demokritou, P. (2017). Nanotechnology to the rescue: Using nano-enabled approaches in microbiological food safety and quality. *Current Opinion in Biotechnology*, *44*, 87–93. https://doi.org/10.1016/j.copbio.2016.11.012

Emamifar, A., & Bavaisi, S. (2020). Nanocomposite coating based on sodium alginate and Nano-ZnO for extending the storage life of fresh strawberries (Fragaria×ananassa Duch.). *Journal of Food Measurement and Characterization*, *14*(2), 1012–1024. https://doi.org/10.1007/s11694-019-00350-x

Feynman, R. (2018). There's plenty of room at the bottom. In: Hey, A. (eds.) *Feynman and computation* (pp. 63–76). CRC Press. https://doi.org/10.1201/9780429500459

Gallocchio, F., Belluco, S., & Ricci, A. (2015). Nanotechnology and food: Brief overview of the current scenario. *Procedia Food Science*, *5*, 85–88.

Gómez-Estaca, J., López-de-Dicastillo, C., Hernández-Muñoz, P., Catalá, R., & Gavara, R. (2014). Advances in antioxidant active food packaging. *Trends in Food Science and Technology*, *35*(1), 42–51.

Guerreiro, A. C., Gago, C. M., Faleiro, M. L., Miguel, M. G., & Antunes, M. D. (2015). The use of polysaccharide-based edible coatings enriched with essential oils to improve shelf-life of strawberries. *Postharvest Biology and Technology*, *110*, 51–60. https://doi.org/10.1016/J.POSTHARVBIO.2015.06.019

Hermawan, D., Lai, T. K., Jafarzadeh, S., Gopakumar, D. A., Hasan, M., Owolabi, F. T., & Khalil, H. A. (2019). Development of seaweed-based bamboo microcrystalline cellulose films intended for sustainable food packaging applications. *BioResources, 14*(2), 3389–3410.

Issa, A. T. (2017). *Biodegradable nanocomposite films based on sweet potato starch: Development, characterization, and their antimicrobial effects on the shelf-life of packed baby spinach leaves* (Doctoral Dissertation). North Carolina Agricultural and Technical State University. ProQuest Dissertations Publishing; ProQuest LLC.

Jafarzadeh, S., Alias, A. K., Ariffin, F., & Mahmud, S. (2018). Physico-mechanical and microstructural properties of semolina flour films as influenced by different sorbitol/glycerol concentrations. *International Journal of Food Properties, 21*(1), 983–995.

Jafarzadeh, S., Ariffin, F., Mahmud, S., Hosseini, S. F., Ahmad, M., & Ahmad, M. (2017). Improving the physical and protective functions of semolina films by embedding a blend nanofillers (ZnO-nr and nano-kaolin). *Food Packaging and Shelf Life, 12*, 66–75.

Jafarzadeh, S., & Jafari, S. M. (2021). Impact of metal nanoparticles on the mechanical, barrier, optical and thermal properties of biodegradable food packaging materials. *Critical Reviews in Food Science and Nutrition, 61*(16), 2640–2658.

Jafarzadeh, S., Salehabadi, A., & Jafari, S. M. (2020). Metal nanoparticles as antimicrobial agents in food packaging. In: Jafari, S. M. (eds.) *Handbook of food nanotechnology* (pp. 379–414). Academic Press.

Jamróz, E., Kopel, P., Juszczak, L., Kawecka, A., Bytesnikova, Z., Milosavljevic, V., & Makarewicz, M. (2019). Development of furcellaran-gelatin films with Se-AgNPs as an active packaging system for extension of mini kiwi shelf life. *Food Packaging and Shelf Life, 21*, 100339. https://doi.org/10.1016/j.fpsl.2019.100339

Kaewklin, P., Siripatrawan, U., Suwanagul, A., & Lee, Y. S. (2018). Active packaging from chitosan-titanium dioxide nanocomposite film for prolonging storage life of tomato fruit. *International Journal of Biological Macromolecules, 112*, 523–529. *Food Packaging and Shelf Life, 18*, 140–146.

Kanikireddy, V., Varaprasad, K., Rani, M. S., Venkataswamy, P., Reddy, B. J. M., & Vithal, M. (2020). Biosynthesis of CMC-Guar gum-Ag0 nanocomposites for inactivation of food pathogenic microbes and its effect on the shelf life of strawberries. *Carbohydrate Polymers, 236*, 116053.

Kaur, M., Kalia, A., & Thakur, A. (2017). Effect of biodegradable chitosan–rice-starch nanocomposite films on post-harvest quality of stored peach fruit. *Starch-Stärke, 69*(1–2), 1600208. https://doi.org/10.1002/star.201600208

Khan, J. H., Lin, J., Young, C., Matsagar, B. M., Wu, K. C., Dhepe, P. L., & Hossain, M. S. A. (2018). High surface area nanoporous carbon derived from high quality jute from Bangladesh. *Materials Chemistry and Physics, 216*, 491–495.

Khan, S. T., Al-Khedhairy, A. A., & Musarrat, J. (2015). ZnO and TiO_2 nanoparticles as novel antimicrobial agents for oral hygiene: A review. *Journal of Nanoparticle Research, 17*(6), 1–16. https://doi.org/10.1007/s11051-015-3074-6

Kim, I., Viswanathan, K., Kasi, G., Thanakkasaranee, S., Sadeghi, K., & Seo, J. (2020). ZnO nanostructures in active antibacterial food packaging: Preparation methods, antimicrobial mechanisms, safety issues, future prospects, and challenges. *Food Reviews International, 38*(4), 537–565. https://doi.org/10.1080/87559129.2020.1737709

Kumar, A., Pratush, A., & Bera, S. (2021). Significance of nanoscience in food microbiology: Current trend and future prospects. In: Maddela N. R., Chakraborty, S., Prasad, R. (eds.) *Nanotechnology for advances in medical microbiology* (pp. 249–267).

Kumar, L., Ramakanth, D., Akhila, K., & Gaikwad, K. K. (2022). Edible films and coatings for food packaging applications: A review. *Environmental Chemistry Letters*, 20(4):1–26.

Kumar, S., Boro, J. C., Ray, D., Mukherjee, A., & Dutta, J. (2019). Bionanocomposite films of agar incorporated with ZnO nanoparticles as an active packaging material for shelf life extension of green grape. *Heliyon*, *5*(6), e01867. https://doi.org/10.1016/J.HELIYON.2019.E01867

Kurt, A. S., & Cekmecelioglu, D. (2021). Bacterial cellulase production using grape pomace hydrolysate by shake-flask submerged fermentation. *Biomass Conversion and Biorefinery*, *13*(8), 6981–6988.

Lara, G., Yakoubi, S., Villacorta, C. M., Uemura, K., Kobayashi, I., Takahashi, C., Nakajima, M., & Neves, M. A. (2020). Spray technology applications of xanthan gum-based edible coatings for fresh-cut lotus root (*Nelumbo nucifera*). *Food Research International*, *137*, 109723. https://doi.org/10.1016/j.foodres.2020.109723

Lee, S. Y., Lee, S. J., Choi, D. S., & Hur, S. J. (2015). Current topics in active and intelligent food packaging for preservation of fresh foods. *Journal of the Science of Food and Agriculture*, *95*(14), 2799–2810. https://doi.org/10.1002/jsfa.7218

Li, D., Ye, Q., Jiang, L., & Luo, Z. (2017). Effects of Nano-TiO_2-LDPE packaging on postharvest quality and antioxidant capacity of strawberry (*Fragaria ananassa* Duch.) stored at refrigeration temperature. *Journal of the Science of Food and Agriculture*, *97*(4), 1116–1123.

Li, J., Zhou, M., Cheng, G., Cheng, F., Lin, Y., & Zhu, P. X. (2019). Fabrication and characterization of starch-based nanocomposites reinforced with montmorillonite and cellulose nanofibers. *Carbohydrate Polymers*, *210*, 429–436. https://doi.org/10.1016/j.carbpol.2019.01.051

Lipin, A. A., & Lipin, A. G. (2022). Prediction of coating uniformity in batch fluidized-bed coating process. *Particuology*, *61*, 41–46. https://doi.org/10.1016/J.PARTIC.2021.03.010

Maftoonazad, N., & Ramaswamy, H. (2019). Design and testing of an electrospun nanofiber mat as a pH biosensor and monitor the pH associated quality in fresh date fruit (Rutab). *Polymer Testing*, *75*, 76–84. https://doi.org/10.1016/j.polymertesting.2019.01.011

Mahato, R. P., Kumar, S., & Singh, P. (2021). Optimization of growth conditions to produce sustainable polyhydroxyalkanoate bioplastic by Pseudomonas aeruginosa EO1. *Frontiers in Microbiology*, *12*, 711588. https://doi.org/10.3389/fmicb.2021.711588

Mahato, R. P., Kumar, S., & Singh, P. (2023). Production of polyhydroxyalkanoates from renewable resources: A review on prospects, challenges and applications. *Archives of Microbiology*, *205*(5), 172. https://doi.org/10.1007/s00203-023-03499-8

Mangaraj, S., & Goswami, T. K. (2009). Modified atmosphere packaging of fruits and vegetables for extending shelf-life - A review. *Fresh Produce*, *3*(1), 1–31.

Martínez-Abad, A., Ocio, M. J., Lagarón, J. M., & Sánchez, G. (2013). Evaluation of silver-infused polylactide films for inactivation of *Salmonella* and feline calicivirus *in vitro* and on fresh-cut vegetables. *International Journal of Food Microbiology*, *162*(1), 89–94.

Meindrawan, B., Suyatma, N. E., Wardana, A. A., & Pamela, V. Y. (2018). Nanocomposite coating based on carrageenan and ZnO nanoparticles to maintain the storage quality of mango. *Food Packaging and Shelf Life*, *18*, 140–146.

Mohamed, S. A., El-Sakhawy, M., & El-Sakhawy, M. A. M. (2020). Polysaccharides, protein and lipid-based natural edible films in food packaging: A review. *Carbohydrate Polymers*, *238*, 116178. https://doi.org/10.1016/j.carbpol.2020.116178

Mohammad, Z. H., Ahmad, F., Ibrahim, S. A., & Zaidi, S. (2022). Application of nanotechnology in different aspects of the food industry. *Discover Food*, *2*(1), 12.

Myszka, K., & Czaczyk, K. (2011). Bacterial biofilms on food contact surfaces - A review. *Polish Journal of Food and Nutrition Sciences*, *61*(3), 173–180.

Naknaen, P. (2014). Utilization possibilities of antimicrobial biodegradable packaging produced by poly (butylene succinate) modified with zinc oxide nanoparticles in fresh-cut apple slices. *International Food Research Journal*, *21*(6), 2413–2420.

Nasrollahzadeh, M., Sajjadi, M., Iravani, S., & Varma, R. S. (2020). Trimetallic nanoparticles: Greener synthesis and their applications. *Nanomaterials*, *10*(9), 1784.

Nile, S. H., Baskar, V., Selvaraj, D., Nile, A., Xiao, J., & Kai, G. (2020). Nanotechnologies in food science: Applications, recent trends, and future perspectives. *Nano-Micro Letters*, *12*(1), 1–34.

No, H. K., Meyers, S. P., Prinyawiwatkul, W., & Xu, Z. (2007). Applications of chitosan for improvement of quality and shelf life of foods: A review. *Journal of Food Science*, *72*(5), R87–R100. https://doi.org/10.1111/j.1750-3841.2007.00383.x

Oloye, M. T., Jabar, J. M., Adetuyi, A. O., & Lajide, L. (2021). Extraction and characterization of pectin from fruit peels of *Irvingia gabonensis* and pulp of *Cola milleni* and *Theobroma cacao* as precursor for industrial applications. *Biomass Conversion and Biorefinery*, *11*, 1–9. https://doi.org/10.1007/S13399-021-01366-4

Ortiz-Duarte, G., Pérez-Cabrera, L. E., Artés-Hernández, F., & Martínez-Hernández, G. B. (2018). Ag-chitosan nanocomposites in edible coatings affect the quality of fresh-cut melon. *Postharvest Biology and Technology*, *147*, 174–184. https://doi.org/10.1016/j.postharvbio.2018.09.021

Pandey, J. K., Nakagaito, A. N., & Takagi, H. (2013). Fabrication and applications of cellulose nanoparticle-based polymer composites. *Polymer Engineering and Science*, *53*(1), 1–8. https://doi.org/10.1002/pen.23242

Parisi, C., Vigani, M., & Rodríguez-Cerezo, E. (2015). Agricultural nanotechnologies: What are the current possibilities? *Nano Today*, *10*(2), 124–127.

Pereira de Abreu, D. A., Cruz, J. M., & Paseiro Losada, P. (2012). Active and intelligent packaging for the food industry. *Food Reviews International*, *28*(2), 146–187.

Pirozzi, A., Del Grosso, V., Ferrari, G., & Donsì, F. (2020). Edible coatings containing oregano essential oil nanoemulsion for improving postharvest quality and shelf life of tomatoes. *Foods*, *9*(11), 1605.

Rana, R. A., Siddiqui, M., Skalicky, M., Brestic, M., Hossain, A., Kayesh, E., Popov, M., Hejnak, V., Gupta, D. R., Mahmud, N. U., & Islam, T. (2021). Prospects of nanotechnology in improving the productivity and quality of horticultural crops. *Horticulturae*, *7*(10), 332.

Resende, N. S., Gonçalves, G. A. S., Reis, K. C., Tonoli, G. H. D., & Boas, E. V. B. V. (2018). Chitosan/cellulose nanofibril nanocomposite and its effect on quality of coated strawberries. *Journal of Food Quality*, *2018*, 1–13. https://doi.org/10.1155/2018/1727426

Rezaee, P., Shahriari, S., & Mostaghim, T. (2018). The effect of polylactic acid packaging modified with clay nanoparticles on quality and shelf life of mushroom. *Journal of Food Biosciences and Technology*, *8*(2), 29–40.

Rocha, M. A. M., Coimbra, M. A., & Nunes, C. (2017). Applications of chitosan and their derivatives in beverages: A critical review. *Current Opinion in Food Science*, *15*, 61–69. https://doi.org/10.1016/j.cofs.2017.06.008

Saba, M. K., & Amini, R. (2017). Nano-ZnO/carboxymethyl cellulose-based active coating impact on ready-to-use pomegranate during cold storage. *Food Chemistry*, *232*, 721–726.

Saekow, M., Naradisorn, M., Tongdeesoontorn, W., & Hamauzu, Y. (2019). Effect of carboxymethyl cellulose coating containing ZnO-nanoparticles for prolonging shelf life of persimmon and tomato fruit. *Journal of Food Science and Agricultural Technology (JFAT)*, *5*, 41–48.

Salinas-Roca, B., Soliva-Fortuny, R., Welti-Chanes, J., & Martín-Belloso, O. (2016). Combined effect of pulsed light, edible coating and malic acid dipping to improve fresh-cut mango safety and quality. *Food Control*, *66*, 190–197. https://doi.org/10.1016/j.foodcont.2016.02.005

Satalkar, P., Elger, B. S., & Shaw, D. M. (2016). Defining Nano, nanotechnology and nanomedicine: Why should it matter? *Science and Engineering Ethics*, *22*(5), 1255–1276.

Servin, A. D., & White, J. C. (2016). Nanotechnology in agriculture: Next steps for understanding engineered nanoparticle exposure and risk. *NanoImpact*, *1*, 9–12.

Shi, S., Wang, W., Liu, L., Wu, S., Wei, Y., & Li, W. (2013). Effect of chitosan/nano-silica coating on the physicochemical characteristics of longan fruit under ambient temperature. *Journal of Food Engineering*, *118*(1), 125–131.

Shiekh, R. A., Malik, M. A., Al-Thabaiti, S. A., & Shiekh, M. A. (2013). Chitosan as a novel edible coating for fresh fruits. *Food Science and Technology Research*, *19*(2), 139–155. https://doi.org/10.3136/fstr.19.139

Singh, P., Maheshwari, A., Dharmesh, V., Anand, V., Kaur, J., Srivastava, S., & Srivastava, S. (2022). Microbial Biofilm in Remediation of Environmental Contaminants from Wastewater: Mechanisms, Opportunities, Challenges, and Future Perspectives. *Omics for Environmental Engineering and Microbiology Systems*, 469–492.

Singh, A., & Prasad, S. M. (2017). Nanotechnology and its role in agro-ecosystem: A strategic perspective. *International Journal of Environmental Science and Technology*, *14*(10), 2277–2300.

Singh, D. P., & Packirisamy, G. (2022). Biopolymer based edible coating for enhancing the shelf life of horticulture products. *Food Chemistry: Molecular Sciences*, *4*, 100085.

Singh, G., Singh, S., Kumar, B., & Gaikwad, K. K. (2021). Active barrier chitosan films containing gallic acid based oxygen scavenger. *Journal of Food Measurement and Characterization*, *15*(1), 585–593. https://doi.org/10.1007/s11694-020-00669-w

Singh, V., Yadav, V. K., & Mishra, V. (2020). Nanotechnology: An application in biofuel production. In M. Srivastava, N. Srivastava, P. Mishra, & V. Gupta (Eds.), *Nanomaterials in biofuels research. Clean energy production technologies*. Springer. https://doi.org/10.1007/978-981-13-9333-4_6

Siripatrawan, U., & Kaewklin, P. (2018). Fabrication and characterization of chitosan-titanium dioxide nanocomposite film as ethylene scavenging and antimicrobial active food packaging. *Food Hydrocolloids*, *84*, 125–134. https://doi.org/10.1016/j.foodhyd.2018.04.049

Soares, N. M., Fernandes, T. A., & Vicente, A. A. (2016). Effect of variables on the thickness of an edible coating applied on frozen fish–Establishment of the concept of safe dipping time. *Journal of Food Engineering*, *171*, 111–118. https://doi.org/10.1016/j.jfoodeng.2015.10.016

Sobhan, A., Muthukumarappan, K., Wei, L., Van Den Top, T., & Zhou, R. (2020). Development of an activated carbon-based nanocomposite film with antibacterial property for smart food packaging. *Materials Today Communications*, *23*, 101124. https://doi.org/10.1016/j.mtcomm.2020.101124

Song, H., Yuan, W., Jin, P., Wang, W., Wang, X., Yang, L., & Zhang, Y. (2016). Effects of chitosan/nano-silica on postharvest quality and antioxidant capacity of loquat fruit during cold storage. *Postharvest Biology and Technology*, *119*, 41–48.

Sorrentino, A., Gorrasi, G., & Vittoria, V. (2007). Potential perspectives of bio-nanocomposites for food packaging applications. *Trends in Food Science and Technology*, *18*(2), 84–95. https://doi.org/10.1016/j.tifs.2006.09.004

Suhag, R., Kumar, N., Petkoska, A. T., & Upadhyay, A. (2020). Film formation and deposition methods of edible coating on food products: A review. *Food Research International*, *136*, 109582. https://doi.org/10.1016/j.foodres.2020.109582

Sundqvist-Andberg, H., & Åkerman, M. (2021). Sustainability governance and contested plastic food packaging–An integrative review. *Journal of Cleaner Production*, *306*, 127111. https://doi.org/10.1016/j.jclepro.2021.127111

Swaroop, C., & Shukla, M. (2019). Development of blown polylactic acid-MgO nanocomposite films for food packaging. *Composites – Part A: Applied Science and Manufacturing*, *124*, 105482.

Tavakoli, S. A., Mirzaei, S., Rahimi, M., & Tavassolian, I. (2019). Assessment of peppermint, clove, cumin essential oils and silver Nano particles on biochemical and shelf life of *Citrus limon* (L.). *Indian Journal of Biochemistry and Biophysics*, *56*, 269–275.

Tripathi, S., Mehrotra, G., & Dutta, P. (2009). Physicochemical and bioactivity of cross-linked chitosan–PVA film for food packaging applications. *International Journal of Biological Macromolecules*, *45*(4), 372–376. https://doi.org/10.1016/j.ijbiomac.2009.07.006

Wang, K., Jin, P., Shang, H., Li, H., Xu, F., Hu, Q., & Zheng, Y. (2010). A combination of hot air treatment and nano-packing reduces fruit decay and maintains quality in postharvest Chinese bayberries. *Journal of the Science of Food and Agriculture*, *90*(14), 2427–2432.

Xing, Y., Yang, H., Guo, X., Bi, X., Liu, X., Xu, Q., Wang, Q., Li, W., Li, X., Shui, Y., & Chen, C. (2020). Effect of chitosan/Nano-TiO_2 composite coatings on the postharvest quality and physicochemical characteristics of mango fruits. *Scientia Horticulturae*, *263*, 109135.

Yang, D., Zhu, L., & Jiang, X. (2010). Electrochemical reaction mechanism and determination of Sudan I at a multi wall carbon nanotubes modified glassy carbon electrode. *Journal of Electroanalytical Chemistry*, *640*(1–2), 17–22. https://doi.org/10.1016/j.jelechem.2009.12.022

Yotova, L., Yaneva, S., & Marinkova, D. (2013). Biomimetic nanosensors for determination of toxic compounds in food and agricultural products. *Journal of Chemical Technology and Metallurgy*, *48*(3), 215–227.

Youssef, A. M., Abdel-Aziz, M. S., & El-Sayed, S. M. (2014). Chitosan nanocomposite films based on Ag-NP and Au-NP biosynthesis by *Bacillus subtilis* as packaging materials. *International Journal of Biological Macromolecules*, *69*, 185–191. https://doi.org/10.1016/j.ijbiomac.2014.05.047

Yuan, G., Chen, X., & Li, D. (2016). Chitosan films and coatings containing essential oils: The antioxidant and antimicrobial activity, and application in food systems. *Food Research International*, *89*(1), 117–128. https://doi.org/10.1016/j.foodres.2016.10.004

Zhang, C., Li, W., Zhu, B., Chen, H., Chi, H., Li, L., Qin, Y., & Xue, J. (2018). The quality evaluation of postharvest strawberries stored in Nano-Ag packages at refrigeration temperature. *Polymers*, *10*(8), 894.

13 Nanofilms for Enhancing the Shelf-life of Harvested Products

*Galahitigama GAH, Isuri Thakshila, Salinda Sandamal, Tharindi PWM, Suthajini T, Abeysinghe NPM, Asanka Tennakoon, Ayman El Sabagh and Disna Ratnasekera**

13.1 INTRODUCTION

Horticultural crops, especially fruits and vegetables, are important in providing calories and essential micro- and macronutrients for the human body (Ma et al., 2017). Consumers' preference for consuming fruits and vegetables depends on extrinsic (visual attributes such as shape, size, colour, flavour, etc.) and intrinsic (nutritional and functional attributes) factors (Ziv & Fallik, 2021). As a nutritional characteristic, the presence of vitamins, minerals, antioxidants and fibre in fruits and vegetables plays a significant role in a healthy diet. Though relatively large quantities of fruits and vegetables are being produced globally, many regions still face hunger and malnutrition, which has become a serious global issue in the present scenario (Li et al., 2020). Meanwhile, according to the latest prophecies, world food demand has been increasing rapidly in the last few decades, and the total population is estimated to reach nine billion by 2050 (Duro et al., 2020). This exponential incensement should ensure food security by enriching food sovereignty. However, global food supplies are highly impacted by the expenditure of bioresources, manufacturing and application of agrochemicals, higher post-harvest losses, minimum value addition and inefficient distribution and marketing factors (Barrett, 2021). Among these factors, post-harvest loss of fruits and vegetables is a crucial factor to be considered. In contrast, approximately 20%–40% were lost due to poor shelf-life extension approaches (El-Ramady et al., 2015). Thus, a massive amount of wasted food places a heavy burden on society, the environment and the economy, which drives the need for innovative methods of waste reduction as well as improvements to existing processes (Ziv & Fallik, 2021).

Depending on storage conditions, the quality and shelf-life of fresh agricultural products can be dramatically impacted by several physiological processes, including respiration, ripening and senescence (Ziv & Fallik, 2021). Especially with the ripening,

* Corresponding author.

DOI: 10.1201/9781003376446-13

ethylene gas would release, stimulating chlorophyll degradation and tissue softening, accelerating the post-harvesting losses (Ziv & Fallik, 2021). An accelerated respiration rate encourages senescence and excessive softening (Basumatary et al., 2022). In addition, microbial contamination also causes fruit and vegetable deterioration and directly impacts quality reduction at higher incidents (Kumar et al., 2020).

In recent years, several freshness-keeping techniques have been examined to perpetuate the shelf-life of fresh foodstuffs (Minas et al., 2012). Cold storing (Ziv & Fallik, 2021), irradiation techniques (Guan et al., 2016), modified atmosphere storing (Korićanac et al., 2020), application of different chemicals and essential oils (Tzortzakis et al., 2019), physical treatments (Mpai & Sivakumar 2020), correct post-harvest handling (Conversa et al., 2020) and edible coating (Tzortzakis et al., 2019) are some of the techniques being used widely in the food industry. Among them, edible coatings have become more popular after the introduction of nanotechnology (Ma et al., 2017).

Nanotechnology is a newly introduced method that is used in different aspects of agriculture. Nanotechnology with edible coating has proven a strategy to enhance the shelf-life of harvestable products by making barriers for gas and moisture exchange and increasing the antimicrobial properties (Zambrano-Zaragoza et al., 2018; Singh et al., 2020). Furthermore, recent studies have declared that nanocoatings relatively reduced the cost of production of edible coatings preparation (Zambrano-Zaragoza et al., 2018). Therefore, this chapter will discuss the various nanomaterials used to coat fruits and vegetables, how nanocoatings contribute to a reduction in post-harvest losses, the additional benefits of nanofilms and the potential applications of nanotechnology in this field.

13.2 NANOFILMS USED IN POST-HARVEST LOSS MANAGEMENT OF FRESH PRODUCTS

Edible coatings, or translucent films manufactured from food-grade ingredients, are an efficient post-harvest technology to extend the shelf-life of fresh vegetables (Hasan et al., 2020). Nanotechnology-related applications in creating edible coatings are becoming popular due to the development of nanoscale particles (1–100 nm) which provide good protection, barrier qualities for gases and moisture and facilitate the regulated release of bioactive elements such as antioxidants, antimicrobial compounds, enzymes, flavours and colouring agents by serving as a vehicle for their delivery (Vanza et al., 2020). Nanocoatings include nanoemulsions, nanotubes, lipid nanocarriers, solid lipid nanoparticles and polymeric nanoparticles (Zaragoza et al., 2018). Research on reducing post-harvest losses emphasizes nanosystems (Gundewadi et al., 2018; Sami et al., 2021), and the potential benefits of developing them into effective edible coatings are enormous (Mahela et al., 2020).

13.2.1 Nanosystems in Edible Coatings

Nanosystems have made the development of modified edible coatings feasible by integrating nanosized inorganic and organic components (Mahela et al., 2020). The uses of the different nanosystems are discussed in the following sections. Different systems contribute in diverse ways to interactions with the coating matrix. Figure 13.1 illustrates a schematic diagram of the functionality of nanoedible coatings.

13.2.1.1 Nanoemulsions

Nanoemulsions are dispersed in an aqueous solution (the radius of the particles is less than 100 nm) (McClements, 2012). The nanoparticles containing aqueous solution alter the coating's physical characteristics by minimizing the growth of pathogens (micro-organisms), oxidative processes, movement of water, respiration, suppressing item degradation and boosting regulation of physiological problems (Hasan et al., 2020). A thin layer of emulsifying molecules surrounds every drop of oil in nanoemulsions at the interface (Zaragoza et al., 2018). There is tremendous potential for using nanoemulsions as active substance delivery systems for edible coatings. Nanoemulsion particle sizes range from 50 to 500 nm (Ranjan et al., 2014). Nanoemulsions can either be water-in-oil (w/o) or oil-in-water (o/w), depending on the phases involved (Jin et al., 2016). Oil-in-water systems produce edible coatings because they can include antioxidant and antibacterial lipophilic compounds in a hydrophilic polymeric matrix (Zaragoza et al., 2018). The bioactive lipids (essential oils, vitamins, oily flavours, oily colours, etc.) may make up the entire lipid phase of nanoemulsions or be combined with other oils to form oily solutions (olive, corn, sunflower, soybean, etc.). Bioactive lipids, flavouring, antibacterial, antioxidant and nutraceuticals are a few examples of lipophilic substances included in nanoemulsions (Silva et al., 2012).

13.2.1.2 Polymeric Nanoparticles

Particle sizes in these colloidal formations range from 100 nm to 1000 nm. Two prominent types of polymeric nanoparticles are frequently studied in terms of morphology: nanospheres, which are made of a thick polymeric matrix, and other types of nanocapsules, which are produced of an oiled interior and a polymeric covering (Zaragoza et al., 2018). Essential oils have better antibacterial action for polymeric nanoparticles and edible coatings (Zaragoza et al., 2018). Alginate-chitosan nanocapsules containing lemongrass and turmeric oils have been reported (Natrajan et al., 2015). Although numerous formed polymers are widely available to synthesize polymeric nanoparticles, high priority is given to synthesizing chitosan nanoparticles in the food industry (Zaragoza et al., 2018). It has improved mechanical qualities, and the capacity to form films is non-toxic, biocompatible, has fungicidal and antibacterial capabilities and has permeability for selective gases. Chitosan nanoparticles successfully prevented weight loss, retained firmness and reduced respiratory rate by three weeks on fresh strawberries by ionotropic gelation (Pilon et al., 2015).

13.2.1.3 Solid Lipid Nanoparticles (SLNs)

This system is made up of solid lipid particles that are spherical in shape and range in size of range from 50 to 1000 nm (Gastaldi et al., 2014), where the water-insoluble core is disseminated and then sustained by surfactants (Katouzian et al., 2017). The bioactive substance is either disseminated or dissolved in the fatty acid chains or the solid high melting point matrix. Because of their excellent stability and greater loading capacity SLNs represent a promising system as an element of edible coatings (Zaragosa et al., 2018).

13.2.1.4 Inorganic Nanocomposites

Nanocomposites are two-phase materials with a dispersed phase at the nanometer level in the continuous phase (Abd-Elsalam, 2021). This new class of innovative materials provides good macro-scopic characteristics by using the distinct attributes that fillers with nanometer-sized dimensions provide to an appropriate matrix compared to those achieved with identical fillers at micro-metric dimensions (Clarizia & Bernardo, 2019).

Polysaccharides, lipids and proteins have created edible coatings because of their origin and unique qualities (Zaragoza et al., 2018). This technique, which combines nanosystems with inorganic materials to create nanocomposites, is attractive. These mixtures represent a more contemporary choice for enhancing the characteristics of edible coatings since they enable more excellent mechanical resistance, regulated release, transparency and more efficient gas barrier features (Dang et al., 2017).

Inorganic nanoparticles are made of metal, or metal oxides, such as iron oxide (Fe_3O_4), aluminium oxide, zinc oxide (ZnO), copper oxide (CuO), titanium oxide (TiO_2), and calcium carbonate (carbon-based compounds), cerium dioxide hydroxides, and other metals like silver and gold (He & Hwang, 2016). The zinc oxide nanoparticle (ZnONP) has attracted much interest because of its superior mechanical qualities, biocompatibility, barrier abilities and broad-spectrum antibacterial behaviours (Yusof & Zain, 2019). ZnONP can be added to lipids, protein-based biopolymers and polysaccharides to increase the mechanical, barrier and physicochemical characteristics of edible coatings (Wu et al., 2019). Compared to edible chitosan coatings, chitosan (CS) nanoparticles performed better in antibacterial and antioxidant activity (Odetayo et al., 2022). CS-NPs demonstrate higher biological activity because their particle size is smaller and has more surface area than CS (Odetayo et al., 2022). For the same reasons, CS-NPs can migrate through biofilms and eliminate harmful bacteria (Huang & Li, 2014). Silver nanoparticles are efficient against numerous bacteria while safe for humans (Aadil et al., 2018).

13.2.1.5 Nanotubes and Nanofibers

Active compounds (antioxidants or antimicrobials) have recently been predicted to be transported through nanotubes and nanofibers in potentially advantageous ways (Zaragoza et al., 2018). These are considered novel strategies to improve the regulated release and use of active substances to preserve freshly harvested products. These are beneficial in carrying compounds with a favourable impact on fresh produce storage and functionality during consumption, including their application in edible coatings (Aytac et al., 2017).

Another kind of nanosystem is nanofiber. Nanofibers refer to fibrous structures with less than 100 nm or even 500 nm diameters (Fahami & Fathi, 2018). They have been applied to encapsulate active compounds, modify film characteristics and immobilize enzymes. They are a novel option for edible coatings (Fahami & Fathi, 2018). This is possible due to their strong crystallinity and negatively charged structure, which permits a greater bonding with the food surface. Additionally, these nanosystems are incorporated into protein and polysaccharide-based polymeric matrices, which can be functionalized with different EOs and/or plant extracts (Deng et al., 2017). Table 13.1 represents an example list of nanosystems showing their functionalities and applications.

TABLE 13.1
Different nanosystems use in nanoedible films or coatings

Harvestable product	Bioactive substance/ nanoparticle component	Biopolymer/lipid matrix/other ingredients	Functionality	Reference
Cucumber	Carvacrol	Chitosan	Antimicrobial effect	Tastan et al., 2017
Strawberry	Olive oil	Chitosan	Inhibited the fruit decay and antifungal effect	Khalifa et al., 2016
Pomegranate	Clove essential oil	Chitosan	Extended aril shelf-life	Hasheminejad & Khodaiyan, 2020
Grape berry	Lemongrass essential oil	Chitosan	Antimicrobial and antioxidant effects	Oh et al., 2017
Fresh-cut apples	Lemongrass essential oil	Sodium alginate	Antimicrobial effect	Salvia et al., 2015
Fresh-cut apples	Alpha- tocopherol	Nopal mucilage	Extend shelf-life and enzyme-based activities	Zaragoza et al., 2014
Green beans	Mandarin essential oil	Chitosan	Antimicrobial effect	Severino et al., 2014
Okra	Basil essential oil	Sodium alginate	Increased the antimicrobial activity	Gundewadi et al., 2018
Strawberry	Cinnamon essential oil	Pullulan	Increased the antimicrobial activity	Chu et al., 2020
Plums	Lemongrass essential oil	Carnauba wax	Antimicrobial effect	Kim et al., 2013
Fuji apple	Extract of tarbush	Candelilla wax	Increased the wettability of the surface	De León-Zapata et al., 2017
Guava		Wax candeuba	Minimized the senescence process	Zaragoza et al., 2013
Eggplant		Wax candeuba	Extended the shelf-life of white eggplant genotypes	Singh et al., 2016
Pear		Wax candeuba	Maintained the natural ageing process	Aditya et al., 2017
Strawberry		Bees wax	Antifungal, decreases decay rates and physiological damage	
Lemon		Bees wax	Reduced loss of water	Guerrero et al., 2014
Green grapes	ZnO NPs	Agar	Up to 21 days longer shelf-life	Kumar et al., 2019
Okra	ZnO NPs	Chitosan	The antimicrobial and antifungal effects extended the shelf-life	Al-Naamani et al., 2018
Mango	ZnO NPs	Carrageenan	Up to 19 days of shelf-life	Meindrawan et al., 2018
Mango	TiO_2 NPs	Chitosan	Delayed decay rate	Xing et al., 2020
Green tomato	Nano SiO_x	Chitosan	Delayed weight loss, firmness and senescence	Zhu et al., 2019
Apple	Nano SiOx	Soy protein isolate	Prolonged shelf-life and enhanced quality of preservation	Liu et al., 2017

(Continued)

TABLE 13.1 (CONTINUED)
Different nanosystems use in nanoedible films or coatings

Harvestable product	Bioactive substance/ nanoparticle component	Biopolymer/lipid matrix/other ingredients	Functionality	Reference
White button mushroom	Nano SiOx	Chitosan	Extended the shelf-life by efficiently reducing the yeast, mould and aerobic plate micro-organisms	Sami et al., 2021
Pomegranate	Nano ZnO_2	Carboxymethyl cellulose	Antimicrobial effect and maintain bioactive compounds	Saba et al., 2017
Shiitake mushroom	Ag NPs	Alginate	Inhibited the weight loss, softening and browning Reduced the yeast, mould and micro-organism counts	Jiang et al., 2013
Fresh-cut melon	Ag NPs	Chitosan	Reduced ethylene synthesis, growth of micro-organisms and respiration rates	Ortiz-Duarte et al., 2019
Kinnow	Ag NPs	Guar gum$^+$	Retention of volatile compounds	Shah et al., 2016
Cherry tomato	Mg NPs	Pectin	Increased up to 24 days of shelf-life at 10°C	Kumar et al., 2020
Chinese yam	Nano $CaCO_3$		Delayed the ripening process and antimicrobial effect	Luo et al., 2015
Guava	Chitosan	Xanthan gum and tween	Improve overall quality of cold storage	Gad & Zag Zog, 2017
Apple	Chitosan	Acetic acid	Reduced weight loss and fruit softening	Gardesh et al., 2016
Carrot	ZnO NPs		Extended shelf-life of up to 50 days and antimicrobial effect	Xu et al., 2017
Rose	Ag NPs		Increased vase life by up to 19 days with no symptoms of senescence	Alekasir et al., 2017
Chrysanthemum	Ag NPs		Reduced bacterial proliferation and biofilm formation at the cut end	Lin et al., 2019
Gerbera (*Gerbera jamesonii*)	Ag NPs		Improved vase life up to eight days by reducing bacterial count	Shafiee-Masouleh, 2018
Carrot	Microfibrilla-ted carrot (MC)	Starch	Improved mechanical properties	Guimarães et al., 2016
Apples	Zein nanofibers	Curcumin	Antimicrobial effect	Aytac et al., 2017

13.3 BASIC PRINCIPLES OF NANOFILMS RELATED SHELF-LIFE EXPANSION

Most of the standard shelf-life expansion techniques are targeted at three basic principles of preservation objectives: (1) mitigating the senescence process through reducing respiration, (2) controlling micro-organisms' activities involved in food deterioration, (3) balancing internal water evaporation (Zambrano-Zaragoza et al., 2018). Incorporating nanoparticles into edible films would result in greater protection from micro-organisms and other factors involved in post-harvest losses of fresh, harvestable products (Liu et al., 2020; Neme et al., 2021). This is owing to a series of unique characteristics of nanomaterials in nanofilms (Liu et al., 2020). Since nanoparticles consist of a large surface area, only a tiny quantity of addition is needed for a solid interfacial interaction to occur with the polymer; thus, nanofilms or coatings act as an effective barrier/filler for gas exchange (Zambrano-Zaragoza et al., 2018; Liu et al., 2020). Disrupting gas movement prevents cell respiration and preserves volatile aroma compounds, which aid in minimizing microbes' activities in the product (Neme et al., 2021). For instance, Cheng et al., (2016) observed that guar gum (GG)-nanocrystalline cellulose (NCG) film has reduced the oxygen permeability of the product and enhanced the shelf-life. In addition, Mihindukulasuriya & Lim, (2014) reported that mixing nanoparticles with macro-molecule materials encourages the stability of the micro-climate of the storage house where fresh vegetables and fruits are stored. In addition, nanoparticles increased the mechanical properties of the nanocomposite presence in the film by making hydrogen bonds and ionic interactions between other components (Abdollahi et al., 2013). For example, Khan et al., (2012) reported that adding a small quantity of nanocrystalline cellulose improved the forte of chitosan film by 26%, compared to non-nanocrystalline filler. Simultaneously, the interaction between nanoparticles and polymer has encouraged the thermal properties of the nanocomposite available in the edible coating or film, hindering the post-processing treatments of harvested products (Cheng et al., 2016).

Some nanomaterials contain photocatalytic properties, which are essential to the extent of the shelf-life of harvestable products through mitigation of both microbes and ethylene production. As a result of a series of chemical reactions, they produce active substances such as hydroxyl radicals ($OH\cdot$) and superoxide radicals (O_2^-). These substances react with organics and degrade ethylene and other gas production through the acceleration of oxidative decomposition, and result in delaying of post-harvest fruits and vegetable decay (Mihindukulasuriya & Lim, 2014). Nano-TiO_2 is the commonly used photocatalyst material, which is used to mitigate the ethylene synthesis of fresh products (Liu et al., 2020). In addition, antibacterial nanomaterials available in the edible film would prolong the shelf-life of harvested products by hindering the growth of the microbial population (Liu et al., 2020). These antibacterial nanomaterials can be divided into two types, namely, metal ionic nano-antimicrobial particles (ex: Ag, Cu, Zn, Ni and Al) and metal oxide types (ex: TiO_2, SiO_2, ZnO, etc.) (Liu et al., 2020). For example, chitosan film-based nano-SiO_2 and nano-ZnO coating ensure the durability of fruit freshness by reducing the microbial population (Yu et al., 2012). Furthermore, antimicrobial metal ions can enter bacterial cells and

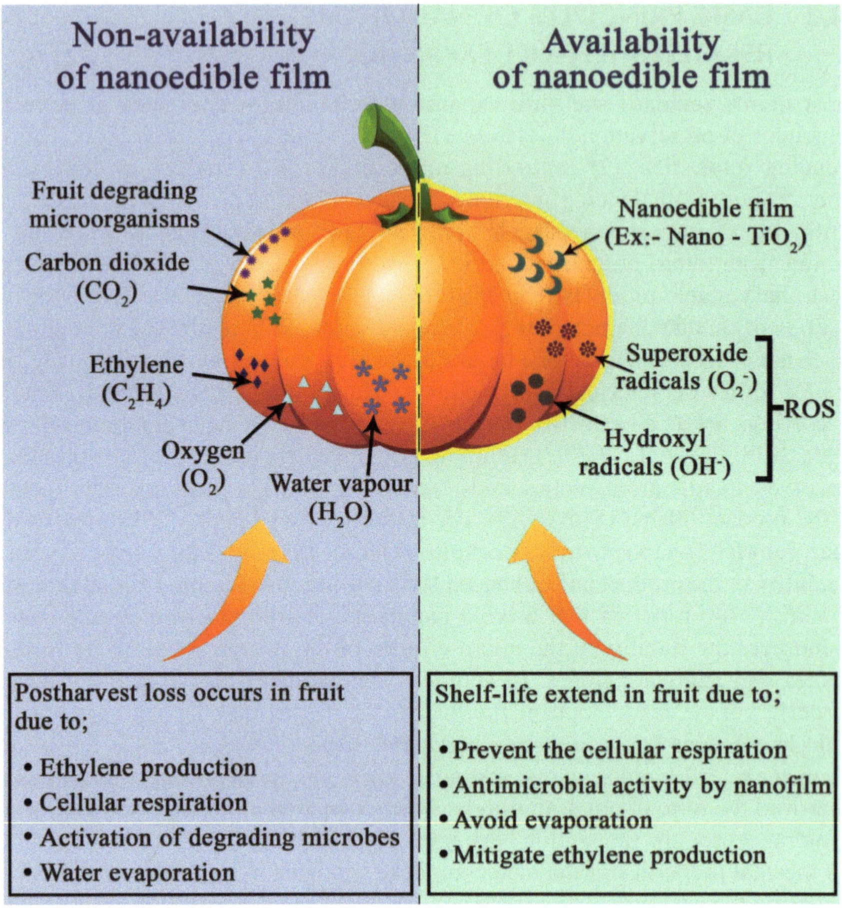

FIGURE 13.1 Impact of availability of nanoedible films in fresh fruits

denature the proteins; hence, bacterial cells' functions would be disrupted (Liu et al., 2020). Apart from nanomaterials, other antimicrobial compounds can be included in edible coatings, including organic acids, polypeptides, mineral salts and essential oils (Flores-López et al., 2016). For instance, Hasheminejad & Khodaiyan, (2020) reported that clove oil presence in the nanocomposite of edible coating of pomegranate improved the shelf-life by eliminating aerobic microbes. Therefore, nanoedible films would be essential to ensure fresh harvestable products' physical properties. Figure 13.1 illustrates the effect of the availability of nanoedible films in fresh fruits.

13.4 DIFFERENT NANOPACKAGING TO PROTECT FROM POST-HARVESTING LOSSES

Food packaging is the principal area for nanotechnology application in the food sector (Pradhan et al., 2015). Notably, the use of nanotechnology has significantly increased recently. The fundamental goal of nanopackaging is to reduce the need for preservatives and antibiotics as well as decrease environmental contamination (Fuertes et al., 2016). Various nanomaterials, including silver nanoparticles (Ag NPs), titanium nitride nanoparticles (TiN-NPs), nanotitanium dioxide (TiO_2-NPs), nanozinc oxide (ZnO NPs) and nanoclay are introduced as functional nanoadditives to food packaging (Pal, 2017). This section focuses on novel active nanopackaging that protects food from post-harvest losses.

13.4.1 Active Packaging

Active packaging serves as a barrier between a product and the external environment by using active nanopackaging material (Rai et al., 2019), thus encouraging to enhance and sustain food quality by absorbing inert substances from the packaging and its surroundings (Wyser et al., 2016). Mainly, the elements of active food packaging comprises (a) antimicrobial film, (b) nanocomposites (metal nanoparticles such as silver (Ag NPs), copper (CuO-NPs) and oxides such as titanium dioxide (Kuswandi, 2017) and magnesium oxide (MgO-NPs) and (c) gas scavengers (Lee et al., 2015). For example, silicate nanoparticle (Si-NPs)-packed film can lower oxygen flow into the package allowing moisture to escape and keeping the food fresh (Kasaai, 2015). Ag NPs enhance shelf-life by absorbing ethylene gas (Rana et al., 2021). Active packaging materials with oxygen scavengers can be used to stop oxidation reactions, maintain a suitable atmosphere, maintain quality and extend the food's shelf-life (Echegoyen, 2015). Palladium nanoparticle-based oxygen-scavenging film was developed by Cherpinski et al. (2018) using poly-3-hydroxybutyrate.

Additionally, carbon nanotubes and chitosan nanoparticles are utilized in nanopackaging as active components due to their potential antibacterial properties. Magnesium oxide and zinc dots floating in or mixed into polystyrene foam antimicrobial packaging made of polyvinylpyrrolidone can protect food from germs like *Listeria*, *Salmonella enteritidis* and *E. coli* (Joshi, 2012). The effect of chitosan-zinc oxide nanocomposite-coated low-density polyethylene (LDPE) packaging on the quality and shelf-life of okra were examined by Al-Naamani et al. (2018), and it was found that LDPE film preserves the moisture content of the preserved okra samples while also preventing bacterial and fungal growth on samples of okra.

13.5 FUTURE ASPECTS OF NANOFILM ON ENHANCING OF SHELF-LIFE OF HARVESTABLE PRODUCTS

Nanotechnology has already entered the food and dairy industries, research facilities have been constructed, and potential uses are being investigated (Suthar et al., 2021). However, the application of nanoscience in agriculture is still limited due to a lack of scientific and technical advances in regulating health hazards, environmental impact (toxicity, the effect of exposure, bioaccumulation in the food chain, etc.), public concern and consumer confidence, legal aspects and economics (Kirdar, 2015; Prasad et al., 2017). Therefore, future aspects of nanofilms/nanosciences should focus on risk assessment, policy development and cost-benefit analysis for their effective and efficient application in smart and sustainable agriculture.

13.5.1 Research and Development

Previous studies have confirmed that adding additives to edible nanofilms improves functionality and provides benefits to human health (Radev & Pashova, 2020; Romanazzi & Moumni, 2022). However, further studies are required to develop specific nanofilms for each product with the optimum preservation qualities (Rojas et al., 2015). Understanding the behaviour or effect of adding bioactive chemicals capable of extending shelf-life requires extensive investigation (de Oliveira Filho et al., 2021). Numerous elements must be considered and investigated to link effects on coating performance to specific commodities, such as solution, emulsion and/or film (composition, concentration, viscosity, density, mechanical, permeability, thermal, thickness, temperature, atmospheric conditions) and the product on which it is applied (fresh or processed, water activity, respiration, etc.). The ultimate goal is to develop coatings that can be marketed; hence, future research should be interdisciplinary to supply comprehensive information to build tailored coatings for specific products favourable to the environment (Ashfaq et al., 2022). The nanofilm utilization for extending the shelf-life of harvestable products is still in the preliminary and research stage, with apparent obstacles to overcome before the pit reaches the industrial scale (Bajpai et al., 2018). Through absorption, dissolution and diffusion, nanoparticles may migrate into fruits, vegetables and other harvestable fresh products when in contact with the surfaces. However, the behaviour of nanomaterials inside biological systems has yet to be entirely understood (Chellaram et al., 2014). Additionally, modelling research may help predict how nanotechnology affects the preservation and product quality; thus, it is essential to focus more on pertinent studies (Liu et al., 2020; Basumatary et al., 2022).

13.5.2 Legal Framework and Safety Regulations

Different major and minor edible coating regulations have been introduced by countries including Europe, the United States, Canada, Japan, New Zealand, etc. (Ummi & Siddiquee, 2019; Cloete et al., 2022). A well-defined national or global nanotech policy and an inflexible institutional structure might be a stumbling barrier to the long-term success of the application of nanoscience and nanotechnology in most countries (Patel et al., 2018). Several researchers have highlighted a few gaps in the regulations of these nanocoatings. There is no internationally recognized standard for the additives used in the coating and the labelling of the harvested products (Saxena et al., 2015). It is critical to regulate labelling requirements to ensure consumers receive all essential information about the presence of animal-derived products and any allergen components added (Davari et al., 2017). Before incorporating nanotechnology into the food sector, all responsible authorities, institutions and governmental bodies representing all the countries should discuss the common difficulties and risks connected with its application and should initiate legislative actions (e.g., rules and regulations) (Singh et al., 2017). Moreover, it is essential to create standards for assessing food safety, packaging and other applications of new nanomaterials (Devi et al., 2021).

13.5.3 Environmental Risk Assessment

Several prominent investors in the food business are investigating millions of dollars in introducing nanofilms/nanocoatings which would improve the preservation of harvested products (Jafarzadeh et al., 2021). However, nanoparticles might occasionally have irreversible and adverse impacts on ecosystems (Yousuf et al., 2018). For instance, some chemicals used to construct the nanosystems should be replaced with natural nanosystems to develop nanofilms and design eco-friendly nanoparticles safe for environmental sustainability (Hamad et al., 2018). Moreover, research and industrial collaborations should be strengthened for risk-free product development. (Pradhan et al., 2015; Ghosh et al., 2017). Before releasing of any nanocoated product, a risk assessment of the ecosystem should be carried out and legal aspects of such procedures should be entertained (Odetayo et al., 2022).

13.5.4 Consumer Preference and Economics of Nanofilms

When introducing novel nanofilms into the global fruit and vegetable trade, the cost-effectiveness and industrial applicability should be considered and studied thoroughly (Nile et al., 2020). It is essential to determine at which stage the respective nanofilm would apply to harvestable products to reduce post-harvest losses at the retail or farmer levels (Cloete et al., 2022). The capability of edible coatings to protect the fruits' nutritional value, sensory integrity and freshness while extending their shelf-life should also be considered with a cost-benefit analysis (Vijayakumar et al., 2022). Consumer acceptability is strongly influenced by a need for more knowledge and awareness regarding the safety of edible coatings and the limited marketing efforts to highlight their advantages (Rana et al., 2021). Public understanding and attitudes towards the use of nanotechnology must be developed. Since this technology is still very new and has a promising future, serious efforts must be made to successfully deliver high-quality, fresh food to consumers (Rai et al., 2019).

BIBLIOGRAPHY

Aadil, K. R., Mussatto, S. I., & Jha, H. (2018). Synthesis and characterization of silver nanoparticles loaded poly (vinyl alcohol)-lignin electrospun nanofibers and their antimicrobial activity. *International Journal of Biological Macromolecules, 120*(Part A), 763–767.

Abd-Elsalam, K. A. (2021). Chapter 1. Silver-based nanomaterials for sustainable applications in agroecology: A note from the editor. In K. A. Abd-Elsalam (Ed.), *Silver nanomaterials for agri-food applications* (pp. 1–14). Elsevier.

Abdollahi, M., Alboofetileh, M., Rezaei, M., & Behrooz, R. (2013). Comparing physico-mechanical and thermal properties of alginate nanocomposite films reinforced with organic and/or inorganic nanofillers. *Food Hydrocolloids, 32*(2), 416–424.

Aditya, N. P., Espinosa, Y. G., & Norton, I. T. (2017). Encapsulation systems for the delivery of hydrophilic nutraceuticals: Food application. *Biotechnology Advances, 35*(4), 450–457.

Aguirre-Joya, J. A., Pastrana-Castro, L., Nieto-Oropeza, D., Ventura-Sobrevilla, J., Rojas-Molina, R., & Aguilar, C. N. (2018). The physicochemical, antifungal and antioxidant properties of a mixed polyphenol based bioactive film. *Heliyon, 4*(12), e00942.

Alkasir, K., Naghshiband Hassani, R., & Motalebi Azar, A. (2017). Effects of silver nanoparticles (SNPs) pulsing treatment and sucrose holding on flower and leaf senescence of cut rose. *Journal of Ornamental Plants*, *7*(2), 103–113.

Ashfaq, A., Khursheed, N., Fatima, S., Anjum, Z., & Younis, K. (2022). Application of nanotechnology in food packaging: Pros and cons. *Journal of Agriculture and Food Research*, *7*, 100270.

Aytac, Z., Ipek, S., Durgun, E., Tekinay, T., & Uyar, T. (2017). Antibacterial electrospun zein nanofibrous web encapsulating thymol/cyclodextrin-inclusion complex for food packaging. *Food Chemistry*, *233*, 117–124.

Bagchi, A., & He, T. (2012). Intelligent sensing and packaging of foods for enhancement of shelf life: Concepts and applications. *International Journal of Scientific and Engineering Research*, *3*(10), 1–13.

Bajpai, V. K., Kamle, M., Shukla, S., Mahato, D. K., Chandra, P., Hwang, S. K., Kumar, P., Huh, Y. S., & Han, Y.-K. (2018). Prospects of using nanotechnology for food preservation, safety, and security. *Journal of Food and Drug Analysis*, *26*(4), 1201–1214.

Barrett, C. B. (2021). Overcoming global food security challenges through science and solidarity. *American Journal of Agricultural Economics*, *103*(2), 422–447.

Basumatary, I. B., Mukherjee, A., Katiyar, V., & Kumar, S. (2022). Biopolymer-based nanocomposite films and coatings: Recent advances in shelf-life improvement of fruits and vegetables. *Critical Reviews in Food Science and Nutrition*, *62*(7), 1912–1935.

Chellaram, C., Murugaboopathi, G., John, A. A., Sivakumar, R., Ganesan, S., Krithika, S., & Priya, G. (2014). Significance of nanotechnology in food industry. *APCBEE Procedia*, *8*, 109–113.

Chen, C., Chaudhary, A., & Mathys, A. (2020). Nutritional and environmental losses embedded in global food waste. *Resources, Conservation and Recycling*, *160*, 104912.

Cheng, S., Zhang, Y., Cha, R., Yang, J., & Jiang, X. (2016). Water-soluble nanocrystalline cellulose films with highly transparent and oxygen barrier properties. *Nanoscale*, *8*(2), 973–978.

Cherpinski, A., Gozutok, M., Sasmazel, H. T., Torres-Giner, S., & Lagaron, J. M. (2018). Electrospun oxygen scavenging films of poly(3-hydroxybutyrate) containing palladium nanoparticles for active packaging applications. *Nanomaterials (Basel, Switzerland)*, *8*(7), 469.

Chu, Y., Gao, C., Liu, X., Zhang, N., Tian, X., Feng, X., Yang, Y., Shen, X., & Tang, X. (2020). Improvement of storage quality of strawberries by pullulan coatings incorporated with cinnamon essential oil nanoemulsion. *LWT*, *122*, 109054.

Clarizia, G., & Bernardo, P. (2019). *Diverse applications of organic-inorganic nanocomposites: Emerging research and opportunities*. IGI Global. https://services.igi-global.com/resolvedoi/resolve.aspx?doi=10.4018/978-1-7998-1530-3

Cloete, L., Picot-Allain, C., Ramasawmy, B., Neetoo, H., Ramful-Baboolall, D., & Emmambux, M. N. (2022). Drivers and barriers for commercial uptake of edible coatings for fresh fruits and vegetables industry: A review. *Food Reviews International*, *39*(6), 3481–3514.

Conversa, G., Lazzizera, C., Bonasia, A., & Elia, A. (2020). Harvest season and genotype affect head quality and shelf-life of ready-to-use broccoli. *Agronomy*, *10*(4), 527. https://doi.org/10.3390/agronomy10040527

Dan, N. (2016). Compound release from nanostructured lipid carriers (NLCs). *Journal of Food Engineering*, *171*, 37–43.

Dang, X., Cao, X., Ke, L., Ma, Y., An, J., & Wang, F. (2017). Combination of cellulose nanofibers and chain-end-functionalized polyethylene and their applications in nanocomposites. *Journal of Applied Polymer Science*, *134*(42), 45387.

Davari, M. R., Bayat Kazazi, S., & Akbarzadeh Pivehzhani, O. (2017). Nanomaterials: Implications on agroecosystem. In R. Prasad, M. Kumar, & V. Kumar (Eds.), *Nanotechnology* (pp. 59–71). Springer.

De León-Zapata, M. A., Pastrana-Castro, L., & Barbosa-Pereira, L. Rua-Rodríguez, M. L., Saucedo, S., Ventura-Sobrevilla, J. M., Salinas-Jasso, T. A., Rodríguez-Herrera, R., & Aguilar, C. N. (2017). Nanocoating with extract of tarbush to retard Fuji apples senescence. *Postharvest Biology and Technology, 134*, 67–75.

de Oliveira Filho, J. G., Miranda, M., Ferreira, M. D., & Plotto, A. (2021). Nanoemulsions as edible coatings: A potential strategy for fresh fruits and vegetables preservation. *Foods, 10*(10), 2438. https://doi.org/10.3390/foods10102438

Deng, Z., Jung, J., Simonsen, J., Wang, Y., & Zhao, Y. (2017). Cellulose nanocrystal reinforced chitosan coatings for improving the storability of postharvest pears under both ambient and cold storages. *Journal of Food Science, 82*(2), 453–462.

Devi, L. S., Purkayastha, M. D., Mukherjee, A., & Kumar, S. (2021). Biopolymer-based films and coatings: Emerging technologies to extend shelf-life of fruits and vegetables.

Duro, J. A., Lauk, C., Kastner, T., Erb, K.-H., & Haberl, H. (2020). Global inequalities in food consumption, cropland demand and land-use efficiency: A decomposition analysis. *Global Environmental Change, 64*, 102124.

Dutta, J., & Dobretsov, S. (2018). Nanocomposite zinc oxide-chitosan coatings on polyethylene films for extending storage life of okra (*Abelmoschus esculentus*). *Nanomaterials, 8*(7), 149. https://doi.org/10.3390/nano8070479

Echegoyen, Y. (2015). Nano-developments for food packaging and labeling applications. In M. Rai, C. Ribeiro, L. Mattoso, & N. Duran (Eds.), *Nanotechnologies in food and agriculture* (pp. 141–166). Springer International Publishing.

El-Ramady, H. R., Domokos-Szabolcsy, É., Abdalla, N. A., Taha, H. S., & Fári, M. (2015). Postharvest management of fruits and vegetables storage. In E. Lichtfouse (Ed.), *Sustainable agriculture reviews* (Vol. 15, pp. 65–152). Springer International Publishing.

Esmaeili, H., Nourafkan, E., Nakisa, M., & Ahmed, W. (2021). Application of nanotechnology for biofuel production. *Emerging nanotechnologies for renewable energy*, 149–172). https://doi.org/10.1016/B978-0-12-821346-9.00005-5

Fahami, A., & Fathi, M. (2018). Fabrication and characterization of novel nanofibers from cress seed mucilage for food applications. *Journal of Applied Polymer Science, 135*(6), 45811. https://doi.org/10.1002/app.45811

FAO, E. (2019). Moving forward on food loss and waste reduction. *The State of Food and Agriculture*.Rome. https://doi.org/10.4060/CA6030EN

Flores-López, M. L., Cerqueira, M. A., de Rodríguez, D. J., & Vicente, A. A. (2016). Perspectives on utilization of edible coatings and nano-laminate coatings for extension of postharvest storage of fruits and vegetables. *Food Engineering Reviews, 8*(3), 292–305.

Food Wastage Footprint (Project). (2013). *Food wastage footprint: Impacts on natural resources: Summary report.* Food & Agriculture Org.

Fuertes, G., Soto, I., Carrasco, R., Vargas, M., Sabattin, J., & Lagos, C. (2016). Intelligent packaging systems: Sensors and nanosensors to monitor food quality and safety. *Journal of Sensors, 2016*, e4046061.

Gad, M., & Zagzog, O. (2017). Mixing xanthan gum and chitosan Nano particles to form new coating for maintaining storage life and quality of elmamoura guava fruits. *International Journal of Current Microbiology and Applied Sciences, 6*(11), 1582–1593.

Garavand, F., Rouhi, M., Razavi, S. H., Cacciotti, I., & Mohammadi, R. (2017). Improving the integrity of natural biopolymer films used in food packaging by crosslinking approach: A review. *International Journal of Biological Macromolecules, 104*(Part A), 687–707.

Gardesh, A. S. K., Badii, F., Hashemi, M., Ardakani, A. Y., Maftoonazad, N., & Gorji, A. M. (2016). Effect of nanochitosan based coating on climacteric behavior and postharvest shelf-life extension of apple cv. Golab Kohanz. *LWT: Food Science and Technology, 70*, 33–40.

Gastaldi, L., Battaglia, L., Peira, E., Chirio, D., Muntoni, E., Solazzi, I., Gallarate, M., & Dosio, F. (2014). Solid lipid nanoparticles as vehicles of drugs to the brain: Current state of the art. *European Journal of Pharmaceutics and Biopharmaceutics: Official Journal of Arbeitsgemeinschaft Fur Pharmazeutische Verfahrenstechnik e.V, 87*(3), 433–444.

Geszke-Moritz, M., & Moritz, M. (2016). Solid lipid nanoparticles as attractive drug vehicles: Composition, properties and therapeutic strategies. *Materials Science and Engineering: C, 68*, 982–994.

Ghosh, A., Dey, K., Bauri, F., Mani, A., Dey, A., & Das, B. (2017). *Implication of Nano technology in horticulture.* NEW INDIA PUBLISHING AGENCY (NIPA) (pp. 141–150).

Guan, W., Zhang, J., Yan, R., Shao, S., Zhou, T., Lei, J., & Wang, Z. (2016). Effects of UV-C treatment and cold storage on ergosterol and vitamin D2 contents in different parts of white and brown mushroom (*Agaricus bisporus*). *Food Chemistry, 210*, 129–134.

Guerrero, D. Q., & Zaragoza, M. de la L. Z., Cárdenas, A. A., & Silva, E. M. (2014). *Composition of solid lipid nanoparticles for the long-term conservation of fruits, vegetables, seeds, cereals and/or fresh foodstuffs using a coating* (European Union Patent No. EP2698066A2).

Guimarães, I. C., dos Reis, K. C., Menezes, E. G. T., Rodrigues, A. C., da Silva, T. F., de Oliveira, I. R. N., & de Boas, E. V. B. V. (2016). Cellulose microfibrillated suspension of carrots obtained by mechanical defibrillation and their application in edible starch films. *Industrial Crops and Products, 89*, 285–294.

Gundewadi, G., Rudra, S. G., Sarkar, D. J., & Singh, D. (2018). Nanoemulsion based alginate organic coating for shelf life extension of okra. *Food Packaging and Shelf Life, 18*, 1–12.

Gustavsson, J. (Ed.). (2011). *Global food losses and food waste: Extent, causes and prevention; study conducted for the international congress save food! At interpack 2011, [16–17 May].* Food and Agriculture Organization of the United Nations.

Hamad, A. F., Han, J.-H., Kim, B.-C., & Rather, I. A. (2018). The intertwine of nanotechnology with the food industry. *Saudi Journal of Biological Sciences, 25*(1), 27–30.

Hasan, S. M. K., Ferrentino, G., & Scampicchio, M. (2020). Nanoemulsion as advanced edible coatings to preserve the quality of fresh-cut fruits and vegetables: A review. *International Journal of Food Science and Technology, 55*(1), 1–10.

Hasheminejad, N., & Khodaiyan, F. (2020). The effect of clove essential oil loaded chitosan nanoparticles on the shelf life and quality of pomegranate arils. *Food Chemistry, 309*, 125520.

He, X., & Hwang, H.-M. (2016). Nanotechnology in food science: Functionality, applicability, and safety assessment. *Journal of Food and Drug Analysis, 24*(4), 671–681.

Huang, Y.-C., & Li, R.-Y. (2014). Preparation and characterization of antioxidant nanoparticles composed of chitosan and fucoidan for antibiotics delivery. *Marine Drugs, 12*(8), 4379–4398.

Jafarzadeh, S., Mohammadi Nafchi, A., Salehabadi, A., Oladzad-abbasabadi, N., & Jafari, S. M. (2021). Application of bio-nanocomposite films and edible coatings for extending the shelf life of fresh fruits and vegetables. *Advances in Colloid and Interface Science, 291*, 102405.

Jiang, T., Feng, L., & Wang, Y. (2013). Effect of alginate/Nano-Ag coating on microbial and physicochemical characteristics of shiitake mushroom (*Lentinus edodes*) during cold storage. *Food Chemistry, 141*(2), 954–960.

Jin, W., Xu, W., Liang, H., Li, Y., Liu, S., & Li, B. (2016). Nanoemulsions for food: Properties, production, characterization, and applications. In A. M. Grumezescu (Ed.), *Emulsions* (Vol. 1, pp. 1–36). Academic Press.

Joshi, S. (2012). Solid lipid nanoparticle: A review. *IOSR Journal of Pharmacy (IOSRPHR), 2*(6), 34–44. https://doi.org/10.9790/3013-26103444

Kabir, J., Kore, V., & Tawade, S. (2016). Application of edible coatings on fruits and vegetables. *Imperial Journal of Interdisciplinary Research (IJIR), 3*, 591–603.

Kasaai, M. R. (2015). Nanosized particles of silica and its derivatives for applications in various branches of food and nutrition sectors. *Journal of Nanotechnology, 2015*, e852394.

Katouzian, I., Faridi Esfanjani, A., Jafari, S. M., & Akhavan, S. (2017). Formulation and application of a new generation of lipid nano-carriers for the food bioactive ingredients. *Trends in Food Science and Technology, 68*, 14–25.

Khalifa, I., Barakat, H., El-Mansy, H., & Soliman, S. (2016). Effect of chitosan–olive oil processing residues coatings on keeping quality of cold-storage strawberry (*Fragaria ananassa*. Var. festival). *Journal of Food Quality, 1*, 12.

Khan, A., Khan, R. A., Salmieri, S., Le Tien, C., Riedl, B., Bouchard, J., Chauve, G., Tan, V., Kamal, M. R., & Lacroix, M. (2012). Mechanical and barrier properties of nanocrystalline cellulose reinforced chitosan based nanocomposite films. *Carbohydrate Polymers, 90*(4), 1601–1608. https://doi.org/10.1016/j.carbpol.2012.07.037

Kim, I.-H., Lee, H., Kim, J. E., Song, K. B., Lee, Y. S., Chung, D. S., & Min, S. C. (2013). Plum coatings of lemongrass oil-incorporating carnauba wax-based nanoemulsion. *Journal of Food Science, 78*(10), E1551–E1559.

Kim, J. U., Ghafoor, K., Ahn, J., Shin, S., Lee, S. H., Shahbaz, H. M., Shin, H. H., Kim, S., & Park, J. (2016). Kinetic modeling and characterization of a diffusion-based time-temperature indicator (TTI) for monitoring microbial quality of non-pasteurized angelica juice. *LWT - Food Science and Technology, 67*, 143–150.

Kirdar, S. S. (2015). Current and future applications of nanotechnology in the food industry. Conference paper, ISITES2015, Valencia, Vol. 23.

Kore, V. T., Tawade, S. S., & Kabir, J. (2017). Application of edible coatings on fruits and vegetables. *Imperial Journal of Interdisciplinary Research, 3*(1), 591–603.

Korićanac, A., Miletić, N., Popović, B., Mitrović, O., Lukić, M., Pešaković, M., & Tomić, J. (2019). The effect of ULO and NA storage on changes in the quality of apple fruit (*Malus domestica* Borkh.) during shelf life. *Agronomy, 10*(1), 25.

Koushesh Saba, M., & Amini, R. (2017). Nano-ZnO/carboxymethyl cellulose-based active coating impact on ready-to-use pomegranate during cold storage. *Food Chemistry, 232*, 721–726.

Kumar, N., Kaur, P., Devgan, K., & Attkan, A. K. (2020). Shelf life prolongation of cherry tomato using magnesium hydroxide reinforced bio-nanocomposite and conventional plastic films. *Journal of Food Processing and Preservation, 44*(4), e14379.

Kumar, S., Boro, J. C., Ray, D., Mukherjee, A., & Dutta, J. (2019). Bionanocomposite films of agar incorporated with ZnO nanoparticles as an active packaging material for shelf life extension of green grape. *Heliyon, 5*(6), e01867.

Kumar, S., Mukherjee, A., & Dutta, J. (2020). Chitosan based nanocomposite films and coatings: Emerging antimicrobial food packaging alternatives. *Trends in Food Science and Technology, 97*, 196–209.

Kuswandi, B. (2017). Environmental friendly food nano-packaging. *Environmental Chemistry Letters, 15*(2), 205–221.

Lee, S. Y., Lee, S. J., Choi, D. S., & Hur, S. J. (2015). Current topics in active and intelligent food packaging for preservation of fresh foods. *Journal of the Science of Food and Agriculture, 95*(14), 2799–2810.

Li, X., Yadav, R., & Siddique, K. H. M. (2020). Neglected and underutilized crop species: The key to improving dietary diversity and fighting hunger and malnutrition in Asia and the Pacific. *Frontiers in Nutrition*, *7*, 593711.

Lin, X., Li, H., Lin, S., Xu, M., Liu, J., Li, Y., & He, S. (2019). Improving the postharvest performance of cut spray 'Prince' carnations by vase treatments with nano-silver and sucrose. *Journal of Horticultural Science and Biotechnology*, *94*(4), 513–521.

Liu, R., Liu, D., Liu, Y., Song, Y., Wu, T., & Zhang, M. (2017). Using soy protein SiOx nanocomposite film coating to extend the shelf life of apple fruit. *International Journal of Food Science and Technology*, *52*(9), 2018–2030.

Liu, W., Zhang, M., & Bhandari, B. (2020). Nanotechnology – A shelf life extension strategy for fruits and vegetables. *Critical Reviews in Food Science and Nutrition*, *60*(10), 1706–1721.

Luo, Z., Wang, Y., Jiang, L., & Xu, X. (2015). Effect of Nano-$CaCO_3$-LDPE packaging on quality and browning of fresh-cut yam. *LWT: Food Science and Technology*, *60*(2), 1155–1161.

Ma, L., Zhang, M., Bhandari, B., & Gao, Z. (2017). Recent developments in novel shelf life extension technologies of fresh-cut fruits and vegetables. *Trends in Food Science and Technology*, *64*, 23–38.

Mahela, U., Rana, D. K., Joshi, U., & Tariyal, Y. S. (2020). Nano edible coatings and their applications in food preservation. *Journal of Postharvest Technology*, *8*(4), 52–63.

Mallakpour, S., & Sadaty, M. A. (2016). Thiamine hydrochloride (vitamin B1) as modifier agent for TiO_2 nanoparticles and the optical, mechanical, and thermal properties of poly(vinyl chloride) composite films. *RSC Advances*, *6*(95), 92596–92604.

McClements, D. J. (2012). Nanoemulsions versus microemulsions: Terminology, differences, and similarities. *Soft Matter*, *8*(6), 1719–1729.

Meindrawan, B., Suyatma, N. E., Wardana, A. A., & Pamela, V. Y. (2018). Nanocomposite coating based on carrageenan and ZnO nanoparticles to maintain the storage quality of mango. *Food Packaging and Shelf Life*, *18*, 140–146.

Mihindukulasuriya, S. D. F., & Lim, L.-T. (2014). Nanotechnology development in food packaging: A review. *Trends in Food Science and Technology*, *40*(2), 149–167.

Minas, I., Tanou, G., Belghazi, M., Job, D., Manganaris, G., Molassiotis, A., & Vasilakakis, M. (2012). Physiological and proteomic approaches to address the active role of ozone in kiwifruit post-harvest ripening. *Journal of Experimental Botany*, *63*(7), 2449–2464.

Momin, J. K., & Joshi, B. H. (2015). Nanotechnology in foods. In M. Rai, C. Ribeiro, L. Mattoso, & N. Duran (Eds.), *Nanotechnologies in food and agriculture* (pp. 3–24). Springer International Publishing. https://doi.org/10.1007/978-3-319-14024-7_1

Mpai, S., & Sivakumar, D. (2020). Stimulation of light-emitting diode treatment on defence system and changes in mesocarp metabolites of avocados cultivars (Hass and Fuerte) during simulated market shelf conditions. *Agronomy*, *10*(11), 1654. https://doi.org/10.3390/agronomy10111654

Muraleedaran, K., & Mujeeb, V. A. (2015). Applications of chitosan powder with in situ synthesized nano ZnO particles as an antimicrobial agent. *International Journal of Biological Macromolecules*, *77*, 266–272.

Natrajan, D., Srinivasan, S., Sundar, K., & Ravindran, A. (2015). Formulation of essential oil-loaded chitosan-alginate nanocapsules. *Journal of Food and Drug Analysis*, *23*(3), 560–568.

Neme, K., Nafady, A., Uddin, S., & Tola, Y. B. (2021). Application of nanotechnology in agriculture, postharvest loss reduction and food processing: Food security implication and challenges. *Heliyon*, *7*(12), e08539.

Nile, S. H., Baskar, V., Selvaraj, D., Nile, A., Xiao, J., & Kai, G. (2020). Nanotechnologies in food science: Applications, recent trends, and future perspectives. *Nano-Micro Letters*, *12*(1), 45.

Odetayo, T., Tesfay, S., & Ngobese, N. Z. (2022). Nanotechnology-enhanced edible coating application on climacteric fruits. *Food Science and Nutrition, 10*(7), 2149–2167.

Oh, Y. A., Oh, Y. J., Song, A. Y., Won, J. S., Song, K. B., & Min, S. C. (2017). Comparison of effectiveness of edible coatings using emulsions containing lemongrass oil of different size droplets on grape berry safety and preservation. *LWT - Food Science and Technology, C*, 75, 742–750.

Ortiz-Duarte, G., Pérez-Cabrera, L. E., Artés-Hernández, F., & Martínez-Hernández, G. B. (2019). Ag-chitosan nanocomposites in edible coatings affect the quality of fresh-cut melon. *Postharvest Biology and Technology, 147*, 174–184.

Pal, M. (2017). Nanotechnology: A new approach in food packaging. *Journal of Food: Microbiology, Safety and Hygiene, 2*(2). DOI: 10.4172/2476-2059.1000121

Patel, A., Patra, F., Shah, N., & Khedkar, C. (2018). Application of nanotechnology in the food industry: Present status and future prospects. In: Grumezescu, A. M. & Holban A. M. (eds) *Impact of nanoscience in the food industry* (pp. 1–27).

Pilon, L., Spricigo, P. C., Miranda, M., de Moura, M. R., Assis, O. B. G., Mattoso, L. H. C., & Ferreira, M. D. (Eds.). (2015). Chitosan nanoparticle coatings reduce microbial growth on fresh-cut apples while not affecting quality attributes. *International Journal of Food Science & Technology, 50*(2), 440–448.

Pradhan, N., Singh, S., Ojha, N., Shrivastava, A., Barla, A., Rai, V., & Bose, S. (2015). Facets of nanotechnology as seen in food processing, packaging, and preservation industry. *BioMed Research International, 2015*, e365672. https://doi.org/10.1155/2015/365672

Prasad, R., Gupta, N., Kumar, M., Kumar, V., Wang, S., & Abd-Elsalam, K. A. (2017). Nanomaterials act as plant defense mechanism. In *Nanotechnology: Food and environmental paradigm* (pp. 253–269). Springer. https://doi.org/10.1007/978-981-10-4678-0_14

Radev, R., & Pashova, S. (Eds.). (2020). Application of edible films and coatings for fresh fruit and vegetables. *Food Safety Management, Nanotechnology, 21*(177), 108–112.

Rai, M., Ingle, A. P., Gupta, I., Pandit, R., Paralikar, P., Gade, A., Chaud, M. V., & dos Santos, C. A. (2019a). Smart nanopackaging for the enhancement of food shelf life. *Environmental Chemistry Letters, 17*(1), 277–290.

Ramachandraiah, K., Han, S. G., & Chin, K. B. (2015). Nanotechnology in meat processing and packaging: Potential applications—A review. *Asian-Australasian Journal of Animal Sciences, 28*(2), 290–302.

Rana, R. A., Siddiqui, M. N., Skalicky, M., Brestic, M., Hossain, A., Kayesh, E., Popov, M., Hejnak, V., Gupta, D. R., Mahmud, N. U., & Islam, T. (2021). Prospects of nanotechnology in improving the productivity and quality of horticultural crops. *Horticulturae, 7*(10), 332. https://doi.org/10.3390/horticulturae7100332

Ranjan, S., Dasgupta, N., Chakraborty, A. R., Samuel, S. M., Ramalingam, C., Shanker, R., & Kumar, A. (2014). Nanoscience and nanotechnologies in food industries: Opportunities and research trends. *Journal of Nanoparticle Research, 16*(6), 1–23.

Read, Q. D., Brown, S., Cuéllar, A. D., Finn, S. M., Gephart, J. A., Marston, L. T., Meyer, E., Weitz, K. A., & Muth, M. K. (2020). Assessing the environmental impacts of halving food loss and waste along the food supply chain. *The Science of the Total Environment, 712*, 136255.

Rojas, R., Vicente, A. A., & Aguilar, C. N. (2015). *Advances in the preservation of fruits and vegetables with bioactive coatings.* Boletim do Centro de Pesquisa de Processamento de Alimentos. http://dx.doi.org/10.5380/cep.v33i2.47168

Romanazzi, G., & Moumni, M. (2022). Chitosan and other edible coatings to extend shelf life, manage postharvest decay, and reduce loss and waste of fresh fruits and vegetables. *Current Opinion in Biotechnology, 78*, 102834.

Rubisela, E. (2018). Foliar application of copper nanoparticles increases the fruit quality and the content of bioactive compounds in tomatoes. *Applied Sciences, 8*(7), 1020. https://doi.org/10.3390/app8071020

Sabbah, M., Di Pierro, P., Ruffo, F., Schiraldi, C., Alfano, A., Cammarota, M., & Porta, R. (2020). Glutamic acid as repeating building block for bio-based films. *Polymers*, *12*(7), 1613.

Salvia-Trujillo, L., Rojas-Graü, M. A., Soliva-Fortuny, R., & Martín-Belloso, O. (2015). Use of antimicrobial nanoemulsions as edible coatings: Impact on safety and quality attributes of fresh-cut Fuji apples. *Postharvest Biology and Technology, Complete*, *105*, 8–16.

Sami, R., Elhakem, A., Almushhin, A., Alharbi, M., Almatrafi, M., Benajiba, N., Fikry, M., & Helal, M. (2021). Enhancement in physicochemical parameters and microbial populations of mushrooms as influenced by nano-coating treatments. *Scientific Reports*, *11*(1), 7915.

Saxena, R., Hardainiyan, S., & Nandy, B. C. (2015). Future aspect of antimicrobial nanoproduct in food processing, packaging and safety. *Food Science.*

Severino, R., Vu, K. D., Donsì, F., Salmieri, S., Ferrari, G., & Lacroix, M. (2014). Antibacterial and physical effects of modified chitosan based-coating containing nanoemulsion of mandarin essential oil and three non-thermal treatments against *Listeria innocua* in green beans. *International Journal of Food Microbiology*, *191*, 82–88.

Shafiee-Jood, M., & Cai, X. (2016). Reducing food loss and waste to enhance food security and environmental sustainability. *Environmental Science & Technology*, *50*(16), 8432–8443.

Shafiee-Masouleh, S.-S. (2018). Effects of nano-silver pulsing, calcium sulfate and gibberellin on an antioxidant molecule and vase life of cut gerbera flowers on JSTOR. *Advances in Horticultural Science*, *32*(2), 185. https://doi.org/26525436

Shah, S. W. A., Qaisar, M., Jahangir, M., Abbasi, K. S., Khan, S. U., Ali, N., & Liaquat, M. (2016). Influence of CMC- and guar gum-based silver nanoparticle coatings combined with low temperature on major aroma volatile components and the sensory quality of kinnow (*Citrus reticulata*). *International Journal of Food Science & Technology*, *51*(11), 2345–2352.

Silva, H. D., Cerqueira, M. Â., & Vicente, A. A. (2012). Nanoemulsions for food applications: Development and characterization. *Food and Bioprocess Technology*, *5*(3), 854–867.

Singh, S., Khemariya, P., Rai, A., Rai, A. C., Koley, T. K., & Singh, B. (2016). Carnauba wax-based edible coating enhances shelf-life and retain quality of eggplant (*Solanum melongena*) fruits. *LWT - Food Science and Technology C*, *74*, 420–426.

Singh, T., Shukla, S., Kumar, P., Wahla, V., Bajpai, V. K., & Rather, I. A. (2017). Application of nanotechnology in food science: Perception and overview. *Frontiers in Microbiology*, *8*, 1501.

Singh, V., Yadav, V. K., & Mishra, V. (2020). Nanotechnology: An application in biofuel production. In M. Srivastava, N. Srivastava, P. K. Mishra, & V. K. Gupta (Eds.), *Nanomaterials in biofuels research* (pp. 143–160). Springer.

Springmann, M., Clark, M., Mason-D'Croz, D., Wiebe, K., Bodirsky, B. L., Lassaletta, L., de Vries, W., Vermeulen, S. J., Herrero, M., Carlson, K. M., Jonell, M., Troell, M., DeClerck, F., Gordon, L. J., Zurayk, R., Scarborough, P., Rayner, M., Loken, B., Fanzo, J., … Willett, W. (2018). Options for keeping the food system within environmental limits. *Nature*, *562*(7728), 519–525.

Suhag, R., Kumar, N., Petkoska, A. T., & Upadhyay, A. (2020). Film formation and deposition methods of edible coating on food products: A review. *Food Research International (Ottawa, Ont.)*, *136*, 109582.

Sun, Q., Li, J., & Le, T. (2018). Zinc oxide nanoparticle as a novel class of antifungal agents: Current advances and future perspectives. *Journal of Agricultural and Food Chemistry*, *66*(43), 11209–11220.

Suthar, T. R., Gavane, A. B., Shah, A. Y., & Devkatte, A. N. (2021a). Nanotechnology, a revolutionary technique in the food industry: Systematic review. *Sustainability, Agri, Food and Environmental Research*, *9*(1), 127–141.

Taştan, Ö., Pataro, G., Donsì, F., Ferrari, G., & Baysal, T. (2017). Decontamination of fresh-cut cucumber slices by a combination of a modified chitosan coating containing carvacrol nanoemulsions and pulsed light. *International Journal of Food Microbiology, 260*, 75–80.

Thyberg, K. L., & Tonjes, D. J. (2016). Drivers of food waste and their implications for sustainable policy development. *Resources, Conservation and Recycling, 106*, 110–123.

Toker, R. D., Kayaman-Apohan, N., & Kahraman, M. V. (2013). UV-curable nano-silver containing polyurethane based organic–inorganic hybrid coatings. *Progress in Organic Coatings, 9*(76), 1243–1250.

Tzortzakis, N., Xylia, P., & Chrysargyris, A. (2019). Sage essential oil improves the effectiveness of aloe vera gel on postharvest quality of tomato fruit. *Agronomy, 9*(10), 635. https://doi.org/10.3390/agronomy9100635

Ummi, A. S., & Siddiquee, S. (2019). Nanotechnology applications in food: Opportunities and challenges in food industry. In S. Siddiquee, G. J. H. Melvin, & Md. M. Rahman (Eds.), *Nanotechnology: Applications in energy, drug and food* (pp. 295–308). Springer International Publishing.

Vanham, D., Bouraoui, F., Leip, A., Grizzetti, B., & Bidoglio, G. (2015). Lost water and nitrogen resources due to EU consumer food waste. *Environmental Research Letters, 10*(8), 084008.

Vanza, J. D., Patel, R. B., & Patel, M. R. (2020). Nanocarrier centered therapeutic approaches: Recent developments with insight towards the future in the management of lung cancer. *Journal of Drug Delivery Science and Technology, 60*, 102070.

Vázquez-Rowe, I., Ziegler-Rodriguez, K., Margallo, M., Kahhat, R., & Aldaco, R. (2021). Climate action and food security: Strategies to reduce GHG emissions from food loss and waste in emerging economies. *Resources, Conservation and Recycling, 170*, 105562.

Vijayakumar, M. D., Surendhar, G. J., Natrayan, L., Patil, P. P., Ram, P. M. B., & Paramasivam, P. (2022). Evolution and recent scenario of nanotechnology in agriculture and food industries. *Journal of Nanomaterials, 2022*, 1–17.

Wu, J., Sun, Q., Huang, H., Duan, Y., Xiao, G., & Le, T. (2019). Enhanced physico-mechanical, barrier and antifungal properties of soy protein isolate film by incorporating both plant-sourced cinnamaldehyde and facile synthesized zinc oxide nanosheets. *Colloids and Surfaces. B, Biointerfaces, 180*, 31–38.

Wyser, Y., Adams, M., Avella, M., Carlander, D., Garcia, L., Pieper, G., Rennen, M., Schuermans, J., & Weiss, J. (2016). Outlook and challenges of nanotechnologies for food packaging. *Packaging Technology and Science, 29*(12), 615–648.

Xing, Y., Yang, H., Guo, X., Bi, X., Liu, X., Xu, Q., Wang, Q., Li, W., Li, X., Shui, Y., Chen, C., & Zheng, Y. (2020a). *Effect of chitosan/nano-TiO_2 composite coatings on the postharvest quality and physicochemical characteristics of mango fruits*. Elsevier B.V.

Xing, Y., Yang, S., Xu, Q., Xu, L., Zhu, D., Li, X., Shui, Y., Liu, X., & Bi, X. (2021). Effect of chitosan/nano-TiO_2 composite coating on the postharvest quality of blueberry fruit. *Coatings, 11*(5), 512.

Xu, J., Zhang, M., Bhandari, B., & Kachele, R. (2017). ZnO nanoparticles combined radio frequency heating: A novel method to control microorganism and improve product quality of prepared carrots. *Innovative Food Science & Emerging Technologies, 44*, 46–53.

Yousuf, B., Qadri, O. S., & Srivastava, A. K. (2018). Recent developments in shelf-life extension of fresh-cut fruits and vegetables by application of different edible coatings: A review. *LWT, 89*, 198–209.

Yu, Y., Zhang, S., Ren, Y., Li, H., Zhang, X., & Di, J. (2012). Jujube preservation using chitosan film with nano-silicon dioxide. *Journal of Food Engineering, 113*(3), 408–414.

Yusof, N. A. A., & Zain, N. M. (2019). The effect of ZnO nanoparticles on the physical, mechanical, and antibacterial properties of chitosan/gelatin hydrogel films. *Jurnal Teknologi*, *81*(2), Article 2. https://doi.org/10.11113/jt.v81.12605

Zafar, R., Zia, K. M., Tabasum, S., Jabeen, F., Noreen, A., & Zuber, M. (2016). Polysaccharide based bionanocomposites, properties and applications: A review. *International Journal of Biological Macromolecules*, *92*, 1012–1024.

Zambrano-Zaragoza, M. L., González-Reza, R., Mendoza-Muñoz, N., Miranda-Linares, V., Bernal-Couoh, T. F., Mendoza-Elvira, S., & Quintanar-Guerrero, D. (2018). Nanosystems in edible coatings: A novel strategy for food preservation. *International Journal of Molecular Sciences*, *19*(3), Article 3. https://doi.org/10.3390/ijms19030705

Zhu, Y., Li, D., Belwal, T., Li, L., Chen, H., Xu, T., & Luo, Z. (2019). Effect of nano-SiOx/chitosan complex coating on the physicochemical characteristics and preservation performance of green tomato. *Molecules*, *24*(24), 4552.

Ziv, C., & Fallik, E. (2021). Postharvest storage techniques and quality evaluation of fruits and vegetables for reducing food loss. *Agronomy*, *11*(6), 1133. https://doi.org/10.3390/agronomy11061133

14 Exploiting the Bioprospecting Potential of Nanoparticles with Micro-organisms

*Manaal Zahera, Katyayani Mishra and Faria Fatima**

14.1 INTRODUCTION

Nanotechnology is emerging as a cutting-edge technology involving a wide range of academic disciplines including physics, material science, chemistry, biology and medicine (Castillo-Henríquez et al., 2020). The prefix "nano" in the term nanotechnology is derived from a Greek word nano, which means "dwarf". It refers to any engineered matter that is one billionth (10^{-9} m) in size or at least one of its dimensions. Nanoparticles are naturally formed and present in the environment as a result of processes such as mineralization, natural disasters and geological recycling of matter. As a result, numerous studies have been conducted to investigate the mode of action of nanoparticles on living organisms and their surroundings (Rawat et al., 2021). Because of the ease with which they can be produced in various shapes, nanoparticles are intriguing nanoscale systems (Singhal et al., 2012; Kolahalam et al., 2019). According to their nanoscale sizes, they exhibit a high surface-to-volume ratio, distinctive shapes and dimensions that are close to those of biomolecules, giving them special features for biological applications. The limitations of the chemical and physical approaches for creating nanoparticles can be solved by applying biological technique (X. F. Zhang et al., 2016). Additionally, using micro-organisms in the biogenic synthesis process has provided a trustworthy, long-lasting, secure and environmentally beneficial method for nano-synthesis. It is well-known that bacterial, algal, fungal and yeast cells can take metals from their surroundings and transform them into elemental nanoparticle forms that can then be collected or released. Viral technology has also been used to create strong nanocarriers (Ghosh et al., 2021). Capping agents are frequently released during biosynthesis to inhibit aggregation and encourage stability of the nanoparticles (Javed et al., 2020). Biogenic synthesis of NPs using microbes is a recent trend and a greener approach than physical and chemical methods of synthesis, which require higher costs, more energy consumption and complex reaction conditions with potentially hazardous environmental

* Corresponding author

DOI: 10.1201/9781003376446-14

consequences (Khan et al., 2019). Several micro-organisms have been observed trapping metals in-situ and converting them to elemental NPs forms. They have been discovered to accumulate both inside and outside the cell, as well as in the periplasmic space. Despite the toxicity of NPs, the driving factor for NP production within micro-organisms remains unknown.

14.2 NANOPARTICLES SYNTHESIS BY MICROBES

There are three main ways to create nanoparticles: physically, chemically and biologically. These three methods for synthesizing nanoparticles fall under either the top-down or bottom-up categories (Bahrulolum et al., 2021). One of the primary goals of nanotechnology is to develop an environmentally friendly manufacturing process capable of producing low-toxicity nanoparticles (Hulkoti & Taranath, 2014). As a result, the biological synthesis of nanoparticles includes a diverse range of natural species including viruses, bacteria, fungi, algae and plants (using their enzymes, proteins, DNA, lipids and carbohydrates, etc.) (Figure 14.1).

Microbes that can reduce metals have been discovered to be environmentally friendly catalysts for bioremediation and material synthesis. Indeed, microbes may aid in the synthesis of various metal oxides via respiration processes (Koul et al., 2021). Microbes are cultured in the proper conditions for the extracellular production of NPs. The supernatant containing microbial enzymes is obtained after centrifuging the broth containing

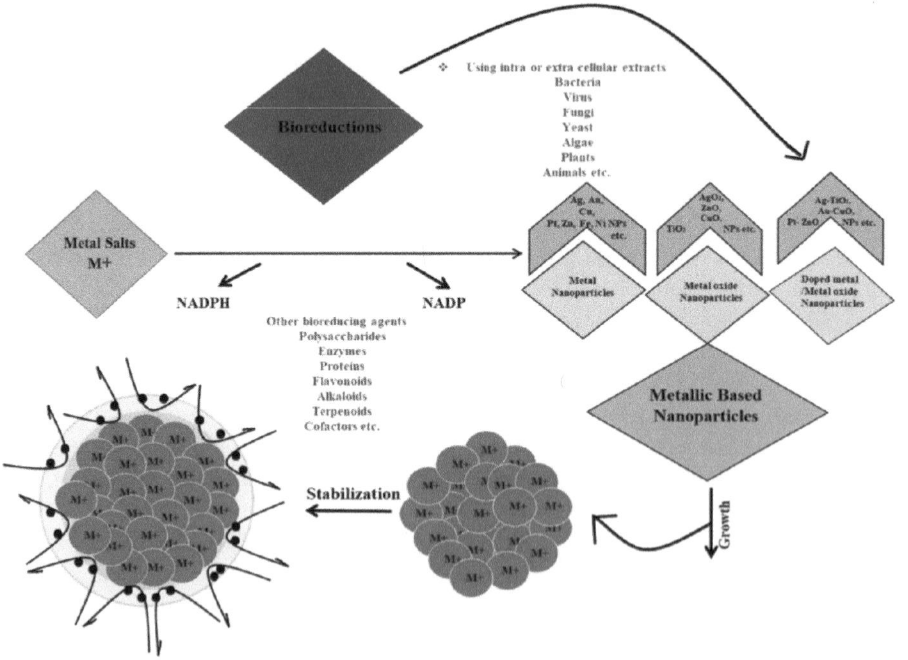

FIGURE 14.1 Mechanism of metal-based nanoparticles synthesis via micro-organisms and other bio reducing agents

TABLE 14.1
Metal nanoparticles synthesized by microbes

Micro-organisms	Products	Location	References
Shewanella oneidensis	Ag	Extracellular	Wu & Ng, 2017
Candida utilis		Intracellular	Waghmare et al., 2015
Rhodopseudomonas capsulate		Extracellular	Malarkodi et al., 2013
Trichoderma viride	Ag	Extracellular	Castro-Longoria, 2022
Yeast	Au/Ag	Extracellular	Shu et al., 2020
Ureibacillus thermosphaericus	Au	Extracellular	Juibari et al., 2015
Alternaria brassicae	Au	Extracellular	Fatima et al., 2016
Enterobacter sp	Hg	Intracellular	El-Shanshoury et al., 2012
Shewanella sp	Se	Extracellular	Mondal et al., 2020
Pyrobaculum islandicum	Cr, Co, Mn	Extracellular	Iravani, 2014
Aeromonas hydrophila, Pseudomonas aeruginosa	Zn	Extracellular	Jayaseelan et al., 2012.
Aspergillus terreus, Candida albicans, Fusarium spp.		Extracellular	
Bacillus mycoides, Lactobacillus sp. Sachharomyces cerevisae	Ti-NPs	Extracellular	Órdenes-Aenishanslins et al., 2014

microbial cells is subsequently applied to the synthesis of NPs. Reductase enzyme-containing supernatant is allowed to interact with the metal ions in a different vessel. NPs are produced as a result of the bioreduction of metal ions in a cell-free supernatant. The features of freshly produced NPs (morphology, uniformity and interaction) are subsequently assessed using XRD (X-ray diffraction), SEM (scanning electron microscopy), TEM (transmission electron microscopy) and FTIR (Fourier transform infrared spectroscopy). The mixture of metals and microbial biomass is then grown under the desired incubation conditions until a particular chromatic shift is noticed. In this procedure, positively charged metal ions are held inside negatively charged microbial cell walls. Enzymes bioreduce these metal ions within the cell wall, forming nanoclusters along the way before the NPs finally diffuse out of the cell wall and into the solution. As a consequence, the microbial cells absorb the metal ions throughout the intracellular process, and in the presence of cellular enzymes, metal ions are changed into NPs within and the metal ions are trapped outside the cell surface in the extracellular process where they are reduced by cellular enzymes to create NPs (Table 14.1). It is known that the kind of live extract, its concentration, the concentration of the metal salt, the pH, temperature and contact duration all have an impact on how quickly the nanoparticles are produced as well as their number and other properties.

14.2.1 Nanoparticles Synthesis by Bacteria

Bacteria have a remarkable ability to reduce heavy metal ions, making them excellent candidates for nanoparticle synthesis. For example, some bacterial species have evolved the ability to use specific defence mechanisms to combat stresses such as heavy metal ion or metal toxicity. It was discovered that some of them could survive and grow even in environments with high metal ion concentrations. There have been reports of additional biomineralization phenomena, including the production of tellurium (Te) by *Escherichia coli K12*, the direct enzymatic reduction of Tc (VII) by resting cells of *Shewanella* (previously *Alteromonas*) *putrefaciens* and *Geobacter metallireducens* (previously known as strain GS-15), and the conversion of selenite to selenium by *Enterobacter cloacae* removal of Ag^{+}, Cd^{2+}, Cu^{2+} and La^{3+} from solutions by *Bacillus cereus*, *B. subtilis*, *E. coli*, and *P. aeruginosa.* They discovered that bacterial cells had the capacity to bind significant amounts of metallic cations. *Deinococcus radiodurans* was used in the biosynthesis of AuNPs in three forms: spherical, triangular and irregular (approximate size 43.75 nm) (Iravani, 2014). Silver NPs-cell-free culture supernatants of five psychrophilic bacteria *Phaeocystis antarctica*, *Pseudomonas proteolytica*, *Pseudomonas meridiana*, *Arthrobacter kerguelensis*, and *Arthrobacter gangotriensis* and two *mesophilic bacteria Bacillus indicus* and *Bacillus cecembensis* have been used to biosynthesize silver NPs (~6–13 nm) (Koul et al., 2021). These NPs remained stable in the dark for eight months. The synthesis and stability of silver nanoparticles appeared to be affected by temperature, pH and the species of bacteria used in the supernatant. It was discovered that the *A. kerguelensis* supernatant could not produce silver NPs at the same temperature that *P. antarctica* could (Iravani, 2014). As a result, this study provided important evidence that between bacterial species.

14.2.2 Nanoparticles Synthesis by Fungi

Fungi can be easily cultured under controlled conditions and are well-known for secreting metabolites and enzymes involved in the formation of nanoparticles or nanominerals. Nanoparticles can form intracellularly or extracellularly, with the latter being more convenient for harvesting, while the cell wall also provides numerous nucleation sites for their formation. Moreover, fungal mediated silver nanoparticles have received a lot of attention recently because of their antimicrobial properties (Fatima et al., 2016). Early research found that the yeasts *Debaryomyces hansenii*, *Candida albicans*, *Saccharomyces cerevisiae* and *Rhodotorula rubra*, as well as the polymorphic fungus *Aureobasidium pullulans*, could grow in silver-containing media, resulting in blackening of colonies and the external medium. *Fusarium oxysporum*, *Bipolaris tetramera* (Fatima et al., 2016) and *Verticillium sp.* can synthesize gold (Au (0)) nanoparticles either extracellularly or intracellularly because they produce significant amounts of extracellular proteins and secondary metabolites that act as a reductase and stabilization agent for the nanoparticles produced. Several *Fusarium oxysporum* strains produced extracellular Ag nanoparticles with sizes ranging from 20 to 50 nm.

14.2.3 Nanoparticles Synthesis by Actinomycetes

Actinomycetes have been investigated as distinct nanomachineries for the extracellular and intracellular synthesis of gold, silver and various types of metallic NPs.

Actinomycetes are an interesting group of microbes for nanomaterials because of their unique ability to produce various bioactive compounds and enzymes that act as strong reductants (Manivasagan et al., 2016). Actinomycetes-derived nanoparticles have been extensively tested for a variety of applications. Characterization of actinomycetes-derived NPs has been carried out using techniques such as FT-IR, TEM, UV-vis, SEM, AFM, XRD and DLS, among others (Kumari et al., 2020). Due to the significant surface to bulk ratio of silver atoms, silver is primarily constituted of silver oxide. A powerful antibacterial, antifungal, antiviral and anti-inflammatory agent is silver nanoparticles. Silver nanoparticles were created using a quick, cheap and environmentally friendly biosynthetic technique.

14.2.4 Nanoparticles Synthesis by Yeast

Yeasts are non-pathogenic, eukaryotic micro-organisms that come in a variety of strains and have economic value in breweries and bakeries as nutrient supplements and for their ability to ferment sugar. A recently discovered use of yeasts is their ability to produce metal (gold, silver, titanium, palladium and selenium) NPs of specific dimensions, either extracellularly or intracellularly, via reducing enzymes involving different mechanisms (Skalickova et al., 2017). These synthesized NPs are also extremely useful as antibacterial, antifungal, antioxidant and anticancer agents. The yeast biomass is then transferred to water to allow the compounds involved in the synthesis to be released (Moghaddam et al., 2015). The strains were cultivated in YNBG liquid medium at 22°C with 160 rpm agitation for 96 hours in order to synthesize nanoparticles using yeast biomass and yeast supernatant. Centrifugation was used to separate the yeast biomass from the culture broth (5000 rpm, 10°C, 10 min). The yeast biomass was then washed three times with sterile distilled water. Actinomycetes can produce both intracellular and extracellular nanoparticles.

14.2.5 Nanoparticles Synthesis by Algae

The capacity of algae to collect heavy metal ions and transform into more malleable forms is well-known (Mutanda et al., 2020). Algae have thus been recommended as model organisms for the synthesis of bio-nanomaterials. Phycocyanin and phycoerythrin contain varying concentrations of carbohydrates, vitamins, nutrients, oil, fats, polyunsaturated fatty acids, bioactive substances such as antioxidants (polyphenols and tocopherols), pigments such as carotenoids (carotene and xanthophyll), chlorophylls and phycobilin's, depending on the species and age of the algae. These active substances have been described as reducing and stabilizing agents for the formation of nanoparticles in theory (Menaa et al., 2021). Nanoparticles can be synthesized intracellularly or extracellularly, depending on the algae species and method of activity. Without any additional processing steps, a live culture of algae is used. There is already evidence for the synthesis of Ag and Au nanoparticles using living cyanobacterial strains of *Anabaena flos-aquae*, *Calothrix pulvinate* and *Leptolyngbya foveolarum*. Two district paths are possible for synthesis. There are many papers that describe different studies that were done by researchers to create silver nanoparticles utilizing living culture. Three different genera of micro-alga live cultures (*Nannochloropsis oculata* (*Ochrophyta*), *Dunaliella salina* and *C. vulgaris* (*Chlorophyta*)) were subjected to three different doses of aqueous AgNO3 solution (1 mM, 2 mM and 5 mM) (Chugh et al., 2021).

14.2.6 Nanoparticles Synthesis by Virus

Because of their well-characterized geometries with surface and size uniformity, viruses have become an ideal example for nanoscale fabrication/nanomaterials. They replicate and produce in living cells/host cells, allowing millions of nanoparticles to be assembled. The plant virus nanoparticles (VNPs) and virus-like nanoparticles (VLPs) produced by viral nanotechnology have emerged as a flexible channel for a number of applications, including attaining optimum selectivity as well as specificity, optics and biosensing, therapeutic agents and targeting, nanocatalysis and next-generation nanoelectronics. Plant viral capsid proteins may self-assemble and generate well-organized icosahedral viruses with variable coat protein components, internal and external size features, and can also help produce unique nanomaterials. The sensitive chemicals are specifically protected and encapsulated by the viral inside, while the surface of the virus may be employed to precisely coat tiny molecules. VNPs/VLPs have been developed as a result of the characteristics of viruses, especially their biocompatibility, to accomplish targeted drug delivery. Plant viruses have been modified to serve as vaccines against different infections since they are naturally immunogenic. In order to provide an efficient and different approach connected to both the fields of medicine and nanotechnology, we address their applications and function in nanoparticle synthesis in this chapter (Srinivasa et al., 2021).

14.3 APPLICATION OF MICROBIAL SYNTHESIZED NANOMATERIALS

One of the most recent technical advancements that shows specific targeted qualities with increased strength is nanoparticle engineering. There are several uses for nanomaterials in medical, environmental research, agriculture and food processing as a result of recent advances in their manufacturing (Fig. 11.2). Precision agriculture, which entails the introduction of wireless networking and the simplification of the sensors for monitoring, evaluating and regulating agricultural activities, has presented intriguing uses for nanotechnology in agriculture. Recent developments in bioengineering and targeted genetic modification (GM) delivery using engineered nanomaterials are notable scientific achievements (Shang et al., 2019).

Additionally, nanotechnology offers good alternatives for a growing variety of environmental problems. Therefore, such ongoing advancements in nanotechnology, with a focus on issue identification and the creation of cooperative ways, have a great potential to bring about a range of social and equitable advantages.

14.3.1 Biocontrol Agent

Recent years have seen an upsurge in the use of silver-based antiseptics due to the presence and growth of bacteria that are resistant to several antibiotics. A fungus, *Trichoderma viride*, was used to biosynthesize silver nanoparticles. The enhanced antibacterial activity of the nanoparticles with different antibiotics against gram-positive as well as gram-negative bacteria were also assessed. In the presence of AgNPs, test strain resistance to the antibacterial effects of erythromycin, ampicillin, kanamycin and chloramphenicol was enhanced (Fatima et al., 2021). Ampicillin had the strongest boosting impact when used against test strains. The findings offered valuable information for the creation of novel antimicrobial drugs and demonstrated that antibiotics and AgNPs together had improved antibacterial properties. The *Aspergillus flavus*-derived AgNPs shown potent antibacterial activity against *Trichoderma spp.*, *Bacillus subtilis*, *Staphylococcus aureus*, and *Bacillus subtilis* (Fátima et al., 2016, 2018). Using *Fusarium oxysporum*, Durán and colleagues demonstrated how extracellularly generated silver nanoparticles may be integrated into textiles to prevent or reduce infection with harmful bacteria like *Staphylococcus aureus*. As a result, new antimicrobial agents like silver, gold copper, poly-£-lysine, zinc, iron and chitosan have been produced that have remarkable potential by proving they can be used as effective antibiotic agents against a variety of pathogenic bacterial species. At concentrations of 3 and 2 mg/mL, respectively, the Ni-EPS nanocomposite demonstrated antibacterial and antibiofilm efficacy against resistant pathogens of *Staphylococcus aureus* and *Pseudomonas* aeruginosa (Garza-Cervantes et al., 2019). According to several studies, altering their surfaces activates their anti-adherent and antimicrobial characteristics, which eliminates both gram-negative and gram-positive bacterial biofilms (Table 14.2). *S. aureus*, *E. coli*, and *Shigella* dysentery were examined at doses of 50 and 100 g/ml for their bactericidal effects. Chemically produced FeNPs are often physiologically compatible and soluble in nature. An enhanced antibacterial impact was seen in *S. pneumoniae* bacterial colonies when the antimicrobial medication erythromycin was attached to FeNPs. Similar to this, *P. aeruginosa*, a gram-negative bacteria that causes nosocomial infections, was able to boost the antibactericidal ability of alginate-capped-FeNP-tobramycin compounds. These conjugates exhibit promise lower-cost replacements in medication solutions, antibacterial treatments and other applications where microbial eradication or invasion avoidance is desirable since they also suppress biofilm development.

14.3.2 Anti-biofilm Agents

A biofilm is a consortium of sessile microbes' groups that may cling to both biotic and abiotic surfaces due to glycocalyx with extracellular polymeric molecules. These biofilms result in biofouling on devices and medical equipment as well as a number of chronic disorders. The NPs can effectively stop bacterial development by penetrating a biofilm matrix's exopolysaccharides (Lahiri et al., 2021). The NPs enter the biofilm and alter the quorum-sensing gene pathways, impairing the process of cell-to-cell communication and preventing the production of the biofilm. Using

TABLE 14.2
Bactericidal potencies of some nanoparticles

Nanoparticles	Test bacteria strains	Mechanism of action	References
Silver nanoparticles	*E. coli* and *S. aureus*	DNA damage and protein denaturation	Ramalingam et al., 2016
Copper nanoparticles	*E. coli*, *B. subtilis* and *S. aureus*	Mechanism not clear	Fatima & Wahid, 2022; Kalaiselvan et al., 2022
Zinc oxide nanoparticles	*Camphylobacter jejuni*	Cell membrane destruction	Ravbar et al., 2020
Iron nanoparticles	*Staphylococcus aureus* and *E. coli*	ROS generation and protein denaturation	Darwish et al., 2015
Copper oxide nanoparticles	*S. aureus*, *B. subtilis*, *Psedomonas aeruginosa* and *E. coli*	Protein denaturation	Thakur et al., 2020
Magnetic iron oxide nanoparticles	*S. aureus*	ROS generation	Ismail et al., 2015
Ciprofloxacin loaded chitosan nanoparticles	*Escherichia coli* and *Staphylococcus aureus*	Growth inhibition and increased penetration of NPs loaded drug within the cell	Sobhani et al. (2017)
Silver oxide nanoparticles	*Streptococcus mutans* and *Lactobacillus* acidophilus	Mechanism nor clear yet	Manikandan et al. (2017)
Nickel ferrite nanoparticles	*Staphylococcus aureus*, *Streptococcus pyogenes*, *Escherichia coli* and *Salmonella typhimurium*	Production of reactive oxidative species (ROS) and oxidative stress lead to cell death	Bhosale et al. (2018)

a green synthesis technique, biocomposites of Zn and Ni metal nanoparticles with exopolysaccharide caps were created, ranging in size from 8 to 26 nm, respectively. Microbial mediated spherical shape AgNPs with an average size of 55 nm were assessed for antimicrobial activity exhibited antibiofilm activity against two MDR strains MF953599 and MF953600. [Siddique et al., 2020].

14.3.3 Anthelminthic Activity

AgNPs and *M. charantia* aqueous extract were used in this efficient, secure, economical and environmentally friendly method, where it was discovered that Ag-extract NPs had outstanding anti-helmintic action (Nayaka et al., 2020). The combination of *M. charantia* and AgNPs will be used to create novel anthelmintic drugs as a result of this work. Similar investigations shown that polymeric (85.7%) and lipid

(14.3%) organic nanoparticles had an impact on *Haemonchus contortus*, an infectious gastrointestinal nematode parasite. The use of encapsulated bromelain against egg, larval and adult worm lethality at 4, 2 and 1 mg/ml, respectively, produced the maximum in vitro effectiveness of 100%. *In vivo*, encapsulated *Cymbopogon citratus* nanoemulsion and *Eucalyptus staigeriana* oil both decreased worm load by 83.1% and 83.75%, respectively. *In vivo*, encapsulated *Cymbopogon citratus* nanoemulsion, encapsulated *Eucalyptus staigeriana* oil and encapsulated bromelain were all safe and non-toxic. The cuticle was harmed by encapsulated bromelain, which also led to paralysis as well as death. To combat the growing problem of anthelmintic resistance in *H. contortus*, nanoparticles may be a source for the creation of brand-new anthelmintic medications. Based on the several outcomes, silver nanoparticles, when conjugated with tannic acid, may be considered as potential agents against *Acanthamoeba spp.* (Padzik et al., 2018).

14.3.4 Bioformulations of PGPR for Growth Promotion

Through the development of more effective fertilizers and pesticides, nanotechnology has created new opportunities for sustainable and precise farming. Enhanced seed germination, stem length, seedling weight and physiological characteristics that boost nitrogen metabolism and photosynthetic capacity in many agricultural plants are all results of using these nanomaterials. For instance, by controlling the demand-based release and improving the plant absorption process, the use of porous nanomaterials, such as zeolites, clay or chitosan considerably minimizes nitrogen losses (Shang et al., 2019). Zeolites that have been ammonium charged may make phosphate minerals more soluble, improving the availability and absorption of phosphorus by plants. A carbon-based nanocomposite, called graphene oxide coatings, can prolong the release of potassium nitrate, extending the period of function and reducing leaching losses. Thus, it seems a great initiative to enhance the use of nanomaterials in place of conventional fertilizers in agricultural production. They showed that the application of nanocalcite $CaCO_3$ (40%) with nano-SiO_2 (4%), MgO (1%) and Fe_2O_3 (1%) boosted the absorption of Ca, Mg and Fe and also considerably increased the intake of P with micro-nutrients Zn and Mn. The application of fertilizers with nanotechnology has been found to significantly boost agricultural output. For instance, using fertilizer and carbon nanoparticles combined can boost the grain yields of leafy greens (12.34–19.76%), winter wheat (28.81%), soybean (16.74%), spring maize (10.93%) and rice (10.29%) (Verma et al., 2019). When compared with the control yield factors, the application of chitosan-NPK fertilizer considerably raises the harvest index, crop index and mobilization index of the chosen wheat yield variables. As a result, the use of nanofertilizers can enhance the absorption of nutrients by plants through these pores, or the procedure can make it easier for molecules to combine with root exudates or molecular transporters by opening up new pores, or by using endocytosis or ion channels (Fatima et al., 2021). Therefore, the fertilizer nanoformulations ensure that crops have a balanced diet throughout the growth cycle, which ultimately increases agricultural output. It should be emphasized that a product's improved efficiency can persuade farmers to utilize it more profitably.

14.3.5 Improve Soil Quality

Nanozeolite is effective in delivering minerals and nutrients over a long period of time that supports soil equilibrium and plant development. Microbially mediated nano-Si fertilizers, on the other hand, are used to regulate soil remediation and are preferred over conventional fertilizers in terms of lowering heavy metal build-up. In addition to promoting plant stress tolerance and soil improvement, nanopesticides may be utilized to protect crops effectively (Bala et al., 2022). The effectiveness of mineral absorption is crucial for plant growth since it affects the majority of developmental phases. Most products and fertilizers are designed to boost production without decontaminating soils or waterways or providing protection from various insect pests and micro-biological illnesses. One excellent illustration is nano-Si, which is used to remove hazardous metals from soil. Because accumulated harmful metals can seriously poison many different plant tissues (Azam et al., 2022), plants cannot develop in an environment with reactive oxygen species as a result. Rice plants exposed to cadmium stress benefit from nano-Si that controls Cd build-up. Various additional pressures on plant species, such those caused by nickel and chromium, are improved by nano-Si. Thus, through enhancing soil structure and planning, nanotechnology has had a significant impact on soil practices.

14.3.6 Nanosensor

To ensure soil and plant health, product quality and general safety for sustainable farming and environmental systems, the development of sensing devices to track and assess crop growth and growing conditions, nutritional deficiencies, toxicity, diseases and the admittance of agrochemicals into the environment would be helpful. By avoiding an overdose of agricultural inputs, this can provide a real-time, thorough monitoring of crop production and effective, high-quality data that creates chances for the best management techniques. Utilizing sensor technologies to automate irrigation systems has the enormous potential to increase water usage effectiveness (Koech & Langat, 2018). When there is a water shortage, real-time soil water tension estimation using nanosensors is combined with autonomous irrigation control. The timely use of pesticides or fertilizers to prevent crops from infestation would also benefit from the quick and precise identification of insects or diseases. A wireless nanosensor for spotting insect attacks. This sensor can distinguish between the sorts of insects and the volatile organic compounds released by a variety of host plant species (Omran, 2020). An immunosensor made of nanogold is efficient in spotting the karnal bunt disease in wheat plants (Qasim et al., 2022). Furthermore, the creation of bionic plants using nanoparticles inserted into live plant cells and chloroplasts to sense or image items in their environment, communicate via infrared devices or even self-power plants as light sources offers enormous promise in precision farming. Developments in nanobionic methods for improving crops and monitoring the environment therefore present a new avenue for research into hybrids of functional plant nanomaterials (Alghuthaymi et al., 2021). Recently, a label-free dual-functional platform utilizing upconversion nanoparticles (UCNPs) and gold nanoparticles (AuNPs) has been developed for real-time detection of hazardous Cd2+. Additionally, a considerable improvement has been achieved in the measurement and monitoring of small-scale contaminants like pesticides. As an illustration, photosystem II with biosensors has the capacity to bind several classes of pesticides and can also keep track of chemical contaminants. Such nanosensors offer a simple and affordable approach for the early detection of a variety of organic contaminants in pesticides before they are released into the agro-environment.

14.3.7 Nanoremediation

One of the key issues for heavy metal pollution in soil and water has also been addressed with nanotechnology. For instance, magnetic konjac glucomannan aerogels have been produced with green step characteristics to remove arsenite pollution from groundwater. However, in recent years, research on graphene nanosheets and graphene oxide, which have been connected to the removal of many types of contaminants from water, has received a lot of interest (Asghar et al., 2022). Heavy metal and organic pollutants have been adsorbent removed from residual water using graphene oxide-titanium oxide nanocomposite. Titanium oxide stands out as a good alternative for residual water treatment due to qualities like cheap cost, non-toxic, effective photo-catalyst and biologically and chemically stability. Additionally, several investigations into the photocatalytic activity of titanium dioxide have revealed that it is capable of eradicating a broad variety of gram-positive and gram-negative bacteria, filamentous and unicellular fungus, algae, protozoa, mammalian viruses and bacteriophages (Foster et al., 2011).

14.3.8 Pond Productivity and Fish Growth

The use of vaccinations in aquaculture has proven essential as a pathogen defence strategy to shield the host from these pathogen diseases. Oral administration or injection are the most dependable and efficient methods of immunization in fisheries. The latter, a conventional adjuvant technique, calls for the preparation of vaccines using oil/water formulations, which has several unfavourable consequences. The scientific community has suggested a nano-delivery system as an alternate method for administering vaccinations to fish that is thought to be both safer and more effective in order to get around these issues (Shah & Mraz, 2020). Alginate particles were chosen among them as the initial candidates for the oral administration of vaccinations to aquatic animals. Alginate is a copolymer of D-mannuronic acid (M) and L-guluronic acid (G) that is present in a variety of brown algae species as well as in certain bacteria such as a polysaccharide. It is very desirable for oral administration due to its mechanical and physical stability as well as mucoadhesive qualities that allow it to make contact with epithelial cell walls. Alginate particles are often created for use in fish by emulsification, one of the quickest and most scalable NP production techniques, and to a lesser extent by alternative techniques such orifice-ionic gelation and the spray method. Chitosan (CS), which is often found in the exoskeleton of crustaceans and insects, is thought to be the second-most prevalent biopolymer in nature.

Due to its unique biological properties, which include being bioadhesive, biodegradable, biocompatible and non-toxic, CS-based formulations are primarily employed as drug delivery systems, edible coatings for bio-nanosensors and in a variety of medical fields (dentistry, surgical procedures etc.). For instance, major capsid protein (MCP) gene of *Lymphocystis Disease Virus (LCDV)*, *Vibrio anguillarum* in Asian sea bass (*Lates calcarifer*), *Philasterides dicentrarchi* in *Scophthalmus maximus*, *Vibrio parahaemolyticus* in *Acanthopagrus schlegelii*, and dietary RNA in *Labeo rohita* have all been effectively encapsulated and delivered in like PLGA, (poly D, L-lactic-co-glycolic acid). PLGA is a polymeric material that has been widely utilized to encapsulate and distribute various chemicals in fish. As the name suggests, this polymer is created from two monomers, lactic and glycolic acid, which come in a variety of shapes and sizes. When *Aeromonas hydrophila* was delivered to rohu via PLGA, it significantly increased these fishes' immune-stimulatory and antibody responses compared to the control group (Behera et al., 2010).

14.3.9 Nanotechnology in Food Science

Numerous scientific and industrial fields, including that of the food industry, food science and food microbiology, have been altered by recent advances in nanotechnology. These fields now include food manufacturing, food packaging, nutritional and functional development, food safety, detection of food-borne pathogens and the extension of food and/or packaged food shelf lives. In food processing, the food ingredients containing nanostructures are being created with the promise of better consistency, taste and texture. The shelf-life of many food products is being extended through nanotechnology, and it has also helped reduce the amount of food that is wasted because of microbial contamination (Thiruvengadam et al., 2018). Presently, food additives are delivered in food items using nanocarriers without affecting their fundamental shape. The creation of encapsulation, emulsions, biopolymer matrices, simple solutions and association colloids using nanotechnology provides effective delivery methods with all the properties listed above. In food packaging, nanopolymers are attempting to take the place of traditional materials (Chaudhary et al., 2020) The presence of pollutants, mycotoxins and microbes in food may be demonstrated using nanosensors. In order to more efficiently distribute nutrients like protein and antioxidants for precisely targeted nutritional and health advantages, numerous methods including nanocomposite, nanoemulsification and nanostructuration have been used. It has been discovered that polymeric nanoparticles are ideal for encasing bioactive substances (such as flavonoids and vitamins) and delivering such substances to their intended functions.

Edible nanocoatings on different food components may operate as a barrier against moisture and gas exchange, supply tastes, colours, antioxidants, enzymes and anti-browning agents and may extend the shelf-life of produced meals even after the packaging has been opened. By manipulating the interfacial layer's characteristics around functional components, it is frequently possible to slow down chemical breakdown processes. Montmorillonite (MMT), zinc oxide (ZnO-NPs) capped silicate, silver NPs (Ag NPs) and titanium dioxide (TiO_2NPs) are the most significant nanomaterials utilized in food packaging because their coatings on film act as a barrier against oxygen, carbon dioxide and other noxious substances (Nasir & Kausar, 2015). Additionally, they have the capacity to scavenge oxygen, have antibacterial properties and are temperature-tolerant. The full dispersion of these nanocomposites inside the polymer matrix and their compatibility are the most challenging aspects of their synthesis. As a result, there is a rising need to enhance the mechanical durability, biodegradability and antibacterial efficiency of nanopackaging materials.

14.4 CONCLUSION

Due to their carefully regulated biocompatible dimensions and distinctive characteristics, nanoparticles produced by micro-organisms show promise for a variety of agricultural, environmental and pharmaceutical purposes. Since nanoparticles are frequently covered with a lipid layer or biomolecules that provide them physiological stability and solubility which are crucial for biomedical applications and the constraint of other synthetic techniques, methods of biosynthesis are also advantageous. There are few significant factors that might be taken into account while synthesizing well-characterized nanoparticles. When crucial factors like organism types, genetically determined and genetic attributes of organisms, ideal conditions for cell growth and enzymatic activities, ideal reaction conditions and biocatalyst state selection have been taken into account, biological methods may be used for the production of highly stable and well-characterized NPs. Additionally, the majority of biological research involving microbial nanoparticles has been conducted *in vitro*, making large-scale clinical trials and safety evaluations crucial to understanding their effects *in vivo*. Thus, it is believed that microbial nanoparticles would offer enormous promise in healthcare, agriculture, environmental and medicine food industry with future in-depth investigations.

BIBLIOGRAPHY

Abdeen, S., Geo, S., Praseetha, P. K., & Dhanya, R. P. (2014). Biosynthesis of silver nanoparticles from *Actinomycetes* for therapeutic applications. *International Journal of Nano Dimension*, *5*(2), 155–162

Abid, N., Khan, A. M., Shujait, S., Chaudhary, K., Ikram, M., Imran, M., Haider, J., Khan, M., Khan, Q., & Maqbool, M. (2022). Synthesis of nanomaterials using various top-down and bottom-up approaches, influencing factors, advantages, and disadvantages: A review. *Advances in Colloid and Interface Science*, *300*, 02597.

Alghuthaymi, M. A., Kalia, A., Bhardwaj, K., Bhardwaj, P., Abd-Elsalam, K. A., Valis, M., & Kuca, K. (2021). Nanohybrid antifungals for control of plant diseases: Current status and future perspectives. *Journal of Fungi*, *7*(1), 48.

Altinsoy, B. D., Karatoprak, G. Ş., & Ocsoy, I. (2019). Extracellular directed ag NPs formation and investigation of their antimicrobial and cytotoxic properties. *Saudi Pharmaceutical Journal*, *27*(1), 9–16.

Anu Thakur, N., Kumar, K., & Sharma, K. K. (2020). Application of Co-doped copper oxide nanoparticles against different multidrug resistance bacteria. *Inorganic and Nano-Metal Chemistry*, *50*(10), 933–943.

Asghar, F., Shakoor, B., Fatima, S., Munir, S., Razzaq, H., Naheed, S., & Butler, I. S. (2022). Fabrication and prospective applications of graphene oxide-modified nanocomposites for wastewater remediation. *RSC Advances*, *12*(19), 11750–11768.

Azam, N., Khan, M. W., Sardar, S., Yousaf, I., Zahid, A., Ismail, A., & Javed, M. S. (2022). New insights for improving agriculture soils through nanotechnology. *Haya Saudi Journal of Life Sciences*, *7*, 244–248.

Bahrulolum, H., Saghi, N., Nahid, J., Hossein, T., Vasighe, S. M., Andrew, J. E., & Gholamreza, A. (2021). Green synthesis of metal nanoparticles using microorganisms and their application in the agrifood sector. *Journal of Nanobiotechnology*, *19*, 1–26.

Bala, M., Kumar Bansal, S., & Fatima, F. (2022). Nanotechnology: A boon for agriculture. *Materials Today: Proceedings*, *73*, 267–270. https://doi.org/10.1016/j.matpr.2022.09.498

Behera, T., Nanda, P., Mohanty, C., Mohapatra, D., Swain, P., Das, B., Routray, P., Mishra, B., & Sahoo, S. (2010). Parenteral immunization of fish, Labeo rohita with Poly d, l-lactide-co-glycolic acid (PLGA) encapsulated antigen microparticles promotes innate and adaptive immune responses. *Fish & Shellfish Immunology*, *28*(2), 320–325. https://doi.org/10.1016/j.fsi.2009.11.009

Bhosale, S. V., Ekambe, P. S., Bhoraskar, S. V., & Mathe, V. L. (2018). Effect of surface properties of $NiFe_2O_4$ nanoparticles synthesized by DC thermal plasma route on antimicrobial activity. *Applied Surface Science*, *441*, 724–733.

Castillo-Henríquez, L., Alfaro-Aguilar, K., Ugalde-Álvarez, J., Vega-Fernández, L., Montes de Oca-Vásquez, G., & Vega-Baudrit, J. R. (2020). Green synthesis of gold and silver nanoparticles from plant extracts and their possible applications as antimicrobial agents in the agricultural area. *Nanomaterials*, *10*(9), 1763.

Castro-Longoria, E. (2022). Biosynthesis of metal-based nanoparticles by trichoderma and its potential applications. In N. Amaresan, A. Sankaranarayanan, M. K. Dwivedi, & I. S. Druzhinina (Eds.), *Advances in trichoderma biology for agricultural applications. Fungal biology*. Springer. https://doi.org/10.1007/978-3-030-91650-3_17

Chaudhary, P., Fatima, F., & Kumar, A. (2020). Relevance of nanomaterials in food packaging and its advanced future prospects. *Journal of Inorganic and Organometallic Polymers and Materials*, *30*(12), 5180–5192.

Chokriwal, A., Sharma, M. M., & , A. (2014). Biological synthesis of nanoparticles using bacteria and their applications. *American Journal of PharmacyTech Research*, *4*, 38–61.

Chugh, D., Viswamalya, V. S., & Das, B. (2021). Green synthesis of silver nanoparticles with algae and the importance of capping agents in the process. *Journal of Genetic Engineering and Biotechnology*, *19*(1), 126.

Chung, Y. H., Cai, H., & Steinmetz, N. F. (2020). Viral nanoparticles for drug delivery, imaging, immunotherapy, and theranostic applications. *Advanced Drug Delivery Reviews*, *156*, 214–235.

Consolo, V. F., Torres-Nicolini, A., & Alvarez, V. A. (2020). Mycosinthetized Ag, CuO and ZnO nanoparticles from a promising *Trichoderma harzianum* strain and their antifungal potential against important phytopathogens. *Scientific Reports*, *10*(1), 1–9.

Dadwal, A., Baldi, A., & Kumar Narang, R. (2018). Nanoparticles as carriers for drug delivery in cancer. *Artificial Cells, Nanomedicine, and Biotechnology*, *46*(Suppl. 2), 295–305.

Darwish, M. S., Nguyen, N. H., Ševců, A., & Stibor, I. (2015). Functionalized magnetic nanoparticles and their effect on *Escherichia coli* and *Staphylococcus aureus*. *Journal of Nanomaterials*, *16*, 89.

El-Shanshoury, A. E. R. R., Elsilk, S. E., Ateya, P. S., & Ebeid, E. M. (2012). Synthesis of lead nanoparticles by *Enterobacter* sp. and avirulent *Bacillus anthracis* PS2010. *Annals of Microbiology*, *62*(4), 1803–1810.

Fatima, F., Bajpai, P., Pathak, N., , S., Priya, S., & Verma, S. R. (2015). Antimicrobial and immunomodulatory efficacy of extracellularly synthesized silver and gold nanoparticles by a novel phosphate solubilizing fungus *Bipolaris tetramera*. *BMC Microbiology*, *15*, 1–10.

Fatima, F., Hashim, A., & Anees, S. (2021). Efficacy of nanoparticles as nanofertilizer production: A review. *Environmental Science and Pollution Research*, *28*(2), 1292–1303.

Fatima, F., Pathak, N., & Verma, S. R. (2018). Toxicity and immunomodulatory efficacy of biosynthesized silver myconanosomes on pathogenic microbes and macrophage cells. *Artificial Cells, Nanomedicine, and Biotechnology*, *46*(8), 1637–1645.

Fatima, F., Pathak, N., Verma, S. R., & Bajpai, P. (2016). In vitro antimicrobicidal and cytotoxicity efficacy of gold nanoparticles synthesized from *Alternaria brassicae* (KF934409). *SOJ Pharmacy & Pharmaceutical Sciences (SOJPPS)*, *3*, 1–6.

Fatima, F., Siddiqui, S., & Khan, W. A. (2021). Nanoparticles as novel emerging therapeutic antibacterial agents in the antibiotics resistant era. *Biological Trace Element Research*, *199*(7), 2552–2564.

Fátima, F., Verma, S. R., Pathak, N., & Bajpai, P. (2016). Extracellular mycosynthesis of silver nanoparticles and their microbicidal activity. *Journal of Global Antimicrobial Resistance*, *7*, 88–92.

Fatima, F., & Wahid, I. (2022). Eco-friendly synthesis of silver and copper nanoparticles by *Shizophyllum commune* fungus and its biomedical applications. *International Journal of Environmental Science and Technology*, *19*(8), 7915–7926.

Figueroa, S. M., Fleischmann, D., & Goepferich, A. (2021). Biomedical nanoparticle design: What we can learn from viruses. *Journal of Controlled Release*, *329*, 552–569.

Foster, H. A., Ditta, I. B., Varghese, S., & Steele, A. (2011). Photocatalytic disinfection using titanium dioxide: Spectrum and mechanism of antimicrobial activity. *Applied Microbiology and Biotechnology*, *90*(6), 1847–1868. https://doi.org/10.1007/s00253-011-3213-7

Garza-Cervantes, J. A., Escárcega-González, C. E., Castro, E. D. B., Mendiola-Garza, G., Marichal-Cancino, B. A., López-Vázquez, M. A., & Morones-Ramirez, J. R. (2019). Antimicrobial and antibiofilm activity of biopolymer-Ni, Zn nanoparticle biocomposites synthesized using *R. mucilaginosa* UANL-001L exopolysaccharide as a capping agent. *International Journal of Nanomedicine*, *14*, 2557.

Ghosh, S., Ahmad, R., Banerjee, K., AlAjmi, M. F., & Rahman, S. (2021). Mechanistic aspects of microbe-mediated nanoparticle synthesis. *Frontiers in Microbiology*, *12*, 638068.

Golinska, P., Wypij, M., Ingle, A. P., Gupta, I., Dahm, H., & Rai, M. (2014). Biogenic synthesis of metal nanoparticles from actinomycetes: Biomedical applications and cytotoxicity. *Applied Microbiology and Biotechnology*, *98*(19), 8083–8097.

Guilger-Casagrande, M., & Lima, R. D. (2019). Synthesis of silver nanoparticles mediated by fungi: A review. *Frontiers in Bioengineering and Biotechnology*, *7*, 287.

Hamida, R. S., Ali, M. A., Redhwan, A., & Bin-Meferij, M. M. (2020). Cyanobacteria – A promising platform in green nanotechnology: A review on nanoparticles fabrication and their prospective applications. *International Journal of Nanomedicine*, *15*, 6033.

Hulkoti, N. I., & Taranath, T. C. (2014). Biosynthesis of nanoparticles using microbes—A review. *Colloids and Surfaces, Part B: Biointerfaces*, *121*, 474–483.

Iravani, S. (2014). Bacteria in nanoparticle synthesis: Current Status and Future Prospects. *International Scholarly Research Notices*, *2014*, Article ID 359316, 18p. https://doi.org/10.1155/2014/359316

Iravani, S., & Zolfaghari, B. (2022). Plant viruses and bacteriophages for eco-friendly synthesis of nanoparticles: Recent trends and important challenges. *Comments on Inorganic Chemistry*, *42*(4), 226–248.

Ismail, R. A., Sulaiman, G. M., Abdulrahman, S. A., & Marzoog, T. R. (2015). Antibacterial activity of magnetic iron oxide nanoparticles synthesized by laser ablation in liquid. *Materials Science and Engineering: C, 53*, 286–297.

Javed, R., Zia, M., Naz, S., Aisida, S. O., & Ao, Q. (2020). Role of capping agents in the application of nanoparticles in biomedicine and environmental remediation: Recent trends and future prospects. *Journal of Nanobiotechnology, 18*, 1–15.

Jayaseelan, C., Rahuman, A. A., Kirthi, A. V., Marimuthu, S., Santhoshkumar, T., Bagavan, A., & Rao, K. B. (2012). Novel microbial route to synthesize ZnO nanoparticles using *Aeromonas hydrophila* and their activity against pathogenic bacteria and fungi. *Spectrochimica Acta. Part A: Molecular and Biomolecular Spectroscopy, 90*, 78–84.

Juibari, M. M., Yeganeh, L. P., Abbasalizadeh, S., Azarbaijani, R., Mousavi, S. H., Tabatabaei, M., & Salekdeh, G. H. (2015). Investigation of a hot-spring extremophilic *Ureibacillus thermosphaericus* strain thermo-BF for extracellular biosynthesis of functionalized Gold nanoparticles. *BioNanoScience, 5*(4), 233–241.

Kalaiselvan, D., Saravanan, K. M., Rangasamy Nandakumar, M. M., Phukan, G. B., & Arun, M. (2022). *The antibacterial efficacy of synthesized bionanomaterials produced by Acalypha indica plant extract.* Society of Education.

Khambhati, K., Bhattacharjee, G., Gohil, N., Braddick, D., Kulkarni, V., & Singh, V. (2019). Exploring the potential of cell-free protein synthesis for extending the abilities of biological systems. *Frontiers in Bioengineering and Biotechnology, 7*, 248.

Khan, I., Saeed, K., & Khan, I. (2019). Nanoparticles: Properties, applications and toxicities. *Arabian Journal of Chemistry, 12*(7), 908–931.

Koech, R., & Langat, P. (2018). Improving irrigation water use efficiency: A review of advances, challenges and opportunities in the Australian context. *Water, 10*(12), 1771.

Kolahalam, L. A., Viswanath, I. K., Diwakar, B. S., Govindh, B., Reddy, V., & Murthy, Y. L. N. (2019). Review on nanomaterials: Synthesis and applications. *Materials Today: Proceedings, 18*, 2182–2190.

Koudelka, K. J., Pitek, A. S., Manchester, M., & Steinmetz, N. F. (2015). Virus-based nanoparticles as versatile nanomachines. *Annual Review of Virology, 2*(1), 379.

Koul, B., Poonia, A. K., Yadav, D., & Jin, J. O. (2021). Microbe-mediated biosynthesis of nanoparticles: Applications and future prospects. *Biomolecules, 11*(6), 886.

Kumari, S., Tehri, N., Gahlaut, A., & Hooda, V. (2020). Actinomycetes mediated synthesis, characterization, and applications of metallic nanoparticles. *Inorganic and Nano-Metal Chemistry, 51*(10), 1386–1395.

Krishnan, S., Bhardwaj, S. K., Liu, S., Xing, R., & Chavali, M. (2021). Virus-assisted biological methods for greener synthesis of nanomaterials. In: Kharisov, B. and Kharissova, O. *Handbook of greener synthesis of nanomaterials and compounds.* Elsevier.(pp. 785–806).

Lahiri, D., Nag, M., Sheikh, H. I., Sarkar, T., Edinur, H. A., Pati, S., & Ray, R. R. (Eds.). (2021). Microbiologically-synthesized nanoparticles and their role in silencing the biofilm signaling cascade. *Frontiers in Microbiology, 12*, 636588.

Malarkodi, C., Rajeshkumar, S., & Vanaja, M. (2013). Eco-friendly synthesis and characterization of gold nanoparticles using *Klebsiella pneumoniae. Journal of Nanostructure in Chemistry, 3*(1), 30.

Manikandan, V., Velmurugan, P., Park, J. H., Chang, W. S., Park, Y. J., Jayanthi, P., Cho, M., & Oh, B. T. (2017). Green synthesis of silver oxide nanoparticles and its antibacterial activity against dental pathogens. *3 Biotech, 7*(1), 72. https://doi.org/10.1007/s13205-017-0670-4

Manivasagan, P., Venkatesan, J., Sivakumar, K., & Kim, S. K. (2016). Actinobacteria mediated synthesis of nanoparticles and their biological properties: A review. *Critical Reviews in Microbiology, 42*(2), 209–221.

Martínez-Castañon, G. A., Nino-Martinez, N., Martinez-Gutierrez, F., Martinez-Mendoza, J. R., & Ruiz, F. (2008). Synthesis and antibacterial activity of silver nanoparticles with different sizes. *Journal of Nanoparticle Research*, *10*(8), 1343–1348.

Menaa, F., Wijesinghe, U., Thiripuranathar, G., Althobaiti, N. A., Albalawi, A. E., Khan, B. A., & Menaa, B. (2021). Marine algae-derived bioactive compounds: A new wave of nanodrugs? *Marine Drugs*, *19*(9), 484.

Moghaddam, A. B, Namvar, F., Moniri, M., Tahir, Md., P., Azizi, S., & Mohamad, R. (2015). Nanoparticles biosynthesized by fungi and yeast: A review of their preparation, properties, and medical applications. *Molecules*, *20*(9), 16540–16565.

Mohamed, N. A., Marei, I., Crovella, S., & Abou-Saleh, H. (2022). Recent developments in nanomaterials-based drug delivery and upgrading treatment of cardiovascular diseases. *International Journal of Molecular Sciences*, *23*(3), 1404.

Molnár, Z., Bódai, V., Szakacs, G., Erdélyi, B., Fogarassy, Z., Sáfrán, G., & Lagzi, I. (2018). Green synthesis of gold nanoparticles by thermophilic filamentous fungi. *Scientific Reports*, *8*, 1–12.

Mondal, A. H., Yadav, D., Mitra, S., & Mukhopadhyay, K. (2020). Biosynthesis of silver nanoparticles using culture supernatant of *Shewanella* sp. ARY1 and their antibacterial activity. *International Journal of Nanomedicine*, *15*, 8295.

Mukherjee, A., Sarkar, D., & Sasmal, S. (2021). A review of green synthesis of metal nanoparticles using algae. *Frontiers in Microbiology*, *12*, 693899.

Mutanda, T., Naidoo, D., Bwapwa, J. K., & Anandraj, A. (2020). Biotechnological applications of microalgal oleaginous compounds: Current trends on microalgal bioprocessing of products. *Frontiers in Energy Research*, *8*, 598803.

Nasir, A., & Kausar, A. (2015). A review on materials derived from polystyrene and different types of nanoparticles. *Polymer-Plastics Technology and Engineering*, *54*(17), 1819–1849.

Nayaka, S., Chakraborty, B., Bhat, M. P., Nagaraja, S. K., Airodagi, D., Swamy, P. S., & Kanakannanavar, B. (2020). Biosynthesis, characterization, and in vitro assessment on cytotoxicity of actinomycete-synthesized silver nanoparticles on *Allium cepa* root tip cells. *Beni-Suef University Journal of Basic and Applied Sciences*, *9*, 1–18.

Omran, E. S. E. (2020). Nano-technology for real-time control of the red palm weevil under climate change. In Ewis Omran, E.S., Negm, & A. (eds), *Climate change impacts on agriculture and food security in Egypt: Land and water resources—Smart farming—Livestock, fishery, and aquaculture* (pp. 321–344). Springer. https://doi.org/10.1007/978-3-030-41629-4_15

Órdenes-Aenishanslins, N. A., Saona, L. A., Durán-Toro, V. M., Monrás, J. P., Bravo, D. M., & Pérez-Donoso, J. M. (2014). Use of titanium dioxide nanoparticles biosynthesized by Bacillus mycoides in quantum dot sensitized solar cells. *Microbial Cell Factories*, *13*(1), 1–10.

Padzik, M., Hendiger, E. B., Chomicz, L., Grodzik, M., Szmidt, M., Grobelny, J., & Lorenzo-Morales, J. (2018). Tannic acid-modified silver nanoparticles as a novel therapeutic agent against Acanthamoeba. *Parasitology Research*, *117*(11), 3519–3525.

Qasim, M., Akhtar, W., Haseeb, M., Sajjad, H., & Rasheed, M. (2022). Potential role of nanoparticles in plants protection. *Life Sciences Journal*, *19*, 31–38.

Ramalingam, B., Parandhaman, T., & Das, S. K. (2016). Antibacterial effects of biosynthesized silver nanoparticles on surface ultrastructure and nanomechanical properties of gram-negative bacteria viz. *Escherichia coli* and *Pseudomonas aeruginosa*. *ACS Applied Materials and Interfaces*, *8*(7), 4963–4976.

Ravbar, M., Kunčič, A., Matoh, L., Možina, S. S., Šala, M., & Šuligoj, A. (2022). Controlled growth of ZnO nanoparticles using ethanolic root extract of Japanese knotweed: Photocatalytic and antimicrobial properties. *RSC Advances*, *12*(48), 31235–31245.

Rawat, M., Yadukrishnan, P., & Kumar, N. (2021). Mechanisms of action of nanoparticles in living systems. In *Research anthology on synthesis, characterization, and applications of nanomaterials* (pp. 1555–1571). IGI Global. DOI: 10.4018/978-1-7998-8591-7.ch06

Roychoudhury, A. (2020). Yeast-mediated green synthesis of nanoparticles for biological applications. *Indian Journal of Pharmaceutical and Biological Research, 8*, 26–31.

Shah, B. R., & Mraz, J. (2020). Advances in nanotechnology for sustainable aquaculture and fisheries. *Reviews in Aquaculture, 12*(2), 925–942.

Shang, Y., Hasan, M. K., Ahammed, G. J., Li, M., Yin, H., & Zhou, J. (2019). Applications of nanotechnology in plant growth and crop protection: A review. *Molecules, 24*(14), 2558.

Shu, M., He, F., Li, Z., Zhu, X., Ma, Y., Zhou, Z., & Zeng, M. (2020). Biosynthesis and antibacterial activity of silver nanoparticles using yeast extract as reducing and capping agents. *Nanoscale Research Letters, 15*, 1–9.

Siddique, M. H., Aslam, B., Imran, M., Ashraf, A., Nadeem, H., Hayat, S., Khurshid, M., Afzal, M., Malik, I. R., Shahzad, M., & Qureshi, U. (2020). Effect of silver nanoparticles on biofilm formation and EPS production of multidrug-resistant *Klebsiella pneumoniae. BioMed Research International, 2020*, 1–9.

Singh, S., Kumar, V., Dhanjal, D. S., Datta, S., Prasad, R., & Singh, J. (2020). Biological biosensors for monitoring and diagnosis. In Singh, J., Vyas, A., Wang, S., & Prasad, R. (eds), *Microbial biotechnology: Basic research and applications Environmental and Microbial Biotechnology* (pp. 317–335). Springer. https://doi.org/10.1007/978-981-15-2817-0_14

Singhal, V., Singhal, A., Jagatheesh, K., Padmavathi, K., Elangoran, N., Bangr, O., Sil, R. (Eds.). (2012). Multifunctional role of green blood therapy to cure for many diseases. *Chronicles of Young Scientists, 3*, 12.

Skalickova, S., Baron, M., & Sochor, J. (2017). Nanoparticles biosynthesized by yeast: A review of their application. *Kvasny Prumysl, 63*(6), 290–292.

Slavin, Y. N., Asnis, J., Häfeli, U. O., & Bach, H. (2017). Metal nanoparticles: Understanding the mechanisms behind antibacterial activity. *Journal of Nanobiotechnology, 15*(1), 1–20.

Sobhani, Z., Samani, S. M., Montaseri, H., & Khezri, E. (2017). Nanoparticles of chitosan loaded ciprofloxacin: Fabrication and antimicrobial activity. *Advanced Pharmaceutical Bulletin, 7*(3), 427.

Srinivasa, C., Kavitha, G. C., Pallavi, M., Shivamallu, C., Sushma, P., Kollur, S. P., Aiyaz, M., Shukla, A. K., Murali, M., & Ansari, M. A. (2021). Role of viruses in nanoparticles synthesis. In Ansari, M.A., & Rehman, S. (eds), *Microbial nanotechnology: Green synthesis and applications* (pp. 103–119). Springer. https://doi.org/10.1007/978-981-16-1923-6_6

Thiruvengadam, M., Rajakumar, G., & Chung, I. M. (Eds.). (2018). Nanotechnology: Current uses and future applications in the food industry. *3 Biotech, 8*, 1–13.

Verma, S. K., Das, A. K., Gantait, S., Kumar, V., & Gurel, E. (2019). Applications of carbon nanomaterials in the plant system: A perspective view on the pros and cons. *Science of the Total Environment, 667*, 485–499.

Waghmare, S. R., Mulla, M. N., Marathe, S. R., & Sonawane, K. D. (2015). Eco-friendly production of silver nanoparticles using *Candida utilis* and its mechanistic action against pathogenic microorganisms. *3 Biotech, 5*(1), 33–38.

Wen, A. M., & Steinmetz, N. F. (2016). Design of virus-based nanomaterials for medicine, biotechnology, and energy. *Chemical Society Reviews, 45*(15), 4074–4126.

Wu, J. W., & Ng, I. S. (2017). Biofabrication of gold nanoparticles by *Shewanella* species. *Bioresources and Bioprocessing, 4*, 1–9.

Ye, B., An, C., Zhang, Y., Song, C., Geng, X., & Wang, J. (2018). One-step ball milling preparation of nanoscale CL-20/graphene oxide for significantly reduced particle size and sensitivity. *Nanoscale Research Letters*, *13*(1), 1–8.

Zahera, M., Khan, S. A., Khan, I. A., Elgorban, A. M., Bahkali, A. H., Alghamdi, S. M., & Khan, M. S. (2018). Enhancing using glucose encapsulation, the efficacy of CDO NPs against multi-drug resistant *Escherichia coli. Microbial Pathogenesis*, *119*, 42–48.

Zhang, X. F., Liu, Z. G., Shen, W., & Gurunathan, S. (2016). Silver nanoparticles: Synthesis, characterization, properties, applications, and therapeutic approaches. *International Journal of Molecular Sciences*, *17*(9), 1534.

Zhang, X. F., Qu, Y., Shen, W., Wang, J., Li, H., Zhang, Z., & Zhou, J. (2016). Biogenic synthesis of gold nanoparticles by yeast Magnusiomyces ingens LH-F1 for catalytic reduction of nitrophenols. *Colloids and Surfaces A: Physicochemical and Engineering Aspects*, *497*, 280–285.

15 Applications of Nanoparticles as Anti-microbial Agents

Aryadeep Roychoudhury and Ankur Singh*

15.1 INTRODUCTION

Nanoparticles or invisible particulate substances, commonly exhibit a diameter of 1 to 100 nm and display huge scientific relevance in the modern era (Song et al., 2021). By virtue of the huge surface area to volume ratio, nanoparticles serve as potent delivery agents, as well as display paramount anti-microbial properties (Ingle et al., 2013). Nanoparticles have been widely exploited for the sake of human welfare; particularly in medical science and technology, optics, electronics, drug delivery, sensor support, wastewater treatment and agriculture, since the last few decades (Kim et al., 2010; Li et al., 2010; Singh & Nalwa, 2011; Scott & Chen, 2013; Abou-Hammad et al., 2019; Nahrawy et al., 2020; Medhi et al., 2020; Shen et al., 2021). However, the prospective application of various nanoparticles is largely dependent on a series of parameters, viz. size, shape, charge, extent of reactivity and high surface area to mass ratio; thereby seeking the attention of the entire scientific community (Kim et al., 2007). Currently, nanobiomedicine has evolved as an emerging field, which extensively exploits the distinctive properties of a range of important nanoparticles, including silver, platinum, gold and other oxide-based nanoparticles (Zhang et al., 2021).

One of the greatest burdens imposed on the global economy and public health is due to the rising incidences of infectious diseases (caused by bacteria, fungi, viruses, parasites and protozoa) (Singh et al., 2014). Occasionally antibiotics are administered as the first choice for the treatment of such infections. Antibiotics act via inhibiting the microbial growth thereby reducing the associated symptoms and signs associated with the microbial infection (Ewald, 1980). However, a rise in microbial resistance against many commonly used antibiotics, along with the environmental release of antibiotics and an upsurge in drug abuse, has led to stalled progress in the treatment of dreaded infections worldwide (Laxminarayan et al., 2013; Atolani et al., 2020). The huge capacity to mutate and a faster rate of morphological changes aid in the acquisition of microbial resistance. The development of a newer form of β-lactamase by bacteria has resulted in the generation of resistance against the β-lactam group of antibiotics (Fishovitz et al., 2014). In order to overcome this wide spectrum microbial resistance

* Corresponding author.

 DOI: 10.1201/9781003376446-15

against antibiotics, treatment strategies employing nanoparticles seem to be a flourishing and promising field. In addition, the conjugated application of nanoparticles with various antibiotics might prove quite effective in overcoming this wide range of resistance developed by microbes against antibiotics (Sobhani et al., 2017). Planktonic cells and the development of microbial biofilms often lead to the development of antibiotic resistance and the failure of different therapeutic approaches. Bacterial and fungal cells often take resort to biofilm formation, characterized by a complex structure rendering resistance against a range of foreign chemical agents. Harrison et al. (2005) demonstrated that aggregates of microbes tend to be about 10 to 10^4 times more resistant towards anti-microbial agents, as compared to the same microbe residing in suspension. Planktonic micro-organisms can induce antibiotic resistance to a quite lesser extent, as compared to the microbes associated with biofilms. Hence, for the achievement of fungicidal and bactericidal activities, it is quite necessary to inhibit the formation of pathogenic biofilms. Nanotechnology has been found to be an effective strategy in this regard. Biogenic selenium nanoparticles have been reported to be quite effective in the prevention of biofilm formation by pathogenic *Pseudomonas aeruginosa* (Cremonini et al., 2016). Also, titanium dioxide nanoparticles were reported to be quite effective in inhibiting fungal biofilm formation (Haghighi et al., 2013). Gold nanoparticles capped with a second-generation antibiotic, cefaclor displayed quite an appreciable extent of anti-microbial activity, largely due to the presence of the nanoparticle surface-associated β-lactam ring (Rai et al., 2010).

By virtue of the inherent charge, large surface area, shape and size, along with the ability to deliver a range of antibiotics, nanoparticles serve as potent anti-microbial weapons (Yien et al., 2012). Most importantly, the size of nanoparticles largely contributes to their anti-microbial efficiency. In addition, the huge surface area of the nanoparticles aids in the efficient penetration and attachment of these nanoparticles on the cellular surfaces (Yien et al., 2012). Metal nanoparticles tend to occur in a range of shapes, viz. tubes, rods, triangular, spherical, hexagonal, pentagonal and tetragonal. The shapes of nanoparticles also govern the extent and nature of interactions with microbial membranes and crucial microbial enzymes. Cha et al. (2015) demonstrated the shape-dependent behaviour of zinc oxide nanoparticles (particularly pyramid, spherical and plate-shaped nanoparticles) on the β-galactosidase enzyme. The nanoparticle mediated microbial inhibition process initiates with the process of cell penetration, via diffusion or adsorption of nanoparticles at the cellular surface. The negatively charged functional groups on proteins serve as attachment sites for nanoparticles, leading to the destruction of cellular proteins and eventually cellular death. In addition, the production and cellular diffusion of reactive oxygen species (ROS), disruption of cellular integrity and DNA damage also serve as potent inhibitory mechanisms for bacteria, viruses and fungi. Therefore, the destruction and inactivation of microbes occur particularly due to the interaction between an array of surface-exposed microbial groups and nanoparticles (Ray et al., 2007). The anti-microbial power of a range of nanoparticles has been employed to tackle an array of micro-organisms, viz. bacteria, viruses and fungi from time to time. Therefore, this chapter summarizes the efficiency of various nanoparticles against microbial agents, thereby aiming to draw the attention of researchers to conduct a lucid exploration of the promising anti-microbial traits of nanoparticles.

15.2 ANTI-BACTERIAL ROLE OF NANOPARTICLES

Metal nanoparticles have immense potential in preventing various bacterial infections. Depending on some factors like concentration, capping method and particle size, metal nanoparticles can act either as bactericidal or bacteriostatic agents (Slavin et al., 2017). From the existing pieces of evidence, it has been observed that nanoparticles render anti-bacterial activity on various strains of gram-positive and gram-negative bacteria. The gram-negative cell wall comprises a thin layer of peptidoglycan, with the surface harbouring negative charges (Halder et al., 2015). Certain bacterial features aid in the anti-bacterial power of metal nanoparticles. For instance, the negative charges associated with the bacterial surface ensure proper electrostatic interaction between metal nanoparticles and the cell surface, thereby leading to the generation of toxic ROS and subsequent oxidative stress and the destruction of bacterial cells. Moreover, the thin bacterial cell wall also accelerates the penetration of nanoparticles into the cells (Chen et al., 2013; Mai-Prochnow et al., 2016). In the case of gram-positive bacteria, only the outer peptidoglycan layer is present, while the polysaccharide membrane harbouring structural lipopolysaccharides is absent. This cell wall composition rather increases the permeability of the wall towards various substances (Cabeen & Jacobs-Wagner, 2005). In addition, the presence of surface-associated smaller negative charges boosts the penetration of nanoparticles, thereby allowing the easy entry of negatively charged superoxide radicals and peroxide ions even at relatively low concentrations, finally leading to bacterial destruction (Reddy et al., 2007; Padmavathy & Vijayaraghavan, 2008). Zerovalent FeO (iron oxide) nanoparticles have been found to induce bacterial DNA and protein damage via ROS-mediated oxidative stress in some tested bacterial (both gram-positive and gram-negative) strains (Mahdy et al., 2012). Similarly, copper nanoparticles also display appreciable affinity towards amine and carboxyl groups on the bacterial cell surface (Ren et al., 2009), and tend to enter the cell and disrupt the helical structure of DNA molecules by cross-linking within the DNA double helix. Fluorescent silver nanoparticles can effectively inhibit the growth of an array of the bacterial population, viz. *Bacillus megaterium* (gram-positive), *Staphylococcus epidermidis* (gram-positive), *Escherichia coli* (gram-negative) and *Pseudomonas aeruginosa* (gram-negative). This anti-bacterial nanoparticle leads to perforation of the bacterial cell wall thereby resulting in cell death, due to the leakage of cytoplasmic contents (Bera et al., 2014). Additionally, silver nanoparticles obtained from the extracts of tulsi leaves demonstrated significant inhibitory effects on the growth of *Staphylococcus aureus* and *Escherichia coli*, with a minimal inhibitory concentration (MIC) value of 1.25 and 0.314 μg/mL, respectively. However, the exact inhibitory mechanism is yet to be deciphered (Singhal et al., 2011). Aljabali et al. (2018) demonstrated that gold nanoparticles significantly inhibited the growth of *Escherichia coli*, displaying a zone of inhibition of around 0.5 mm.

TABLE 15.1
Bactericidal power of different nanoparticles

Nanoparticle type	Size (nm)	Test bacterial strain	Reference
Biogenic selenium	160–170	*Pseudomonas aeruginosa* (INT, CFC20, CFC21, CFCB, CFCA strains)	Cremonini et al. (2016)
Cephalor reduced gold	52–22	*Staphylococcus aureus Escherichia coli*	Rai et al. (2010)
Calcium oxide	16	*Staphylococcus epidermidis* MTCC435 *Pseudomonas aeruginosa* ATCC27853	Roy et al. (2013)
Chitosan	40	*Salmonella choleraesuis*	Qi et al. (2004)
Copper	10	*Escherichia coli* *Bacillus subtilis* *Staphylococcus aureus*	Jayesh et al. (2008)
Copper oxide	20	*Staphylococcus aureus* *Psedomonas aeruginosaBacillus subtilis* *Escherichia coli*	Azam et al. (2012)
Ciprofloxacin loaded chitosan	72	*Escherichia coli* ATCC 25922 *Staphylococcus aureus* ATCC 25923	Sobhani et al. (2017)
Fluorescent silver	1.5	*Staphylococcus epidermidis* NCIM2493 *Bacillus megaterium Pseudomonas aeruginosa* ATCC27853 *Escherichia coli*	Bera et al. (2014)
Gold-chitosan hybrid	16.9	*Staphylococcus aureus* *Pseudomonas aeruginosa*	Hussein et al. (2020)
Magnesium oxide	10.28	*Escherichia coli* *Enterobacter cloacae Acinetobacter baumannii* *Enterococcus faecalis*, *Staphylococcus aureus*, *Staphylococcus epidermidis*	El-Sayyad et al. (2018)
Magnetic iron oxide	50–110	*Staphylococcus aureus*	Ismail et al. (2015)
Nalidixic acid-vanadium	170–330	*Escherichia coli* *Staphylococcus aureus* *Psedomonas aeruginosaBacillus cereus*	Bueloni et al. (2020)
Silver	7	*Escherichia coli* *Staphylococcus aureus*	Martinez castanon et al. (2008)
Silver oxide	42.7	*Streptococcus mutans Lactobacillus acidophilus*	Manikandan et al. (2017)

(Continued)

TABLE 15.1 (CONTINUED)
Bactericidal power of different nanoparticles

Nanoparticle type	Size (nm)	Test bacterial strain	Reference
Zero-valent iron	31.1	*Escherichia coli* *Staphylococcus aureus*	Mahdy et al. (2012)
Zinc oxide	30	*Camphylobacter jejuni*	Xie et al. (2011)
Zirconium oxide	15–21	*Bacillus subtilis* MTCC 1305 *Staphylococcus aureus* MTCC 3160 *Escherichia coli Pseudomonas aeruginosa* MTCC 2453	Fathima et al. (2017)

Nanoparticles constituting different metallic oxides and metallic compounds display efficient anti-bacterial activity. El-Sayyad et al. (2018) proved the potency of metal oxides in combating several dreaded infections mediated by different pathogenic microbes, including multidrug resistant (MDR) bacterial strains. Calcium oxide nanoparticles tend to interfere with the cellular and membrane integrity of *Staphylococcus epidermidis* and *Pseudomonas aeruginosa* (Roy et al., 2013). Additionally, magnesium oxide nanoparticles significantly inhibit the growth of a consortium of MDR gram-positive, as well as gram-negative bacteria, at a concentration of 7.81 µg/mL. Such inhibition particularly occurs via the liberation of ROS and interaction with the bacterial cell wall (El-Sayyad et al., 2018). Silicon dioxide doped iron (III) oxide nanoparticles having 5.89 to 19.89 nm size, display potent inhibitory effects on the growth of *Bacillus subtilis* and *Escherichia coli* (Arshad et al., 2019). Formation of biofilm has been found to be a quite effective strategy acquired by bacterial strains to evade the inhibitory effects of common anti-bacterial agents. However, metal nanoparticles have been proven to effectively prevent the formation and establishment of bacterial biofilms. For instance, biogenic selenium nanoparticles tend to efficiently inhibit the establishment of *Pseudomonas aeruginosa* mediated biofilms (Cremonini et al., 2016). Similar bactericidal roles of different nanoparticles have been summarized in Table 15.1.

15.3 ANTI-FUNGAL ROLE OF NANOPARTICLES

Fungal infections tend to contribute to the progressive rise in mortality and morbidity rate since ancient times (Friedman et al., 2020). The fungistatic and fungicidal potential of nanoparticles has been widely employed to combat pathogenic fungal outbreaks. Silver nanoparticles upon stabilization with surfactants (about 25 nm size) have robust anti-fungal power when administered against four pathogenic *Candida* strains. The marked surfactant activity of these nanoparticles led to the disruption of the above-tested yeast cell wall (Panacek et al., 2009). In addition, gold nanoparticles harbouring acidification and surfactant activity proved to be equally fungicidal against *Candida* strains, like silver nanoparticles. Interestingly, factors like the size and shape of nanoparticles tend to modulate their fungicidal

TABLE 15.2
Anti-fungal power of different nanoparticles

Nanoparticle type	Size (nm)	Test fungal strain	Reference
Chitosan	170–435	*Aspergillus niger* *Candida albicans* *Fusarium solani*	Yien et al. (2012)
Copper	3-10	*Alternaria alternata* MTCC6572 *Curvularia lunata* MTCC2030 *Phoma destructiva* DBT66 *Fusarium oxysporum* MTCC1755	Kanhed et al. (2014)
Gold	25	*Candida sp.*	Wani and Ahmad. (2012)
Iron oxide	10–30	*Aspergillus niger* *Cladosporium herbarum* *Penicillium chrysogenum,* *Alternaria alternata* *Trichothecium roseum*	Parveen et al. (2018)
SDS stabilized silver nanoparticles	25	*Candida albicans* *Candida parapsilosis* *Candida tropicalis*	Panacek et al. (2009)
Silicon	7–14	*Candida albicans*	Cousins et al. (2007)
Zinc oxide	12–32	*Aspergillus niger ITCC7122* *Alternaria alternata ITCC6531* *Botrytis cinerea ITCC6192* *Penicillium expansum ITCC6755* *Fusarium oxysporum ITCC55*	Jamdagni et al. (2018)
Titanium oxide	70–100	*Candida albicans*	Haghighi et al. (2013)

potency. For instance, nanodiscs of small sizes could potentially inhibit the activity of H^+ ATPase, causing intracellular acidification and eventually cellular death (Wani & Ahmad, 2012). Moreover, nanoparticles can lead to deformation in fungal hyphae and prevent the process of hyphal transition, thereby leading to fungicidal activity (He et al., 2011).

Nanoparticles procured from metal oxides also exhibit an inhibitory impact on different fungi. Previous reports indicate that zinc oxide nanoparticles having a size ranging from 12–32 nm lead to disruption of membrane structure and also lead to alteration in the permeability of a series of fungal species, viz., *Aspergillus niger*, *Alternaria alternata* and *Botrytis cinerea* (Jamdagni et al., 2018). It has been found that titanium oxide nanoparticles at a concentration of about 5.14 μg/mL prevent the formation of *Candida albicans* mediated biofilm (Haghighi et al., 2013). Anti-fungal activity of these nanoparticles particularly occurs due to oxidative stress arising from the generation of ROS, ultimately leading to cell death. Similarly, iron oxide nanoparticles trigger the formation of ROS and induce fungal protein and DNA damage due to oxidative stress (Parveen et al., 2018). The anti-fungal roles of different nanoparticles have been summarized in Table 15.2.

15.4 ANTI-VIRAL ROLE OF NANOPARTICLES

Viruses are the causal agents of many serious illnesses irrespective of the fact that they are even smaller than living cells (Singh et al., 2014). Virucidal agents tend to directly interfere with the first phase of the viral replication mechanism (Cagno et al., 2018). Zinc oxide nanoparticles of size ranging from 16 to 20 nm can efficiently suppress the cellular proliferation of the H1N1 influenza virus, due to its potent anti-viral activity (Ghaffari et al., 2019). Similarly, iron oxide nanoparticles display inhibitory effects on the proliferation of the H1N1 influenza virus (Kumar et al., 2019). By virtue of the impact of zeta potential, the small size of nanoparticles effectively contributes to the anti-viral action (Gaikwad et al., 2013). Silver nanoparticles having a size around 10–80 nm have potent anti-viral activity when administered against the monkeypox virus. However, the 10 nm-sized nanoparticles were found to be more potent when applied as a therapeutic anti-viral agent (Rogers et al., 2008). Additionally, small-sized silver nanoparticles tend to append to viral surfaces and prevent the attachment of viruses to host cells and also attenuate the viral replication process (Elechiguerra et al., 2005).

Silver nanoparticles (30–50 nm sized) can prevent the attack of human immunodeficiency virus (HIV), by inhibiting CD-4 dependent virion binding, infectivity, fusion and post-entry stages of the virus into the host cell (Lara et al., 2010). In addition, Haggag et al. (2019) explained the anti-viral action of silver nanoparticles against hepatitis A virus (HAV), herpes simplex virus (HSV) and coxsackie B4 (CoxB4) viral strains. However, the mechanism of the anti-viral action is still obscure. Silver nanoparticles can also prevent the entry of the H1N1 virus into the host cells (Mori et al., 2013). Lack of precision in the delivery of drugs has become a major setback in the generation of potential anti-viral drug candidates. Presently, researchers tend to emphasize the potency of polymeric nanoparticles as anti-viral drug-delivery agents; due to their controlled and sustained drug-delivery mechanisms, better bioavailability, as well as lesser side effects and toxicity (Haggag et al., 2019). Acyclovir-loaded β-cyclodextrin-poly (4-acryloylmorpholine) conjugate nanoparticles can display considerable anti-viral effects against the HSV-1 virus (Cavalli et al., 2009). Such marked virucidal activity has been achieved particularly due to sustained drug delivery and increased drug accumulation in viral cells. The anti-viral roles of different nanoparticles have been summarized in Table 15.3.

15.5 CONCLUSION

People across the world are regularly challenged by deadly infections, caused by pathogenic agents, viz., bacteria, viruses, fungi, parasites and protozoa. Existing medicines and therapies are often unsuccessful in completely eradicating these microbial infections. Therefore, this study highlights the promising scope of nanoparticles in tackling deadly microbial agents. Presently, nanoparticles have been explored in multiple fields for human welfare. In addition, nanoparticles have immense applications in the biomedical and pharmaceutical sectors. Some unique features like shape, size, surface area and charge of nanoparticles govern their optimal functionality.

TABLE 15.3
Anti-viral power of different nanoparticles

Nanoparticle type	Size (nm)	Test viral strain	Reference
Acyclovir loaded into β-cyclodextrin-poly (4-acryloylmorpholine) conjugate	200	HSV-1 BGM, HSV-1 MRC	Cavalli et al. (2009)
Copper iodide	160	Feline Calicivirus	Shionoiri et al. (2012)
Cuprous oxide	Around 45.4	Hepatitis C virus (HCV)	Hang et al. (2015)
Gold	17	HIV1	Vijayakumar and Ganesan (2012)
Silver	8.91–14.48	HAV-10, HSV-1, CoxB4	Haggag et al. (2019)
Silver/chitosan	3.5–12.9	H1N1 influenza A	Mori et al. (2013)
Iron oxide	10–15	H1N1 Influenza A	Kumar et al. (2019)
Zinc oxide	16–20	H1N1 Influenza	Ghaffari et al. (2019)

Moreover, these factors modulate the role of nanoparticles in combating infectious diseases. It is quite obvious that nanoparticles harbour the immense potential to prevent the growth of pathogenic microbes and the subsequent spread of deadly diseases. The surface-associated charges and small particle size aids in the entry of nanoparticles into microbial cells, leading to the disruption of cellular contents like DNA and proteins, finally triggering the process of programmed cell death.

Combinatorial therapies encompassing both nanotherapy and antibiotics have been widely employed as a methodical approach to tackle widespread microbial resistance. However, there are several drawbacks including efficiency, toxicity and selectivity index, which serve as major factors demanding further analysis. The detailed mechanisms of action of many nanoagents have not been completely deciphered and require more research and study. Additionally, there is also a need for lucid in-vivo analysis for the proper evaluation of the safety of application and effectiveness of nanoparticles against different infectious agents. The scope of application of some nanoparticles has become limited by virtue of their toxic effects with the rise in concentration. However, these nanoagents have a potent scope of application in the field of biomedical sciences, in the coming years, if well harnessed. This precise study adds credence to the previously existing reports on nanoparticles and encourages the scientific community to undertake more research and studies for the establishment of the potency of nanoparticles in the prevention of dreaded microbial infections globally.

ACKNOWLEDGEMENTS

Financial assistance from the Council of Scientific and Industrial Research (CSIR), Government of India, through the research grant [38(1387)/14/EMR-II], Science and Engineering Research Board, Government of India through the grant [EMR/2016/004799] and Department of Higher Education, Science and Technology and Biotechnology, Government of West Bengal, through the grant [264(Sanc.)/ST/P/S&T/1G-80/2017] to Dr. Aryadeep Roychoudhury is gratefully acknowledged. The authors are also thankful to Mr. Swarnavo Chakraborty for his assistance in drafting the chapter.

REFERENCES

Abou Hammad, A. B., Elnahrawy, A. M., Youssef, A. M., & Youssef, A. M. (2019). Sol gel synthesis of hybrid chitosan/calcium aluminosilicate nanocomposite membranes and its application as support for CO_2 sensor. *International Journal of Biological Macromolecules, 125*, 503–509.

Aljabali, A. A., Akkam, Y., Al Zoubi, M. S., Al-Batayneh, K. M., Al-Trad, B., Abo Alrob, O., & Evans, D. J. (2018). Synthesis of gold nanoparticles using leaf extract of *Ziziphus zizyphus* and their antimicrobial activity. *Nanomaterials, 8*(3), 174.

Arshad, M., Abbas, M., Ehtisham-ul-Haque, S., Farrukh, M. A., Ali, A., Rizvi, H., & Iqbal, M. (2019). Synthesis and characterization of SiO_2 doped Fe_2O_3 nanoparticles: Photocatalytic and antimicrobial activity evaluation. *Journal of Molecular Structure, 1180*, 244–250.

Atolani, O., Baker, M. T., Adeyemi, O. S., Olanrewaju, I. R., Hamid, A. A., Ameen, O. M., & Usman, L. A. (2020). Covid-19: Critical discussion on the applications and implications of chemicals in sanitizers and disinfectants. *Excli Journal, 19*, 785–799.

Azam, A., Ahmed, A. S., Oves, M., Khan, M. S., & Memic, A. (2012). Size-dependent antimicrobial properties of CuO nanoparticles against Gram-positive and-negative bacterial strains. *International Journal of Nanomedicine, 7*, 3527.

Bera, R. K., Mandal, S. M., & Raj, C. R. (2014). Antimicrobial activity of fluorescent A g nanoparticles. *Letters in Applied Microbiology, 58*(6), 520–526.

Bueloni, B., Sanna, D., Garribba, E., Casto, G. R., Leon, I. E., & Islan, G. A. (2020). Deign of nalidixic acid-vanadium complex loaded into chitosan hybrid nanoparticles as smart strategy to inhibit bacterial growth and quorum sensing. *International Journal of Biological Macromolecules, 161*, 1568–1580.

Cabeen, M. T., & Jacobs-Wagner, C. (2005). Bacterial cell shape. *Nature Reviews in Microbiology, 3*(8), 601–610.

Cagno, V., Andreozzi, P., D'Alicarnasso, M., Silva, P. J., Mueller, M., Galloux, M., & Weber, J. (2018). Broad-spectrum non-toxic antiviral nanoparticles with a virucidal inhibition mechanism. *Nature Materials, 17*(2), 195–203.

Cavalli, R., Donalisio, M., Civra, A., Ferruti, P., Ranucci, E., Trotta, F., & Lembo, D. (2009). Enhanced antiviral activity of acyclovir loaded into β-cyclodextrin-poly (4-acryloylmorpholine) conjugate nanoparticles. *Journal of Controlled Release, 137*(2), 116–122.

Cha, S. H., Hong, J., McGuffie, M., Yeom, B., VanEpps, J. S., & Kotov, N. A. (2015). Shape dependent biomimetic inhibition of enzyme by nanoparticles and their antibacterial activity. *ACS Nano, 9*(9), 9097–9105.

Chen, Q., Li, J., Wu, Y., Shen, F., & Yao, M. (2013). Biological responses of Gram-positive and Gram-negative bacteria to nZVI (Fe 0), Fe^{2+} and Fe^{3+}. *RSC Advances, 3*(33), 13835–13842.

Cousins, B. G., Allison, H. E., Doherty, P. J., Edward, C., Garvery, M. J., Martin, D. S., & Wiliams, R. L. (2007). Effect of a nanoparticulate silica substrate on cell attachment of *Candida abicans*. The society for applied microbiology. *Journal of Applied Microbiology*, *102*(3), 757–765.

Cremonini, E., Zonaro, E., Donini, M., Lampis, S., Boaretti, M., Dusi, S., & Vallini, G. (2016). Biogenic selenium nanoparticles: Characterization, antimicrobial activity and effects on human dendritic cells and fibroblasts. *Microbial Biotechnology*, *9*(6), 758–771.

Elechiguerra, J. L., Burt, J. L., Morones, J. R., Camacho-Bragado, A., Gao, X., Lara, H. H., & Yacaman, M. J. (2005). Interaction of silver nanoparticles with HIV-1. *Journal of Nanobiotechnology*, *3*(1), 1–10.

El-Sayyad, G. S., Mosallam, F. M., & El-Batal, A. I. (2018). One-pot green synthesis of magnesium oxide nanoparticles using *Penicillium chrysogenum* melanin pigment and gamma rays with antimicrobial activity against multidrug-resistant microbes. *Advanced Powder Technology*, *29*(11), 2616–2625.

Ewald, P. W. (1980). Evolutionary biology and the treatment of signs and symptoms of infectious disease. *Journal of Theoretical Biology*, *86*(1), 169–176.

Fathima, J. B., Pugazhendhi, A., & Venis, R. (2017). Synthesis and characterization of ZrO_2 nanoparticles-antimicrobial activity and their prospective role in dental care. *Microbial Pathogenesis*, *110*, 245–251.

Fishovitz, J., Hermoso, J. A., Chang, M., & Mobashery, S. (2014). Penicillin-binding protein 2a of methicillin-resistant *Staphylococcus aureus*. *IUBMB Life*, *66*(8), 572–577.

Friedman, M., Tam, C. C., Cheng, L. W., & Land, K. M. (2020). Anti-trichomonad activities of different compounds from foods, marine products, and medicinal plants: A review. *BMC Complementary Medicine and Therapies*, *20*(1), 271.

Gaikwad, S., Ingle, A., Gade, A., Rai, M., Falanga, A., Incoronato, N., & Galdiero, M. (2013). Antiviral activity of mycosynthesized silver nanoparticles against herpes simplex virus and human parainfluenza virus type 3. *International Journal of Nanomedicine*, *8*, 4303.

Ghaffari, H., Tavakoli, A., Moradi, A., Tabarraei, A., Bokharaei-Salmi, F., Zahmatkeshan, M., Farahmand, M., Javanmard, D., Kiani, S. J., Esghaei, M., Prihajati- Mahabadi, V., Monavari, S. H., & Pirkooh, A. A. (2019). Inhibition of H1N1 influenza virus infection by zinc oxide nanoparticles: Another emerging application of nanomedicine. *Journal of Biomedical Science*, *26*(1), 70.

Haggag, E. G., Elshamy, A. M., Rabeh, M. A., Gabr, N. M., Salem, M., Youssif, K. A., & Abdelmohsen, U. R. (2019). Antiviral potential of green synthesized silver nanoparticles of *Lampranthus coccineus* and *Malephora lutea*. *International Journal of Nanomedicine*, *14*, 6217.

Haghighi, F., Roudba Mohammadir, S., Mohammadi, P., Hossinkhani, S., & Shidpour, R. (2013, Summer). Antifungal acitivity of TiO_2 nanoparticles and EDTA on *Candida abicans* biofilms. *Infection Epidemiology and Medicine*, *1*, 133–138.

Halder, S., Yadav, K. K., Sarkar, R., Mukherjee, S., Saha, P., Haldar, S., & Sen, T. (2015). Alteration of Zeta potential and membrane permeability in bacteria: A study with cationic agents. *SpringerPlus*, *4*(1), 1–14.

Hang, X., Peng, H., Song, H., Qi, Z., Miao, X., & Xu, W. (2015). Antiviral activity of cuprous oxide nanoparticles against hepatitis C virus in vitro. *Journal of Virological Methods*, *222*, 150–157.

Harrison, J. J., Turner, R. J., Marques, L. L., & Ceri, H. (2005). Biofilms: A new understanding of these microbial communities is driving a revolution that may transform the science of microbiology. *American Scientist*, *93*(6), 508–515.

He, L., Liu, Y., Mustapha, A., & Lin, M. (2011). Antifungal activity of zinc oxide nanoparticles against *Botrytis cinerea* and *Penicillium expansum*. *Microbiology Research*, *166*(3), 207–215.

Hussein, M. A. M., Banos, F. G. D., Grinholc, M., Dena, A. S. A., El-Shebiny, I. M., & Megahed, M. (2020). Exploring the physicochemical and antimicrobial properties of gold-chitosan hybrid nanoparticles composed of varying chitosan amounts. *International Journal of Biological Macromolecules*, *162*, 1760–1769.

Ingle, A. P., Duran, N., & Rai, M. (2013). Bioactivity, mechanism of action, and cytotoxicity of copper-based nanoparticles: A review. *Applied Microbiology and Biotechnology*, *98*(3), 1001–1009.

Ismail, R. A., Sulaiman, G. M., Abdulrahman, S. A., & Marzoog, T. R. (2015). Antibacterial activity of magnetic iron oxide nanoparticles synthesized by laser ablation in liquid. *Materials Science and Engineering. Part C*, *53*, 286–297.

Jamdagni, P., Khatri, P., & Rana, J. S. (2018). Green synthesis of zinc oxide nanoparticles using flower extract of Nyctanthes arbor-tristis and their antifungal activity. *Journal of King Saud University – Science*, *30*(2), 168–175.

Jayesh, P. Ruparelia, C. A. P., Duttagupta, S. P. B., Mukherji, S. (2008). Strain specificity in antimicrobial activity of silver and copper nanoparticles. *Acta Biomater*, *4*, 707–716.

Kanhed, P., Birla, S., Gaikwad, S., Gade, A., Seabra, A. B., Rubilar, O., & Rai, M. (2014). In vitro antifungal efficacy of copper nanoparticles against selected crop pathogenic fungi. *Materials Letters*, *115*, 13–17.

Kim, H. C., Park, S. M., & Hinsberg, W. D. (2010). Block copolymer based nanostructures: Materials, processes, and applications to electronics. *Chemical Reviews*, *110*(1), 146–177.

Kim, J. S., Kuk, E., Yu, K. N., Kim, J. H., Park, S. J., Lee, H. J., & Kim, Y. K. (2007). Antimicrobial effects of silver nanoparticles. *Nanomedicine: Nanotechnology, Biology and Medicine*, *3*(1), 95–101.

Kumar, R., Nayak, M., Sahoo, G. C., Pandey, K., Sarkar, M. C., Ansari, Y., & Das, P. (2019). Iron oxide nanoparticles based antiviral activity of H1N1 influenza A virus. *Journal of Infection and Chemotherapy*, *25*(5), 325–329.

Lara, H. H., Ayala-Nu~nez, N. V., Ixtepan-Turrent, L., & Rodriguez-Padilla, C. (2010). Mode of antiviral action of silver nanoparticles against HIV-1. *Journal of Nanobiotechnology*, 8(1), 1–10.

Laxminarayan, R., Duse, A., Wattal, C., Zaidi, A. K., Wertheim, H. F., Sumpradit, N., & Greko, C. (2013). Antibiotic resistance—The need for global solutions. *Lancet Infectious Diseases*, *13*(12), 1057–1098.

Li, S., Meng Lin, M., Toprak, M. S., Kim, D. K., & Muhammed, M. (2010). Nanocomposites of polymer and inorganic nanoparticles for optical and magnetic applications. *Nanotechnology Reviews*, *1*(1), 5214.

Mahdy, S. A., Raheed, Q. J., & Kalaichelvan, P. T. (2012). Antimicrobial activity of zero-valent iron nanoparticles. *International Journal of Modern Engineering Research*, *2*(1), 578–581.

Mai-Prochnow, A., Clauson, M., Hong, J., & Murphy, A. B. (2016). Gram positive and Gram negative bacteria differ in their sensitivity to cold plasma. *Scientific Reports*, *6*, 38610.

Manikandan, V., Velmurugan, P., Park, J. H., Chang, W. S., Park, Y. J., Jayanthi, P., Cho, M., & Oh, B. T. (2017). Green synthesis of silver oxide nanoparticles and its antibacterial activity agains dental pathogens. *3 Biotech*, *7*(1), 72.

Martinez-Castanon, G. A., Martínez-Guti_errez, F., & Ruiz, F. (2008). Synthesis and antimicrobial activity of solver nanoparticles with different sizes. *Journal of Nano Research*, *10*, 1343–1348.

Medhi, R., Srinoi, P., Ngo, N., Tran, H., & Lee, T. R. (2020). Nanoparticle-based strategies to combat COVID-19. *ACS Applied Nano Materials*, *3*(9), 8557–8580.

Mori, Y., Ono, T., Miyahira, Y., Nguyen, V. Q., Matsui, T., & Ishihara, M. (2013). Antiviral activity of silver nanoparticle/chitosan composites against H1N1 influenza A virus. Mori et al. *Nanoscale Research Letters*, *8*(1), 93.

Nahrawy, A. M. E., Bakr, A. M., Hammad, A. B. A., & Hemdan, B. A. (2020). High performance of talented copper/magneso-zinc titanate nanostructures as biocidal agents for inactivation of pathogens during wastewater disinfection. *Applied Nanoscience*, *10*(9), 3585–3601.

Padmavathy, N., & Vijayaraghavan, R. (2008). Enhanced bioactivity of ZnO nanoparticles—An antimicrobial study. *Science and Technology of Advanced Materials*, *9*(3), 035004.

Panacek, A., Kolar, M., Vecerova, R., Prucek, R., Soukupova, J., Krystof, V., & Kvítek, L. (2009). Antifungal activity of silver nanoparticles against Candida spp. *Biomaterials*, *30*(31), 6333–6340.

Parveen, S., Wani, A. H., Shah, M. A., Devi, H. S., Bhat, M. Y., & Koka, J. A. (2018). Preparation, characterization and antifungal activity of iron oxide nanoparticles. *Microbial Pathogenesis*, *115*, 287–292.

Qi, L., Xu, Z., Jiang, X., Hu, C., & Zou, X. (2004). Preparation and antibacterial activity of chitosan nanoparticles. *Carbohydrate Research*, *339*(16), 2693–2700.

Rai, A., Prabhune, A., & Perry, C. C. (2010). Antibiotic mediated synthesis of gold nanoparticles with potent antimicrobial activity and their application in antimicrobial coatings. *Journal of Materials Chemistry*, *20*(32), 6789–6798.

Ray, S., Mohan, R., Singh, J. K., Samantaray, M. K., Shaikh, M. M., Panda, D., & Ghosh, P. (2007). Anticancer and antimicrobial metallopharmaceutical agents based on palladium, gold, and silver N-heterocyclic carbene complexes. *Journal of the American Chemical Society*, *129*(48), 15042–15053.

Reddy, K. M., Feris, K., Bell, J., Wingett, D. G., Hanley, C., & Punnoose, A. (2007). Selective toxicity of zinc oxide nanoparticles to prokaryotic and eukaryotic systems. *Applied Physics Letters*, *90*(21), 213902.

Ren, G., Hu, D., Cheng, E. W., Vargas-Reus, M. A., Reip, P., & Allaker, R. P. (2009). Characterisation of copper oxide nanoparticles for antimicrobial applications. *International Journal of Antimicrobial Agents*, *33*(6), 587–590.

Rogers, J. V., Parkinson, C. V., Choi, Y. W., Speshock, J. L., & Hussain, S. M. (2008). A preliminary assessment of silver nanoparticle inhibition of monkeypox virus plaque formation. *Nanoscale Research Letters*, *3*(4), 129.

Roy, A., Gauri, S. S., Bhattacharya, M., & Bhattacharya, J. (2013). Antimicrobial activity of CaO nanoparticles. *Journal of Biomedical Nanotechnology*, *9*(9), 1570–1578.

Scott, N., & Chen, H. (2013). Nanoscale science and engineering for agriculture and food systems. *Industrial Biotechnology*, *9*(1), 17–18.

Shen, Y., Lu, S., Xu, W., Lv, A., Wang, Z., Wang, H., & Zhang, Y. (2021). Fabrication of composite material with Pd nanoparticles and graphene on nickel foam for its excellent electrocatalytic performance. *Electrocatalysis*, *11*(5), 522–535.

Shionoiri, N., Sato, T., Fujimori, Y., Nakayama, T., Nemoto, M., Matsunaga, T., & Tanaka, T. (2012). Investigation of the antiviral properties of copper iodide nanoparticles against feline calicivirus. *Journal of Bioscience and Bioengineering*, *113*(5), 580–586.

Singh, R., & Nalwa, H. S. (2011). Medical applications of nanoparticles in biological imaging, cell labeling, antimicrobial agents, and anticancer nanodrugs. *Journal of Biomedical Nanotechnology*, *7*(4), 489–503.

Singh, S. R., Krishnamurthy, N. B., & Mathew, B. B. (2014). A review on recent diseases caused by microbes. *Journal of Applied and Environmental Microbiology*, *2*(4), 106–115.

Singhal, G., Bhavesh, R., Kasariya, K., Sharma, A. R., & Singh, R. P. (2011). Biosynthesis of silver nanoparticles using *Ocimum sanctum* (Tulsi) leaf extract and screening its antimicrobial activity. *Journal of Nanoparticle Research, 13*(7), 2981–2988.

Slavin, Y. N., Asnis, J., Hafeli, U. O., & Bach, H. (2017). Metal nanoparticles: Understanding the mechanisms behind antibacterial activity. *Journal of Nanobiotechnology, 15*(1), 1–20.

Sobhani, Z., Samani, S. M., Montaseri, H., & Khezri, E. (2017). Nanoparticles of chitosan loaded ciprofloxacin: Fabrication and antimicrobial activity. *Advanced Pharmaceutical Bulletin, 7*(3), 427.

Song, X., Bayati, P., Gupta, M., Elahinia, M., & Haghshenas, M. (2021). Fracture of magnesium matrix nanocomposites - A review. *International Journal of Lightweight Materials and Manufacture, 4*(1), 67–98.

Vijayakumar, S., & Ganesan, S. (2012). Gold nanoparticles as an HIV entry inhibitor. *Current HIV Research, 10*(8), 643–646.

Wani, I. A., & Ahmad, T. (2012). Size and shape dependant antifungal activity of gold nanoparticles: A case study of *Candida. Colloids and Surfaces, Part B: Biointerfaces, 101*, 162–170.

Xie, Y., He, Y., Irwin, P. L., Jin, T., & Shi, X. (2011). Antibacterial activity and mechanism of action of zinc oxide nanoparticles against *Campylobacter jejuni. Applied and Environment Microbiology, 77*(7), 2325–2331.

Yien, L., Zin, N. M., Sarwar, A., & Katas, H. (2012). Antifungal activity of chitosan nanoparticles and correlation with their physical properties. *International Journal of Biomaterials.* Article ID 632698, 2012, 9. https://doi.org/10.1155/2012/632698

Zhang, Z., Wang, Y., Ma, Q., Zhang, S., Liu, H., Zhao, B., & Kong, D. (2021). Biomimetic carrier-free nanoparticle delivers digoxin and doxorubicin to exhibit synergetic antitumor activity in vitro and in vivo. *Chemical Engineering Journal, 406*, 126801.

16 Nanotechnology
An Exciting Perspective for Nematode Management

Rupak Jena, Manoranjan Das, Arkadeb Chattopadhyay, Mukesh Patra, Temuhi S, Rahul Kumar Tiwari, Milan Kumar Lal, Ravinder Kumar and Sujit Kumar Behera*

16.1 INTRODUCTION

16.1.1 Nanotechnology and Its Basic Principles

The term nanotechnology originates from the fusion of two words, the Greek numerical prefix "nano" referring to a billionth and the term "technology" as an outcome. One billionth of a metre, or 100 nm, is typically considered the size limit for nanotechnology or nanoscale technology. Nanoscale science, often known as nanoscience, is the study of phenomena, characteristics and responses of materials at the atomic, molecular and macro-molecular scales, generally between 1 and 100 nm in size. The properties of matter are very different at this scale, and notably below 5 nm, from those at a larger particulate scale (i.e., quantum-scale effects play a considerable role). Nanotechnology is then the arrangement, manipulation, configuration, production and application, by controlling the size and shape, the properties-responses, functionality of structures and devices, systems of the order or less than 100 nm (Sattler, 2010; Bhusan, 2004). Nanoscience, which deals with developing materials at the atomic, molecular and macro-molecular scales, is a fusion of physics, chemistry, materials science and biology. It promises greater functionality in smaller, more affordable, lighter and faster technologies that use fewer raw materials and use less energy (Table 16.1). The ability of molecules and biological systems to "self-assemble" tiny structures allows manufacturing to reach the nanoscale either from the bottom-up or from the top-down. In short, nanotechnology is the method to assemble micro- and macro-materials with atomic precision. Composites designed from particles of nanosize ceramics or metals smaller than 100 nanometers can suddenly become much stronger than predicted by existing materials science models (Kirtane et al., 2021). For instance, metals with so-called grains can be as much as seven times tougher than their common counterparts with grains that are hundreds of nanometers in size. These abrupt shifts are the result of

* Corresponding author.

DOI: 10.1201/9781003376446-16

TABLE 16.1
Properties of nanoparticle as compared to bulk materials

Sl. No	Properties	Characterization
1.	Structural	Crystal structure of nanomaterials resembles that of bulk one with different lattice parameters. Gold (Au) and aluminium (Al) in the case of nanoform, the crystal structure will be icosahedral.
2.	Thermal	Melting property of nanomaterials decreases as a function of higher surface energy and change in interatomic spacing. Thermal properties will alter as the reduction in size towards nano size.
3.	Chemical	The ionization potential at nanosize is higher than that for the bulk materials.
4.	Mechanical	The distinctive atom structural arrangement provides higher mechanical strength, elastic limit and flexibility. Nanophase ceramics show ductility behaviour at elevated temperatures, whereas coarse-grained ceramics shows brittle character.
5.	Magnetic	Magnetic moment of nanoparticles is very less than their bulk size.
6.	Optical	Optical properties such as photoconductivity, photocatalysis, photoemission and electroluminescence are depended on the size factor.
7.	Electronic	Electronic properties are related to the wave-like property of the electron for all small-scaled materials.

strange phenomena in the field of quantum physics. At nanoscale technology, various physical phenomena shift paradigms, particularly gravity, which behaves negligibly, surface tension and Van der Waals attraction would become significant, etc. Much of the fascination with nanotechnology roots in these unique quantum and surface phenomena which matter exhibits at the nanoscale (Moosa, 2009).

16.1.2 Nanotechnology in Agriculture

In India, agriculture is at crossroads facing a wide range of constraints such as continuous escalation in population and constant decrement in cultivating land areas, crisis in water availability, imbalanced nutritional and soil health and oscillating climate change. To combat these challenges against the backdrop of sustainability an alternate technology such as "nanotechnology" is of pressing priority (Figure 16.1). Information supports that all materials are made up of atoms which are the smallest unit. The word "nano-agriculture" refers to the infusion of nanotechnology concepts and principles in agriculture science to develop products and process outputs that deliver inputs and promote productivity without harming the environment (Qi & Wang, 2002). Nanotechnology (NT) has been recognized as a potential technology for reviving the food and agriculture industries and can considerably improve the lives of the poor. Nanotechnology is implemented in the field of agriculture, dairy industries, food processing, transportation, packaging and monitoring agricultural product quality. It has an enormous prospective in making agriculture more profitable and resourceful by implementing nanoparticles to improve nutrient availability

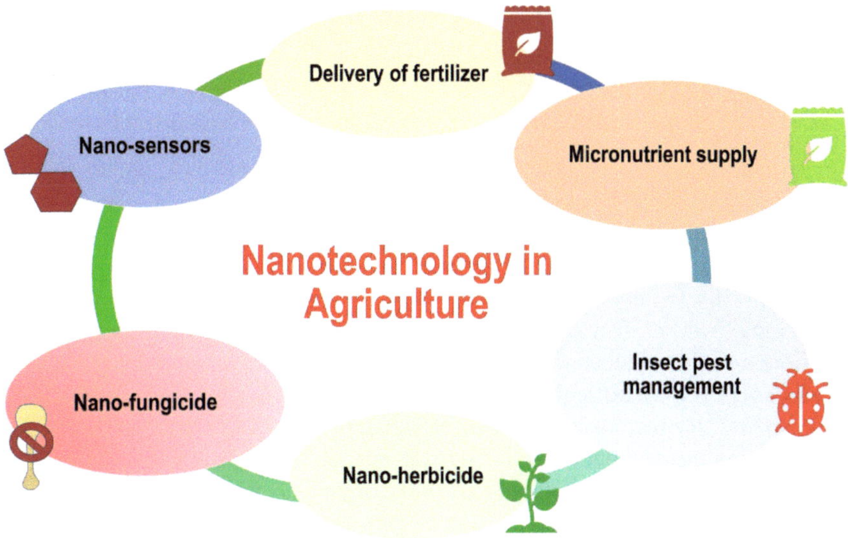

FIGURE 16.1 Representation of the scope of nanotechnology in modern agriculture

(Crucho & Barros, 2017). It makes conventional agriculture proficient and transforms it into scientific agriculture by fusion of nanobased agrochemicals, ceramic devices, filtration methods, etc. The technology improves agricultural practices by exploiting some specific principles in terms of nanopheromone for effective pest monitoring. Enhanced inputs use efficiency, nanofertilizers and balanced crop nutrition, nanoherbicides for effective weed control, nano-insecticide, smart-delivery system, nanodevices for identity preservation, nanobiotechnology, nanofood industry, nanotechnology for environmental safety and sustainable farming, etc (Subramanian & Tarafdar, 2011). Engineered nanomaterials are another important product, having a prominent impact in the industrial and scientific fields and given the global market value for engineered nanomaterials is increasing rapidly (Thomas et al., 2015).

Agriculture aspires to attain a sustainable industry with the help of agro-nanotechnology. The utilization of these technologies for plant disease management, gene delivery and nanopesticide development holds the potential to revolutionize the agricultural sector. These innovations represent agriculture's propitious future and ability to lessen the carbon impact imposed by agricultural operations (Sangeetha et al., 2017). In agriculture, NPs are employed as an intelligent pesticide or fungicide delivery system. There are also 200–400 nm oil- or water-based nanoemulsions that have numerous uses in plant pathology (Goswami et al., 2010). These substances make pesticides more stable, less poisonous and less prone to leaching into the soil. The need for frequent chemical treatment can be decreased by adding NPs with a component for gradual release. The field of nanotechnology as it exists today is constantly developing to incorporate metabolites, nanostructures, enzymes and the cell system to create a fusion that can be practised in pest management and agriculture (Bailey et al., 2010).

16.2 NANOTECHNOLOGY IN THE MANAGEMENT OF ECONOMICALLY IMPORTANT PLANT-PARASITIC NEMATODES

The majority of soil types, roughly all of them, contain nematodes, which can infest both agricultural and horticultural crops and result in huge losses to global agriculture. The International *Meloidogyne* Project has calculated that nematodes cause loss to an extent of 78 billion US dollars in developed nations and more than 100 billion in developing nations per year (Kumar et al., 2020). The infestation of plant roots by nematodes results in a variety of symptoms such as stunting, wilting, yellowing, a reduction in blooming, fruit set, fruit development, dieback and occasionally even plant death. It is a prime example of the severity of these parasites. Besides chemicals plant-parasitic nematodes (PPNs) are difficult to manage because, once they become established in the soil, even soil sterilization may not be effective (Anwar & McKenery, 2012). Mostly, nematode management depends on the toxic chemicals and nematicides that are harmful to the environment, mammals and other organisms (Table 16.2). The impact of these pests was too great for conventional measures to handle, thus new approaches to pest control are required. As an alternative to the chemical nematicides, other approaches like biocontrol agents, growing resistance varieties and following cultural practices are attempting to leave a significant imprint (Table 16.2). Additionally to the following, nanotechnology is an expanding opportunity that dispenses a wide range of applications in the agriculture field including plant protection.

Rajan (2006) stated that nanoparticles improved chemical reactivity, biological activity, catalytic behaviour and high mobility throughout an organism's body, including cellular penetration. Auría-Soro et al. (2019) stated in the case of nanosize, metals can cross semi-permeable membranes and interact with the different biomolecules. And also, Fard et al. (2015) stated that nanoparticles cause direct toxicity to the organisms which causes a lethal impact on the microorganisms including nematodes. Due to their natural encapsulation by mother protein, biosynthesized nanoparticles are now more significant in agricultural sciences and are durable for more than 90 days (Bhau et al., 2016).

16.3 SYNTHESIS OF NANOPARTICLES

16.3.1 Synthesis of Nanoparticles Against Nematodes

Nanotechnology has the potential to revolutionize a wide range of scientific, technological, medical and agricultural fields. By applying contemporary pesticide formulations such as nanocapsules, nanoparticles and nanosuspension insecticides against plant-parasitic nematodes, there is an opportunity to manage nematodes sustainably. The simplest and most straightforward method of shielding plants from economically important nematode invasion is the direct use of nanoparticles (NPs) through foliar spray or drenching around the rhizosphere in required dosages. In this manner, the NPs can suppress infections comparably similar to chemical pesticides. Besides suppressing harmful pests, NPs have a neutral effect on non-target organisms that are

TABLE 16.2
Traditional plant-parasitic nematode management and its drawbacks

Method	Various practices	Nematodes controlled	Disadvantages	Reference
Cultural method	• Selection of healthy seed material • Adjusting the time of planting • Fallowing • Deep summer ploughing • Manuring • Flooding • Trap cropping • Antagonistic crops • Removal and destruction of infected plants • Use of resistant varieties	*Anguinatritici, Aphelenchoidesbesseyi, Meloidogyne incognita*	Time taking and tedious process. Economically not sustainable. Flooding process required high amount of water which leading to water loss. Labour intensive process.	Sivasubramaniam et al., 2020
Physical control	• Heat treatment of soil • Hot water treatment of planting material • Irradiation • Osmotic pressure	*Anguina tritici, Aphelenchoidesbesseyi, Meloidogyne incognita, Rhadopholus similis*	More technical skill required. Scientific knowledge of farmer is essential for this approach. Not practical and economical method in field condition.	TNAU Agri portal: Crop Protection. https://agritech.tnau.ac.in/crop_protection/crop_prot_nematode_nematicide.html.

(Continued)

TABLE 16.2 (CONTINUED)
Traditional plant-parasitic nematode management and its drawbacks

Method	Various practices	Nematodes controlled	Disadvantages	Reference
Biological control	• Fungal bioagents *Paecilomyces lilacinus, Pachonia chlaydosporium, Trichoderma viride* and *Trichoderma harzianum* • Bacterial bioagents *Pseudomonas fluorescens, Bacillus subtilis, Pasteuria penetrans*	*Meloidogyne incognita, Meloidogyne javanica, Meloidogyne graminicola, Rhadopholus similis, Hoplolaimus spp.*	Incurring more cost of cultivation. Culturing of biocontrol agent is tedious. Host specific biocontrol agent is required.	Abd-Elgawad, M.M., 2020
Chemical control	• Seed treatment • Seedling root dip • Soil application	*Anguina tritici, Meloidogyne incognita, Meloidogyne njavanica, Meloidogyne graminicola, Rhadopholus similis, Hoplolaimus spp., Tylecnhulus semipenetrance, Rotylenchulus reniformis, Heterodera avena, Heterodera zea, Globodera rostochiensis*	Increases cost of cultivation. Harmful for the beneficial microorganism present in the soil environment. Persistent effects in soil hamper the soil flora and fauna. Increases threat to the environment.	Chen et al., 2020

particularly important for soil health as they solubilize minerals impact and participate in various rhizospheric activities (Mahawar & Prasanna, 2018). The use of nanomaterials as a carrier for compounds such as inhibitors of polyamine synthesis, pheromones, SAR including chemicals, or even active elements of chemical pesticides, particularly in flood-prone areas, is the second method for defending plants against nematodes (Khan et al., 2022). The production of noble metal NPs can be accomplished using a wide range of methods, including chemical, physical and biological ones. Additionally, biological entities like fungi, bacteria and plants are good sources for the green synthesis of NPs in the correct size and shape (Bhau et al., 2016). Since it has a multi-site activity against nematodes and doesn't cause phytotoxicity, a nanotechnological strategy is efficient. In order to combat PPNs, some nanoproducts or nanomaterials, such as nanosulfur and nanosilver, were used (Hardman, 2006).

Metalloids, metallic oxides, nonmetals and synthetic NPs from plants, algae, fungi, bacteria, etc. are NPs that have been evaluated for their efficacy against plant-parasitic nematodes. Additionally, the usage of natural materials made from plants, algae, fungi and bacteria as NPs against PPNs boosts the effectiveness of these substances. Because of their distinctive nanoshape and altered characteristics, NPs promote plant growth and substantiate excellent plant health. Due to their exceptional physical and chemical properties, nanoparticles are in demand as they easily enter the plant's system through ion channels and membrane proteins because of their nanoshape. Scientists are becoming concerned about the research on this strategy even though there has only been a small amount of work done on the use of this material against plant nematodes.

Agrochemical delivery with controlled release and site-specific targeting is achieved significantly through nanotechnology. Some of these nanoparticles contain nematicidal capabilities, which are effective against a wide range of nematode taxa that parasitize plants as well as bacterial and fungal pathogens (Sabry, 2019). The nanoparticles are an excellent source to prevent nematode infection in plants because their active components have demonstrated potential as effective nematicides. Targeted site-specific delivery of NP-based agrochemicals is crucial for field application and can be accomplished with the help of nanocapsules, which are highly stable and biodegradable structures (Jha et al., 2009). As NPs are increasingly used in agriculture in moderation with chemical pesticides so they are becoming more successful because of safety, disease and pest management, compatibility, and adherence which ultimately lowers the cost of plant protection (Prasad et al., 2014). Chemical agents coupled with polymers or enclosed on metal surfaces provide sound delivery devices. NP's with an efficient water solubility range of 100–250 nm have been developed by numerous businesses.

16.3.2 Recent Development in Nanotechnology Against Nematodes

16.3.2.1 Metal-based Nanoparticles for the Management of Plant-parasitic Nematodes (PPNs)

16.3.2.1.1 Silver Nanoparticles (AgNP's)

Metal-based NPs have recently become more popular among various nanostructures due to their novel physicochemical characteristics and biological activities. Silver nanoparticles (AgNPs), as a result of nanotechnology, resemble an iconic item. Silver has attracted a lot of attention in recent years because of its high conductivity, chemical stability, catalytic and antibacterial activities. AgNPs were initially investigated in the treatment of plant disease due to their historically known antibacterial action (Richards, 1981). El-Deen and El-Deeb (2018) exhibited that using AgNP's against *Meloidogyne incognita* attacking tomato under greenhouse conditions significantly decreased the nematode population while also improving the plant growth indices. At doses of 0.05–0.5 mg/ml for 72 hours and 5–50 mg/ml for 1–3 days, respectively, AgNPs toxicity impaired the development and reproduction of root-knot nematodes (Lim et al., 2012). This shows that the impact of AgNP sprayed in the field at low concentrations may be gradual and long-lasting. AgNPs have nematicidal effects on root-knot nematodes, but they also have effects on other genera of PPNs, as well as plant pathogenic fungi and bacteria. AgNPs are regarded as broad-spectrum antifungal, antibacterial and antinematicidal agents. High-risk synthetic nematicides and extraordinary biological agents may be effectively replaced by AgNPs' potent nematicidal action.

16.3.2.1.2 Copper Nanoparticles (CuNP's)

The antibacterial properties of Cu and a long history of successful management of plant diseases instigated the synthesis of copper nanoparticles (CuNPs) to be tested against the root-knot nematode genus (Elmer et al., 2018). A variety of microorganisms exhibit the efficiency and specificity of CuNPs (Berry et al., 2014; Bholay et al., 2013). Giannousi et al. (2013) conducted successful initial research on CuNPs as a fungicide/bactericide and nematicides while conducting a study on the grape's *Xiphinema index* to look into the impact of CuNPs on nematodes. Given the reduction demonstrated with AgNPs, it would be logical to assume that CuNPs may also have equal efficiency against nematodes. Mohamed et al. (2019) in their studies advocated CuNPs treatment considerably reduced the nematode *Meloidogyne incognita* population under laboratory conditions. According to his investigation, *M. incognita* juveniles mortality was positively correlated with CuNPs concentrations, with a CuNP concentration of 0.2 g/L being sufficient to cause 100% mortality in J_2 *Meloidogyne incognita* juveniles. Chen & Peng (2019) investigated the effect of CuFeNP's against *Meloidogyne incognita*, the root-knot nematodes, as well as documented the enhancement of plant growth characteristics affected by the nematodes. According to their findings, 0.03 g CuFeNPs perform excellently against nematodes. In tomato plants treated with CuFeNPs, an increase in the weight of the fresh shoots and roots was also noticeable.

16.3.2.1.2 Zinc Nanoparticles (ZnNPs)

Several studies have investigated the antibacterial activity of zinc oxide nanoparticles (ZnO NPs) against plant diseases, similar to AgNPs and CuNPs (Graham et al., 2016; Hafez et al., 2014). The majority of *in vitro* investigations have demonstrated that ZnONPs suppress *Meloidogyne incognita* and other nematodes (Kaushik & Dutta, 2017). ZnONPs have a great deal of optical, physical and antibacterial qualities, making them highly likely to improve farming (Sabir et al.,

2014). It can be used to promote plant development, seed germination and seedling vigour which also help boost plant's stem and root growth (Prasad et al., 2012). In comparison to graphene oxide (GO) NPs, Siddiqui et al. (2019) observed that ZnO NPs at 0.10 mg/ml concentration showed the greatest reduction in the nematode (*M. incognita*) population and galling. The distribution pattern of ZnONPs reveals that the intestine of the pest is targeted by the particles for NPs' toxicity against nematodes (Gupta et al., 2015).

16.3.2.1.3 Other Synthetic Nanoparticles

Ardakani (2013) investigated the effects of three nanoparticles (NPs), namely titanium dioxide, silicon oxide and silver, on the second-stage juveniles (J_2) of *Meloidogyne incognita* in a greenhouse condition. The treatments with AgNPs and 0.02% TiO_2NPs entirely decreased the nematode population. Additionally, TiO_2NPs were found to be less hazardous to tomato plants and nematodes than AgNPs. Thakur et al. (2018) have demonstrated the effectiveness of gold nanoparticles (AuNPs) in eradicating the tomato plant parasite *Meloidogyne incognita*. After incubating *Meloidogyne incognita* with AuNPs for three hours, 100% of second-stage juveniles (J_2) of *Meloidogyne incognita* died when they were exposed directly to water. Due to *Meloidogyne incognita* poor survival rate, AuNPs are expected to exhibit a strong nematicidal effect on it in the soil treatment. According to their findings, AuNPs in the pot experiment not only proved fatal to *Meloidogyne incognita* but also stimulated plant development and had no adverse effects on tomato plant growth. AuNPs were shown to be both harmless and harmful to *Meloidogyne incognita* in this investigation.

16.3.2.2 Green Synthesized Nanoparticles for the Management of Plant-parasitic Nematodes (PPNs)

The synthesis of green nanomaterials has gathered much attention from nanotechnologists and has recently established a science known as "green nanotechnology". It adheres to the principles of green chemistry and forms the cornerstone of sustainable development so biosynthesis is regarded as a green pathway (Rai et al., 2021). Green nanoparticle means the synthesis of nanoparticles by biological means *i.e.* living organisms such as bacteria, fungi, algae, yeast, plant or plant products for the bio-reduction of metal ions into their elemental form in the range of 1–100 nm with the help of biotechnological tools. Agents that reduce or interact with other cells' components serve as stabilizing and capping substances, which reduces adding of capping and stabilizing agents from outside for the synthesis of nanoparticles (Paul & Roychoudhury, 2021). Green synthesis of nanoparticles is eco-friendly and less energy consumable than physical and chemical synthesis. Green synthesis is cost-effective, non-toxic and pollution-free (Ying et al., 2022), and has good characteristics such as stability and biodegradability.

16.3.2.2.1 Algae-mediated Nanoparticles Against Nematodes

In the green biosynthesis of nanoparticles, several microorganisms including fungi, yeast microalgae, bacteria and plant extract have been used. Meanwhile, microalgae are considered a promising tool for nanoparticle synthesis as they are having high growth rates and high biomass production (Hamed et al., 2019). Microalgae are known to have all major and minor nutrients and also biocontrol potential. It helps to stimulate plant growth by producing a wide range of organic compounds and also possesses novel metabolites encompassing significant insecticidal and nematicidal properties (Ghareeb et al., 2019a).

Ghareeb et al. (2020) found that the green silver nanoparticles synthesized by using the freshwater alga *Cladophora glomerata* and the marine alga *Ulva fasciata* caused a 100% reduction in egg hatching and 85.7–95.9% juvenile mortality within 48 hours of exposure. Reduction in the number of galls, eggmasses per plant, eggs per eggmass and nematode population in soil by 99.5%, 98%, 50–63% and 86% respectively was observed. Chitinase gene expression is high in the plants treated with *C. glomerata* and *U. fasciata* nanoparticles after 24 hours. Chitinase is an enzyme that degrades the chitin which is an integral part of the nematode eggshell. Cyanobacterium-based silver nanoparticles reduced the egg hatching and induced marked larval mortality of *M. javanica in vitro* conditions. In faba bean, it reduced the root galling and infective juvenile population in infected soil as well as increased the plant growth due to the production of micro-nutrients and many organic compounds such as auxins, gibberellins and precursor of ethylene and betaine by algal cells (Hamed et al., 2019).

16.3.2.2.1 Plant-mediated Nanoparticles Against Nematodes

Deriving plants and plant products for the synthesis of metal nanoparticles is more advantageous due to their effective, rapid, eco-friendly, economical, non-pathogenic protocol and provides a single-step technique for the green synthesis processes (Roy & Das, 2015) (Table 16.3). Plants and plant products are used as reducing and stabilizing or capping agents (to control the size and to prevent agglomeration) as it has a combination of biomolecules such as proteins, amino acids, enzymes, polysaccharides, tannins, phenolics, saponins, alkaloids terpenoids and vitamins. The different parts of a plant such as a stem, root, fruit, seed, callus, peel, leaves and flowers are used to synthesize metallic nanoparticles in various shapes and sizes by biological approaches (Kuppusamy et al., 2016). Kalaiselvi et al. (2019) reported that the silver nanoparticles synthesized by *Euphorbia tirucalli* latex extract are lethal to the juveniles of *M. incognita* and also it inhibits egg hatching *in vitro.* The *in vivo* studies showed the greatest significant decrease in soil nematode juveniles, egg masses on roots, egg numbers per egg mass and root galls. Silver nanoparticles synthesized using bark and leaf extracts of *Eucalyptus officinalis* increased plant heights, more tillers and heavier seed weight of paddy and it reduced the number of *Heterodera sacchari* cysts and the nematode infestation (Fabiyi et al., 2020). Magnesium oxide nanoparticles (MgO-NP) are synthesized from strawberry seed powder and it is proven that they possess antimicrobial and antinemic ability as it reduces the nematode infestation by increasing the ROS reaction which is a defence mechanism against the nematode attack (Khan et al., 2022) (Table 16.3).

TABLE 16.3
Green synthesized nanoparticles for the management of plant-parasitic nematodes

NP	Green materials	Nematodes	Crop	Effect	References
Silver	*Ulva lactuca* and *Turbinaria turbinata*	*M. javanica*	Egg plants	Reduction in the number of galls, number of female nematodes and number of egg masses per root system.	Abdellatif et al., 2016
	Organic extract of *Urtica urens*	*M. incognita*	*In vitro*	Increase the nematode mortality and reduce the egg hatchability.	Nassar, 2016
	Conyza dioscoroides leaf extract	*M. incognita*	*In vitro*	Inhibit the egg hatching and cause mortality of nematodes juveniles by producing nematicidal secondary metabolites.	Abbassy et al., 2017
	Azadirachta indica and Curcuma longa	*M. incognita*	*In vitro*	Increase the nematode mortality and inhibit the egg hatching.	Kalaiselvi et al., 2017
	Zingiber officinale Rhizome extract	*M. incognita*	Tomato	Reduction in galls and egg masses.	El-Deen & El-Deeb, 2018
	Milled corn residue	*H. sacchhari*	Rice	Improved rice plant growth with a reduction in nematode population.	Fabiyi & Olatunji, 2018
	Aqueous extracts of *Ficus mucoso* bark	*M. incognita*	Groundnut	Increase the tolerance to nematodes by enhancing growth and yield. Reduce the nematodes population also.	Fabiyi & Olatunji, 2018
	Ulva fasciata Delile (UF) (green algae), Corallina mediterranea, C.officinalis	*M. incognita*	Tomato	Production of defence enzymes PAL, PO which cause induced systemic resistance.	Ghareeb et al., 2019
	Cyanobacterium, Nostoc sp.	*M. javanica*	Faba been	Reduction in nematode infestation and increased the tolerance towards nematodes by improving the root morphology and growth parameters.	Hamed et al., 2019

(Continued)

TABLE 16.3 (CONTINUED)
Green synthesized nanoparticles for the management of plant-parasitic nematodes

NP	Green materials	Nematodes	Crop	Effect	References
	Euphorbia tirucalli latex	*M. incognita*	Tomato	Reduced gall formation, eggmasses on roots, number of eggs per egg masses and the J2 populationin the soil. It repels the nematodes by altering the root exudates.	Kalaiselvi et al., 2019
	Microcrystalline cellulose	*M. incognita*	*In vitro*	Inhibit the nematode development by causing abnormal embryogenesis at the tadpole stage and abnormal division at the gastrula stage.	Fouda et al., 2020
	Cladophora glomerata and *Ulva fasciata*	*M. incognita*	Tomato	Increase in the chitinase enzyme expression causes deformed eggs and a reduction in the nematode population.	Ghareeb et al., 2020 (a)
	Cladophora glomerata extract	*M. javanica*	Tomato	Induction of systemic acquired resistance Reduction in nematode population and gall formation.	Ghareeb et al., 2020 (b)
	Eucalyptus officinallis	*Heteroderasacchari*	Paddy	Reduced the number of cysts per plant and also soil nematode populations.	Fabiyi et al., 2020
	Senna seamia Leaf Extract	*M. incognita*	*Trachyspermum ammi*	Reduction in gall formation and production of egg masses. Increase in lignin content in roots. The lignified cell wall acts as a physical barrier against nematode attack and induces resistance.	Danish et al., 2021
	Acalypha wilkesiana aqueous leaf extract	*M. incognita*	*In vitro*	Inhibit the egg hatching and nematode movement and also induces nematode mortality.	Darago, 2014
	The agricultural wastes (corn cob, rice husk, and guinea corn chaff)	*M. javanica*	*In vitro*	Increase the nematode mortality and reduce the egg hatching.	Fabiyi et al., 2021

TABLE 16.3 (CONTINUED)
Green synthesized nanoparticles for the management of plant-parasitic nematodes

NP	Green materials	Nematodes	Crop	Effect	References
	Cnidoscolus aconitifolius stem extract	*M. incognita*	Carrot	Decrease in nematode population in soil, egg masses and the number of galls.	Fabiyi et al., 2021
	Colpomenia sinuosa algal extract	*M. incognita*	Tomato	Reduced the number of nematode galls, egg masses per root, and eggs/egg mass, while also improving plant growth parameters by producing nematicidal organic compounds.	Ghareeb et al., 2022
	Lantana camara and *Conyza dioscoridis* leaf extract	*M. incognita*	Tomato	Reduction in the number of galls, egg masses/root, and eggs/egg mass as the migration is disrupted by the organic compounds.	El-Habashy, 2022
	Plantain peel extract	*M. incognita*	Tomato	Reduction in egg hatching and gall index percentage.	Fabiyi & Olatunji, 2022
Silver (Ag) and silver chloride (AgCl)	*Parachlorella kessleri SAG 211-11*	*M. incognita*	*In vitro*	Induces nematode mortality and inhibit the egg hatching.	Hamed et al., 2019
Zinc oxide (ZnO)	*Streptomyces plicatus*	*M. incognita*	*In vitro*	Deformation of the nematode cuticle and causes immobilization leads to mortality.	Kalaba et al., 2021
	Synadenium grantii latex serum	*M. incognita*	Tomato	Promotes plant growth and causes mortality of nematodes by producing proteolytic enzymes.	Duraisamy et al., 2022
Copper oxide (CuO)	*Jatropha curcas* Leaf extract	*M. incognita*	Chickpea	Reduction in nematode population and galling percentage by altering the root exudates which attract the nematodes.	Khan et al., 2022
Copper (Cu)	*Orobanche aegyptiaca* Stem extract	*M. incognita*	*In vitro*	Induces nematode mortality and inhibits the egg hatching.	Akhter et al., 2020

(Continued)

TABLE 16.3 (CONTINUED)
Green synthesized nanoparticles for the management of plant-parasitic nematodes

NP	Green materials	Nematodes	Crop	Effect	References
	Haplophyllum tuberculatum	*M. incognita*	*In vitro*	Induces nematode mortality and inhibits the egg hatching.	Soliman et al., 2017
Magnesium oxide (MgO)	Strawberry seed powder	*M. incognita*	*In vitro*	Induces nematode mortality and inhibits egg hatching by altering the morphology.	Khan et al., 2022
Sulphur	*Rosmarinus officinalis* leaves aqueous extract	*M. javanica*	*In vitro*	Exhibits cytotoxicity on juveniles leads to mortality.	Al Banna et al., 2020

16.3.3 Mechanism Involved in NP Against Nematodes

Management of nematodes with nanoparticles involves various mechanisms. The disruption of numerous cellular pathways, including membrane permeability, is most likely the mechanism AgNPs work to inhibit nematicidal activity, ATP synthesis and response to oxidative stress (Lim et al., 2012; Kalaiselvi et al., 2019). Enhancement of chitinase gene expression in AgNPs treated plants is reported by (Ghareeb et al., 2020a) which is an enzyme that degrades the chitin. The pharynx and eggshell of nematodes both include chitin as a necessary component. Nematode embryonic lethality, the production of faulty eggs, or moulting failure are all caused by disruptions in chitin synthesis or hydrolysis. Thus, the key components in the chitin metabolic process are promising targets for the anti-nematode agent's development. Lignin deposition in the roots plays a vital role in the development of resistance of plants against nematodes. The plants treated with silver nanoparticles are reported to contain more lignin when compared to untreated plants (Danish et al., 2021). The lignifications of root cells prevent nematode penetration and entry into roots as it is a physical barrier.

Green nanoparticles treatment in plants induced systemic resistance which leads to producing plant defence enzymes including peroxidase (POX), polyphenol oxidase (PPO) and phenylalanine ammonia lyase (PAL) production (Ghareeb et al., 2020b). Mouniga et al. (2022) in his study stated that 1% chitosan nanoparticles were applied to tomato plants that had root-knot nematode *M. incognita* infections in order to increase phenol content and the activity of the antioxidant enzymes peroxidase and polyphenol oxidase, which would then induce systemic resistance in the plants. Root-knot nematodes' ability to develop and reproduce may be hampered by AgNPs' toxicity. The mode of action of AgNPs is not specific and is still being investigated, but is linked to the disruption of numerous cellular processes like membrane permeability, ATP synthesis and oxidative stress response in both bacterial and eukaryotic (Roh et al., 2009) and prokaryotic cells (Sondi & Salopek-Sondi, 2004; Lok et al., 2006; Morones et al., 2005; Choi & Hu, 2008). The mechanism of copper NPs involved DNA damage caused by CuNP's severe toxic effects. The ZnO NPs caused the distorted shape of juveniles in the second stage of the *Meloidogyne javanica* species, as observed under a microscope, which may be caused by the potential negative effects on the cuticles and hypodermis of nematodes. These effects could be caused by their influence on lipids, glycogen and mucopolysaccharides.

16.4 FUTURE THRUSTS AND PERSPECTIVE

Nematicides made by chemicals pollute the environment and have persistent toxicity, and also the nematodes become resistant as a result of their extensive use. Therefore, their management requires a different, environmentally friendly economic strategy. Green AgNPs can be employed for this because they have excellent antibacterial potential. A comprehensive strategy is needed to understand the cognitive and functional characteristics of green AgNPs both *in vitro* and *in vivo*. More research is required to determine their effect on other biological systems and crops.

Nanotechnology has enormous potential to completely transform the current situation in pathogen sensing and management and also promises to offer a clear solution to developing complexities in agriculture. The application of nanotechnology can be studied and implemented in pathogen detection, reduced pesticide solubility, target-oriented delivery based on pH release, increased bioavailability, delivery of specific RNAi molecules for nematode control and release of lower active molecules from NP carriers into the environment. The creation of nanomaterials is a crucial process, and variations in the technology used to create them can affect their dimensions and shapes as well as the dangers involved in using them. Since there are no explicit rules for using these formulations on nanomaterials, risk assessment studies are important prior to the use of such materials, and it is necessary to examine how harmful these compounds are to plants and insects. Much work needs to be done on nanopesticide formulations before they are used more frequently in pest management. This work will involve combining analytical techniques that can identify, characterize (e.g., through size, size range, shape or nature and surface properties), and quantify the active ingredient and adjuvants emanating from the formulations. Nanopesticide-based formulations, such as nanoemulsions, nanodispersions and NPs, have a promising future and the potential to develop safer and more effective chemical pesticide formulations for nematode control.

Zerovalent iron nanoparticles could be used to remediate pesticide-contaminated soil to stop the contamination of groundwater and land since they have a strong affinity for absorbing heavy metals and organic molecules. Moreover, FeNPs harbour wonderful soil-binding qualities like CaCo3. More attention should be placed on using agricultural waste products as raw materials for NM manufacturing in addition to reducing the environmental impact. The compatibility of nutrition, fertilizer and pesticide encapsulating materials as well as the workings of pesticide formulations need to be studied.

16.5 CONCLUSIONS

The application of nanoparticles in the detection and treatment of plant diseases has enormous potential. The biosynthesized and synthetic nanoparticles provide interesting and challenging opportunities in the management of economically important plant-parasitic nematodes. India's population is now at an exponential peak and the unavoidable atrocities like the expansion of industrialization and degradation of arable land are supplementing the major problems of improved productivity and management of resurgent pests and pathogens. The need of the hour to meet future challenges is smart agricultural alternatives and nanotechnology fulfils all the criteria. The various chemicals that are available on the market are a direct result of the utilization of nanoproducts with economic value. Nanotechnology is a multi-faceted science that can minimize nutrient loss, create sustainability and enhance crop growth and yield. The technology also promises to be compatible with pesticides and reduce land water degradation and pesticide contamination. The solubility of water-insoluble pesticides is increased through nanoformulation, which also targets just specific pests and prevents the pesticide from degrading too soon. The green nanoparticles are promising propositions for natural and safe agricultural practices, and they can be explored for better efficacy against devastating nematodes. Hence more research and scientific intervention are required regarding the effects of various nanoparticles and their mechanism in plant growth and protection against pests and nematodes under various abiotic and biotic stresses.

REFERENCES

Abbassy, M. A., Abdel-Rasoul, M. A., Nassar, A. M., & Soliman, B. S. (2017). Nematicidal activity of silver nanoparticles of botanical products against root-knot nematode, meloidogyne incognita. *Archives of Phytopathology and Plant Protection*, *50*(17–18), 909–926.

Abd-Elgawad, M. M. (2020). Biological control agents in the integrated nematode management of potato in Egypt. *Egyptian Journal of Biological Pest Control*, *30*, 1–13.

Abdellatif, K. F., Abdelfattah, R. H., & El-Ansary, M. S. M. (2016). Green nanoparticles engineering on root-knot nematode infecting eggplants and their effect on plant DNA modification. *Iranian Journal of Biotechnology*, *14*(4), 250.

Akhter, G., Khan, A., Ali, S. G., Khan, T. A., Siddiqi, K. S., & Khan, H. M. (2020). Antibacterial and nematicidal properties of biosynthesized Cu nanoparticles using extract of holoparasitic plant. *SN Applied Sciences*, *2*, 1–6.

Al Banna, L. S., Salem, N. M., Jaleel, G. A., & Awwad, A. M. (2020). Green synthesis of sulfur nanoparticles using *Rosmarinus officinalis* leaves extract and nematicidal activity against *Meloidogyne javanica*. *Chemistry International*, *6*(3), 137–143.

Anwar, S. A., & McKenry, M. V. (2012). Incidence and population density of plant-parasitic nematodes infecting vegetable crops and associated yield losses in Punjab, Pakistan. *Pakistan Journal of Zoology*, *44*(2).

Ardakani, A. S. (2013). Toxicity of silver, titanium and silicon nanoparticles on the root-knot nematode, meloidogyne incognita, and growth parameters of tomato. *Nematology*, *15*(6), 671–677.

Auría-Soro, C., Nesma, T., Juanes-Velasco, P., Landeira-Viñuela, A., Fidalgo-Gomez, H., Acebes-Fernandez, V., Gongora, R., Almendral Parra, M. J., Manzano-Roman, R., & Fuentes, M. (2019). Interactions of nanoparticles and biosystems: Microenvironment of nanoparticles and biomolecules in nanomedicine. *Nanomaterials*, *9*(10), 1365.

Bailey, K. L., Boyetchko, S. M., & Längle, T. J. B. C. (2010). Social and economic drivers shaping the future of biological control: A Canadian perspective on the factors affecting the development and use of microbial biopesticides. *Biological Control*, *52*(3), 221–229.

Berry, T. D., Filley, T. R., & Blanchette, R. A. (2014). Oxidative enzymatic response of white-rot fungi to single-walled carbon nanotubes. *Environmental Pollution*, *193*, 197–204.

Bhau, B. S., Phukon, P., Ahmed, R., Gogoi, B., Borah, B., Baruah, J., Sharma, D. K., & Wann, S. B. (2016). A novel tool of nanotechnology: Nanoparticle mediated control of nematode infection in plants. In Singh, D., Singh, H., Prabha, R. (eds), *Microbial Inoculants in Sustainable Agricultural Productivity*. Springer. https://doi.org/10.1007/978-81-322-2644-4_16

Bholay, A. D., Nalawade, P. M., & Borkhataria, B. (2013). Fungicidal potential of biosynthesized silver nanoparticles against phyto-pathogens and potentiation of fluconazole. *World*, *1*(1), 12–15.

Bhushan, B. (2004). *Handbook of nanotechnology*. Springer.

Chen, J., Li, Q. X., & Song, B. (2020). Chemical nematicides: Recent research progress and outlook. *Journal of Agricultural and Food Chemistry*, *68*(44), 12175–12188.

Chen, Q., & Peng, D. (2019). Nematode chitin and application. In Yang, Q., & Fukamizo, T. (eds), *Targeting chitin-containing organisms. Advances in Experimental Medicine and Biology, vol 1142* (pp. 209–219). Springer. https://doi.org/10.1007/978-981-13-7318-3_10

Choi, O., & Hu, Z. (2008). Size dependent and reactive oxygen species related nanosilver toxicity to nitrifying bacteria. *Environmental Science and Technology, 42*(12), 4583–4588.

Crucho, C. I., & Barros, M. T. (2017). Polymeric nanoparticles: A study on the preparation variables and characterization methods. *Materials Science and Engineering: Part C, 80*, 771–784. https://doi.org/10.1016/j.msec.2017.06.004

Danish, M., Altaf, M., Robab, M. I., Shahid, M., Manoharadas, S., Hussain, S. A., & Shaikh, H. (2021). Green synthesized silver nanoparticles mitigate biotic stress induced by meloidogyne incognita in *Trachyspermum ammi* (L.) by improving growth, biochemical, and antioxidant enzyme activities. *ACS Omega, 6*(17), 11389–11403.

Daragó, A. (2014). *The distribution of dagger nematodes species in Hungarian wind regions and newest control options* (Doctoral Dissertation, PhD Thesis). University of Pannonia Georgikon, Keszthely, Hungary.

Duraisamy, K., Amirthalingam, M., Govindhan, T., Kim, J. C., Hasegawa, K., & Palanisamy, S. (2022). Fabrication of zinc oxide nanorods using plant latex serum as a green matrix for the sustainable management of root-knot nematodes. *Materials Letters, 317*, 132098.

El-Ansary, M. S. M., Hamouda, R. A., & Elshamy, M. M. (2022). Using biosynthesized zinc oxide nanoparticles as a pesticide to alleviate the toxicity on banana infested with parasitic-nematode. *Waste and biomass valorization*, 13, 405–415). https://doi.org/10.1007/s12649-021-01527-6

El-Deen, A. N., & El-Deeb, B. A. (2018). Effectiveness of silver nanoparticles against root-knot nematode, meloidogyne incognita infecting tomato under greenhouse conditions. *Journal of Agricultural Science, 10*(2), 148–156.

El-Habashy, D. E. (2022). Effectiveness of nanoparticles of some plant extracts against root-knot nematode, meloidogyne incognita on tomato plants. *SVU-International Journal of Agricultural Sciences, 4, 3*(3), 46–57.

Elmer, W., De La Torre-Roche, R., Pagano, L., Majumdar, S., Zuverza-Mena, N., Dimkpa, C., Gardea-Torresdey, J., & White, J. C. (2018). Effect of metalloid and metal oxide nanoparticles on *Fusarium* wilt of watermelon. *Plant Disease, 102*(7), 1394–1401.

Fabiyi, O. A., Claudius-Cole, A. O., Olatunji, G. A., Abubakar, D. O., & Adejumo, O. A. (2021). Response of *Meloidogyne javanica* to silver nanoparticle liquid from agricultural wastes. *Agrivita, Journal of Agricultural Science, 43*(3), 507–517.

Fabiyi, O. A., & Olatunji, G. A. (2018). Application of green synthesis in Nano particles preparation: Ficus mucoso extracts in the management of meloidogyne incognita infecting groundnut *Arachis hypogea. Indian Journal of Nematology, 48*(1), 13–17.

Fabiyi, O. A., & Olatunji, O. G. A. (2022). Application of green synthesis in silver nanoparticles for root knot nematode management on tomato. *Indian Journal of Nematology, 52*(2), 164–173.

Fabiyi, O., Olatunji, O. G. A., Atolani, O., & Olawuyi, O. (2020). Preparation of bio-nematicidal nanoparticles of *Eucalyptus officinalis* for the control of cyst nematode (Heterodera sacchari). *Journal of Animal and Plant Sciences, 30*(5), 1172–1177.

Fard, J. K., Jafari, S., & Eghbal, M. A. (2015). A review of molecular mechanisms involved in toxicity of nanoparticles. *Advanced Pharmaceutical Bulletin, 5*(4), 447.

Fouda, M. M., Abdelsalam, N. R., Gohar, I. M. A., Hanfy, A. E., Othman, S. I., Zaitoun, A. F., Allam, A. A., Morsy, O. M., & El-Naggar, M. (2020). Utilization of high throughput microcrystalline cellulose decorated silver nanoparticles as an eco-nematicide on root-knot nematodes. *Colloids and Surfaces B: Biointerfaces, 188*, 110805.

Ghareeb, R. Y., Adss, I. A., Bayoumi, S. R., & El-Habashy, D. E. (2019). The nematicidal potentiality of some algal extracts and their role in enhancement the tomato defense genes against root knot-nematodes. *Egyptian Journal of Biological Pest Control, 29*, 1–10.

Ghareeb, R. Y., Alfy, H., Fahmy, A. A., Ali, H. M., & Abdelsalam, N. R. (2020). Utilization of Cladophora glomerata extract nanoparticles as eco-nematicide and enhancing the defense responses of tomato plants infected by *Meloidogyne javanica. Scientific Reports, 10*(1), 19968.

Ghareeb, R. Y., EL-Saedy, M. A., Fathy, M. M., Hafez, E. E., & Basyony, A. B. (2020). Green biosynthesized silver nanoparticles using macroalgae and their usage as a bioagent on meloidogyne incognita. *Plant Archives, 20*, 9682–9692.

Ghareeb, R. Y., Shams El-Din, N. G. E. D., Maghraby, D. M. E., Ibrahim, D. S., Abdel-Megeed, A., & Abdelsalam, N. R. (2022). Nematicidal activity of seaweed-synthesized silver nanoparticles and extracts against meloidogyne incognita on tomato plants. *Scientific Reports, 12*(1), 3841.

Giannousi, K., Avramidis, I., & Dendrinou-Samara, C. (2013). Synthesis, characterization and evaluation of copper based nanoparticles as agrochemicals against *Phytophthora infestans. RSC Advances, 3*(44), 21743–21752.

Goswami, A., Roy, I., Sengupta, S., & Debnath, N. (2010). Novel applications of solid and liquid formulations of nanoparticles against insect pests and pathogens. *Thin Solid Films, 519*(3), 1252–1257.

Graham, J. H., Johnson, E. G., Myers, M. E., Young, M., Rajasekaran, P., Das, S., & Santra, S. (2016). Potential of nano-formulated zinc oxide for control of citrus canker on grapefruit trees. *Plant Disease, 100*(12), 2442–2447.

Gupta, S., Kushwah, T., Vishwakarma, A., & Yadav, S. (2015). Optimization of ZnO-NPs to investigate their safe application by assessing their effect on soil nematode *Caenorhabditis elegans. Nanoscale Research Letters, 10*(1), 303.

Hafez, E. E., Hassan, H. S., Elkady, M., & Salama, E. (2014). Assessment of antibacterial activity for synthesized zinc oxide nanorods against plant pathogenic strains. *International Journal of Sciences and Technology Research (IJSTR), 3*(9), 318–324.

Hamed, S. M., Hagag, E. S., & El-Raouf, N. A. (2019). Green production of silver nanoparticles, evaluation of their nematicidal activity against *Meloidogyne javanica* and their impact on growth of faba bean. *Beni-Suef University Journal of Basic and Applied Sciences, 8*(1), 1–12.

Hardman, R. (2006). A toxicologic review of quantum dots: Toxicity depends on physicochemical and environmental factors. *Environmental Health Perspectives, 114*(2), 165–172.

Jha, A. K., Prasad, K., & Prasad, K. (2009). A green low-cost biosynthesis of Sb_2O_3 nanoparticles. *Biochemical Engineering Journal, 43*(3), 303–306.

Kalaba, M. H., Moghannem, S. A., El-Hawary, A. S., Radwan, A. A., Sharaf, M. H., & Shaban, A. S. (2021). Green synthesized ZnO nanoparticles mediated by *Streptomyces plicatus*: Characterizations, antimicrobial and nematicidal activities and cytogenetic effects. *Plants, 10*(9), 1760.

Kalaiselvi, D., Mohankumar, A., Shanmugam, G., Nivitha, S., & Sundararaj, P. (2019). Green synthesis of silver nanoparticles using latex extract of *Euphorbia tirucalli*: A novel approach for the management of root knot nematode, meloidogyne incognita. *Crop Protection, 117*, 108–114.

Kalaiselvi, D., Sundararaj, P., Premasudha, P., & Hafez, S. L. (2017). Nematicidal activity of green synthesized silver nanoparticles using plant extracts against root-knot nematode *Meloidogyne incognita. International Journal of Nematology, 27*(1), 81–94.

Kaushik, H., & Dutta, P. (2017, August). *Chemical synthesis of zinc oxide nanoparticle: Its application for antimicrobial activity and plant health management.* 2017 APS Annual Meeting. APSNET. San Antonio, Texas.

Khan, A. U., Bani Mfarrej, M. F., Danish, M., Shariq, M., Khan, M. F., Ansari, M. S., Hashem, M., Alamri, S., & Ahmad, F. (2022). Synthesized copper oxide nanoparticles via the green route act as antagonists to pathogenic root-knot nematode, *Meloidogyne incognita. Green Chemistry Letters and Reviews, 15*(3), 491–507.

Khan, A. U., Khan, M., Khan, A. A., Parveen, A., Ansari, S., & Alam, M. (2022). Effect of phyto-assisted synthesis of magnesium oxide nanoparticles (MgO-NPs) on bacteria and the root-knot nematode. *Bioinorganic Chemistry and Applications.* vol. 2022, Article ID 3973841, 11 pages, 2022. https://doi.org/10.1155/2022/3973841

Kirtane, A. R., Verma, M., Karandikar, P., Furin, J., Langer, R., & Traverso, G. (2021). Nanotechnology approaches for global infectious diseases. *Nature Nanotechnology, 16*(4), 369–384.

Kumar, V., Khan, M. R., & Walia, R. K. (2020). Crop loss estimations due to plant-parasitic nematodes in major crops in India. *National Academy Science Letters, 43*(5), 409–412.

Kuppusamy, P., Yusoff, M. M., Maniam, G. P., & Govindan, N. (2016). Biosynthesis of metallic nanoparticles using plant derivatives and their new avenues in pharmacological applications–An updated report. *Saudi Pharmaceutical Journal, 24*(4), 473–484.

Lim, D., Roh, J. Y., Eom, H. J., Choi, J. Y., Hyun, J., & Choi, J. (2012). Oxidative stress-related PMK-1 P38 MAPK activation as a mechanism for toxicity of silver nanoparticles to reproduction in the nematode *Caenorhabditis elegans. Environmental Toxicology and Chemistry, 31*(3), 585–592.

Lok, C. N., Ho, C. M., Chen, R., He, Q. Y., Yu, W. Y., Sun, H., Tam, P. K. H., Chiu, J. F., & Che, C. M. (2006). Proteomic analysis of the mode of antibacterial action of silver nanoparticles. *Journal of Proteome Research, 5*(4), 916–924.

Mahawar, H., & Prasanna, R. (2018). Prospecting the interactions of nanoparticles with beneficial microorganisms for developing green technologies for agriculture. *Environmental Nanotechnology, Monitoring and Management, 10*, 477–485.

Mohamed, E. A., Elsharabasy, S. F., & Abdulsamad, D. (2019). Evaluation of in vitro nematicidal efficiency of copper nanoparticles against root-knot nematode *Meloidogyne incognita. South Asian Journal of Parasitology, 2*(1), 1–6.

Morones, J. R., Elechiguerra, J. L., Camacho, A., Holt, K., Kouri, J. B., Ramírez, J. T., & Yacaman, M. J. (2005). The bactericidal effect of silver nanoparticles. *Nanotechnology, 16*(10), 2346.

Mouniga, R., Anita, B., Shanthi, A., Lakshmanan, A., & Karthikeyan, G. (2022). Phenol and antioxidant enzymatic activity in root knot nematode, *Meloidogyne incognita* infected tomato plants treated with chitosan nanoparticles. *The Pharma Innovation Journal, 11*(4), 241–245.

Moosa, A. A. (2009). *Nanotechnology: principles and applications. Department of production engineering and metallurgy university of technology Baghdad, Iraq.* Dar Djlah- Amman-Jordan.

Nassar, A. M. (2016). Research article effectiveness of silver nano-particles of extracts of *Urtica urens* (Urticaceae) against root-knot nematode meloidogyne incognita. *Asian Journal of Nematology, 5*(1), 14–19.

Paul, A., & Roychoudhury, A. (2021). Go green to protect plants: Repurposing the antimicrobial activity of biosynthesized silver nanoparticles to combat phytopathogens. *Nanotechnology for Environmental Engineering, 6*(1), 10.

Prasad, K. S., Gandhi, P., & Selvaraj, K. (2014). Synthesis of green Nano iron particles (GnIP) and their application in adsorptive removal of As (III) and As (V) from aqueous solution. *Applied Surface Science, 317*, 1052–1059.

Prasad, R., Bhattacharyya, A., & Nguyen, Q. D. (2017). Nanotechnology in sustainable agriculture: Recent developments, challenges, and perspectives. *Frontiers in Microbiology*, *8*, 1014. https://doi.org/10.3389/fmicb.2017.01014

Prasad, R., Kumar, V., & Prasad, K. S. (2014). Nanotechnology in sustainable agriculture: Present concerns and future aspects. *African Journal of Biotechnology*, *13*(6), 705–713.

Prasad, T. N. V. K. V., Sudhakar, P., Sreenivasulu, Y., Latha, P., Munaswamy, V., Reddy, K. R., Sreeprasad, T. S., Sajanlal, P. R., & Pradeep, T. (2012). Effect of nanoscale zinc oxide particles on the germination, growth and yield of peanut. *Journal of Plant Nutrition*, *35*(6), 905–927.

Qi, W. H., & Wang, M. P. (2002). Size effect on the cohesive energy of nanoparticle. *Journal of Materials Science Letters*, *21*(22), 1743–1745.

Rai, M., Ingle, A. P., Trzcińska-Wencel, J., Wypij, M., Bonde, S., Yadav, A., Kratošová, G., & Golińska, P. (2021). Biogenic silver nanoparticles: What we know and what do we need to know? *Nanomaterials*, *11*(11), 2901.

Rajan, M. S. (2006). *Nano: The next revolution*. National Book Trust.

Richards, R. M. (1981). Antimicrobial action of silver nitrate. *Microbios*, *31*(124), 83–91.

Roh, J. Y., Sim, S. J., Yi, J., Park, K., Chung, K. H., Ryu, D. Y., & Choi, J. (2009). Ecotoxicity of silver nanoparticles on the soil nematode *Caenorhabditis elegans* using functional ecotoxicogenomics. *Environmental Science and Technology*, *43*(10), 3933–3940.

Roy, S., & Das, T. K. (2015). Plant mediated green synthesis of silver nanoparticles-A. *International Journal of Plant Biology and Research*, *3*, 1044–1055.

Sabir, S., Arshad, M., & Chaudhari, S. K. (2014). Zinc oxide nanoparticles for revolutionizing agriculture: Synthesis and applications. *Scientific World Journal. 2014*, Article ID 925494. https://doi.org/10.1155/2014/925494

Sabry, A. K. H. (2019). Role of nanotechnology applications in plant-parasitic nematode control. In Abd-Elsalam, K., & Prasad, R. (eds), *Nanobiotechnology applications in plant protection. Nanotechnology in the Life Sciences* (Vol. 2, pp. 223–240). Springer. https://doi.org/10.1007/978-3-030-13296-5_12

Sangeetha, J., Thangadurai, D., Hospet, R., Harish, E. R., Purushotham, P., Mujeeb, M. A., Shrinivas, J., David, M., Mundaragi, A. C., Thimmappa, S. C., & Arakera, S. B. (2017). Nanoagrotechnology for soil quality, crop performance and environmental management. In Prasad, R., Kumar, M., & Kumar, V. (eds) *Nanotechnology: An Agricultural Paradigm*, 73–97. Springer. https://doi.org/10.1007/978-981-10-4573-8_5

Sattler, K. D. (2010). *Handbook of nanophysics, principles and methods*. CRC.

Siddiqui, Z. A., Parveen, A., Ahmad, L., & Hashem, A. (2019). Effects of graphene oxide and zinc oxide nanoparticles on growth, chlorophyll, carotenoids, proline contents and diseases of carrot. *Scientia Horticulturae*, *249*, 374–382.

Sivasubramaniam, N., Hariharan, G., & Zakeel, M. C. M. (2020). Sustainable management of plant-parasitic nematodes: An overview from conventional practices to modern techniques. In R. Ansari, R. Rizvi, & I. Mahmood (Eds.), *Management of phytonematodes: Recent advances and future challenges*. Springer. https://doi.org/10.1007/978-981-15-4087-5_16

Soliman, B., Abbassy, M., Abdel-Rasoul, M., & Nassar, A. (2017). Efficacy of silver nanoparticles of extractives of *Artemisia judaica* against root-knot nematode. *Journal of Environmental Studies and Researches*, *7*(2), 832–844.

Sondi, I., & Salopek-Sondi, B. (2004). Silver nanoparticles as antimicrobial agent: A case study on *E. coli* as a model for gram-negative bacteria. *Journal of Colloid and Interface Science*, *275*(1), 177–182.

Subramanian, K. S., & Tarafdar, J. C. (2011). Prospects of nanotechnology in Indian farming. *Indian Journal of Agricultural Sciences*, *81*(10), 887–893.

Thakur, R. K., Dhirta, B., & Shirkot, P. (2018). Studies on effect of gold nanoparticles on meloidogyne incognita and tomato plants growth and development. *BioRxiv*, 428144. doi: https://doi.org/10.1101/428144

Thomas, S. C., Kumar Mishra, P., Talegaonkar, S., & Talegaonkar, S. (2015). Ceramic nanoparticles: Fabrication methods and applications in drug delivery. *Current Pharmaceutical Design*, *21*(42), 6165–6188. https://doi.org/10.2174/1381612821666151027153246.

Ying, S., Guan, Z., Ofoegbu, P. C., Clubb, P., Rico, C., He, F., & Hong, J. (2022). Green synthesis of nanoparticles: Current developments and limitations. *Environmental Technology and Innovation*, *26*, 102336.

17 Agricultural Applications of Nanobiosensors in Disease Diagnosis and Food Toxicity Analysis

*Avdhut Jadhav, Kunal Kumar, Subhash Rathod, Ramireddy Boppella, Ylias M. Sabri, Samuel J. Ippolito, Bhagwan Toksha, Aniruddha Chatterjee, Manorama V. Sunkara and Shravanti Joshi**

17.1 INTRODUCTION

Advances in farming and food supply are the leading reasons for revenue and occupation for a higher percentage of the population around the globe. Agriculture is an essential pillar for every nation as it supports the country's ability to maintain its economic self-sufficiency, providing raw goods to feed its people and other industrial sectors, and even going to a greater ability in strengthening relations with other nations. Interestingly, current estimates show that more than 2,500 million (roughly 31.25% of the current world's population) are engaged in the agriculture sector, which is highly diverse and has seen rapid technological growth in the last few years (Mathivanan, 2021). In recent times, several biotic and abiotic aspects are observed to be stimulating agricultural efficiency, productivity and yield. However, in the last few decades, almost all countries are observing a rapid decline in agricultural prospects and productivity primarily due to a reduction in the availability of fertile land, a drastic increase in soil and water pollution and the hazardous impact of climate changes. Additionally, the matter is worsened by the alarming use of fertilizers, pesticides and fungicides, subsequently resulting in heavy metal absorption in the farming land. To retain food security for the ever-growing world population, addressing vital concerns such as enhancement of crop yield, improvement in quality assurance, identification and prevention of diseases in the first stage, and proper storage techniques are imperative (Fang & Ramasamy, 2015; Mousavi & Rezaei, 2011). The most effective way to deal with such concerns is the employment of smarter, sustainable technologies. So far, nanoscience and nanotechnology have been viewed as fast-evolving fields that have shown greater capabilities to revolutionize many different sectors in the earlier period and persistence

* Corresponding author.

DOI: 10.1201/9781003376446-17

just as in the future as well. Nanotechnology has shown its unique abilities even in the agriculture and food industries. It has bolstered the countryside economy by encouraging sustainable technology in organic farming, improving the quality of products, reducing yield losses, enhancing tool life and so on. Thus, to enhance farmers' revenue and livelihood, cost-effective technologies that are fast and portable are required for real-time and accurate monitoring of disease-causing agents in the agriculture sector (Mustafa & Andreescu, 2018; Gallocchio et al., 2015; Farahi e al., 2012). Numerous technological advancements have led to factors that have affected crop life and yield. Factors reducing crop yield mainly include unwanted plants, weeds, chemicals, loss of soil nutrients, insects, water quality and pathogens that can be meticulously controlled by the farmers by using fertilizer, weed killers, insecticide, etc. Many times, unintentional usage of such chemical products in higher concentrations leads to the deterioration of soil, water and future crop quality (Chauhan et al., 2016). Consumption of such food items results in ill effects on living beings in the form of cancer, allergies, lower life expectancy, lower productivity, etc. due to the food items being less nutritious or contaminated. In addition, the unintentional establishment of micro-organisms and bacterial variants that are toxic or produce harmful chemical compounds during storage and transportation can also lead to further health concerns. Such concerns can be effectively addressed by monitoring the chemical and biological food contaminants before eating to ensure food safety and avoid food poisoning hazards. Earlier reports on advanced yet standard techniques, namely chromatography, molecular biology and spectroscopy, have been discussed (Mathivanan, 2021; Fang & Ramasamy, 2015). But the methods were found to have poor resolution, reduced sensitivity, a lower limit of detection, were tedious and required highly skilled technical staff, thus limiting their use and further commercialization (Farahi et al., 2012). Despite, many such challenges, there lay prospects to develop ultrasensitive and highly selective nanobiosensors for applications in farming and food segments.

Although farming and food divisions have been revolutionized by nanoscience in terms of basic research and understanding, practical results and commercialization are still in the nascent stages. Agricultural practices in the current era are far superior compared to a few decades ago, let alone centuries ago, but greater advancements in typical farming methods at a rapid pace are required if the security and well-being of future generations are to be safeguarded. In general, farming efficiency or crop yield can be improved firstly by lower the losses caused by insects, weeds, harsh weather conditions, loss of valuable nutrients, etc. Secondly, it can be improved by undertaking progressive crop production methodologies in terms of lowering the input cost and expanding the product yield with higher price profit proportion. In both cases, nanotechnology can be employed in the agriculture sector and food industry to increase farming efficiency, production rate, quality assurance, crop yield, food safety and storage (Sekhon, 2014; Dasgupta et al., 2015). In this context, biosensors integrated with nanomaterials, that is, nanobiosensors, are a self-contained advanced tool facilitating extremely precise and specific detection of biological variants to predict diseases and food toxicity in real-time. Nanobiosensors utilize expertise from numerous disciplines such as biochemistry, engineering, physics, nanotechnology, sensor technology and materials science. It involves all the principles of nanoscience and nanotechnology such as mounting techniques, sensor device fabrication, electrode design, development of facile chemical routes, nanomaterials synthesis, miniaturization techniques, micro-fluidics and nano- and micro-electromechanical systems (NEMS/MEMS) (Bhatia, 2016; Ettenauer et al., 2015; Malhotra et al., 2005). Comprehensive understanding of nanobiosensors are still limited and it is anticipated that only improving their inherent characteristics in terms of nanomaterials synthesis routes, morphology, aspect ratio, physicochemical and electrical characteristics etc., will help in popularizing their widespread applications in the farming sector and food industry (Reddy & Ratna, 2013; Cui et al., 2015; Weiss, 2010). In this review, the different types of biosensors and various nanomaterials used for sensor device fabrication are discussed with a focus on the importance of nanobiosensing from viewpoints of agricultural applications such as disease diagnosis and food toxicity analysis.

17.2 CLASSIFICATION OF BIOSENSORS

An electronic device that detects changes taking place in its surrounding environment and sends the details to other devices that are part of an electrical circuit for analysis is called a sensor. For example, a sensor employing an appropriate semiconducting metal oxide sensitive layer that changes its electrical characteristics in the presence of a given gas analyte (i.e. oxidizing, reducing or any other volatile organic compound) can be used as a sensing element. Typically, these types of sensors are referred to as solid-state chemoresistive gas sensors and can provide a readout in the form of a change in resistance, impedance, capacitance or voltage depending on the design of the supporting electronic circuit (Kulkarni et al., 2022; Joshi et al., 2017, 2022). Similarly, a biosensor is a diagnostic electronic tool used for monitoring organic components based on the physicochemical detection principle. The biological components can be in the form of cells, deoxyribonucleic acid (DNA), ribonucleic acid (RNA), enzymes, tissues, antibodies etc., interacting with the target analyte under consideration through a transducer (Khalilian et al., 2017). The sensitive layer in biosensors can be tailored and synthesized using engineering principles. The transduction mechanism can work in many forms such as optical, electronic, electrochemiluminescence, pyroelectric, piezoelectric, gravimetric, magnetic or even electrochemical for ease of measurement and quantification (Dincer et al., 2019). **Figure 17.1** displays various components in a conventional biosensor. Biosensors can detect even the slightest change taking place in cells or molecules in samples of exceptionally low concentration range. The basic working principle is that upon binding the matter with the biocomponent, the transducing platform emits a readout signal which is in direct proportion with the quantity of the matter (You et al., 2009; Velasco, 2009; Fan et al., 2008; Khanna, 2008). Biosensors integrated with nanomaterials can conquer all the demerits possessed by the typical detection techniques by providing several merits such as fast response kinetics, high specificity, and by being portable, non-destructive and economic (Ozoglu et al., 2017; Neethirajan et al., 2018). **Figure 17.2** illustrates the broader classification of biosensors based on the bioreceptor, transducer, mode of technology and detection systems.

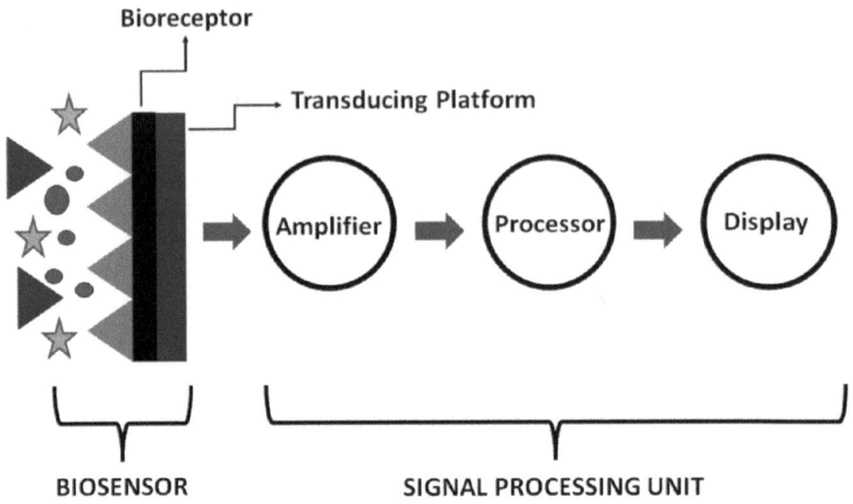

FIGURE 17.1 Different components present in a conventional biosensor

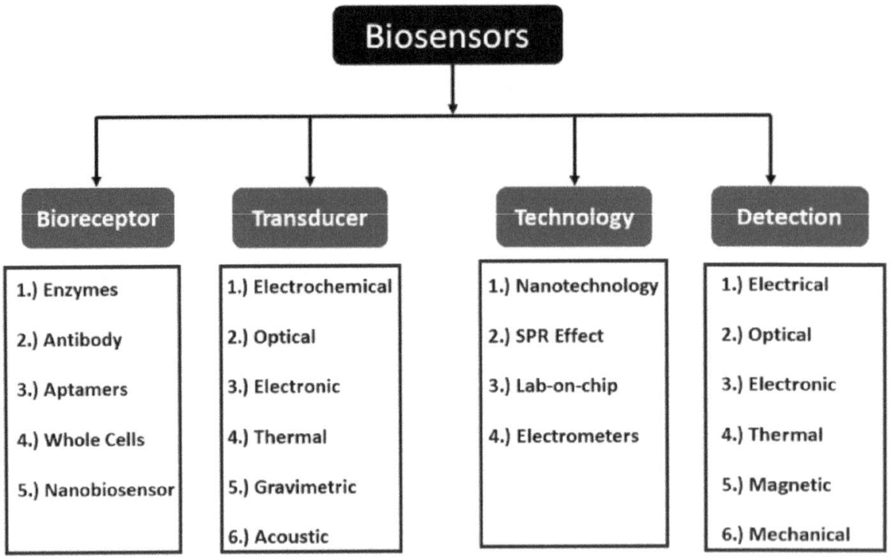

FIGURE 17.2 Classification of a biosensor based on bioreception, transducing platforms, mode of technology and detection systems

17.3 CHARACTERISTICS OF NANOMATERIALS IN BIOSENSORS

Nanomaterials are basic features in nanobiosensors for the detection of biological variants at the atomic scale with greater precision. In a typical biosensor, bioreceptor and transducing platforms are on a macro- and micro-scale. However, in the case

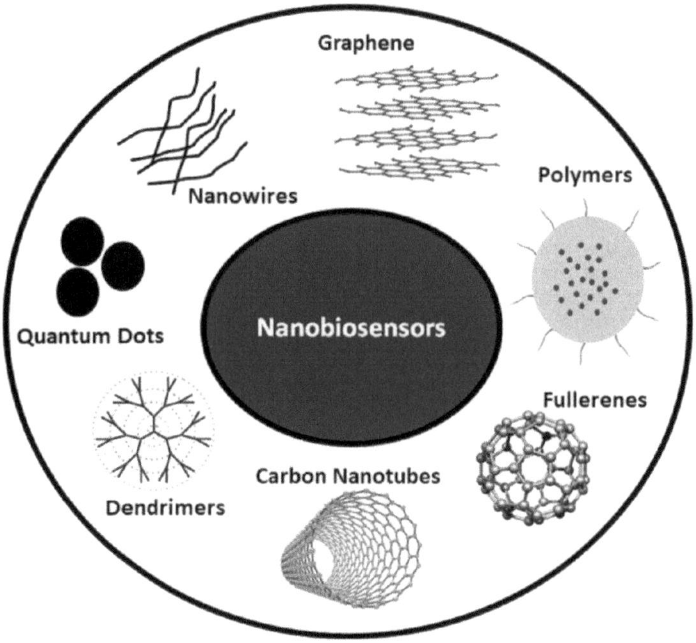

FIGURE 17.3 Various types of nanostructures commonly used for nanobiosensor fabrication

of a nanobiosensor, either bioreceptors or transducing platforms or in some cases even both need to have at least one dimensional parameter in nanoscale, that is less than 100 nm (Doria et al., 2012; Rackauskas et al., 2017). Nanostructures integrated with the biosensor display unique features in terms of optical, mechanical, physical, chemical and electrical properties in comparison to standard materials. Such features provide advantages namely, ultrasensitivity, selectivity, repeatability, long-term stability, low coefficient of variation etc. Researchers have reported several novel nanomaterials showcasing excellent enhancement in biosensing characteristics (Anam et al., 2019; Maduraiveeran & Jin, 2017). In recent times, tailored nanomaterials used in biosensors for different applications included quantum dots, carbon nanotubes, graphene, magnetic nanoparticles, mesoporous nanostructures, photoelectric conversion nanoparticles, upconversion nanoparticles, nanopores, noble metals and metal oxide semiconductors (Wang et al., 2014). **Figure 17.3** showcases various nanomaterials used for biosensor fabrication. In the case of nanobiosensors, the attachment of molecules is a complex part requiring extensive research and a basic understanding of nanomaterials engineering for controlling specific parameters, structural orientation or surface properties in the nanoscale region (Sharma et al., 2021; Lim & Ramakrishna, 2006). Recent progress in nanoscience and nanotechnology has made it possible to prepare nanomaterials with unique material properties and high surface areas using facile chemical methodologies for biosensing applications in the agriculture and food sectors (Shipway et al., 2000). **Figure 17.4** depicts various applications of nanomaterials-integrated biosensors in the agriculture sector.

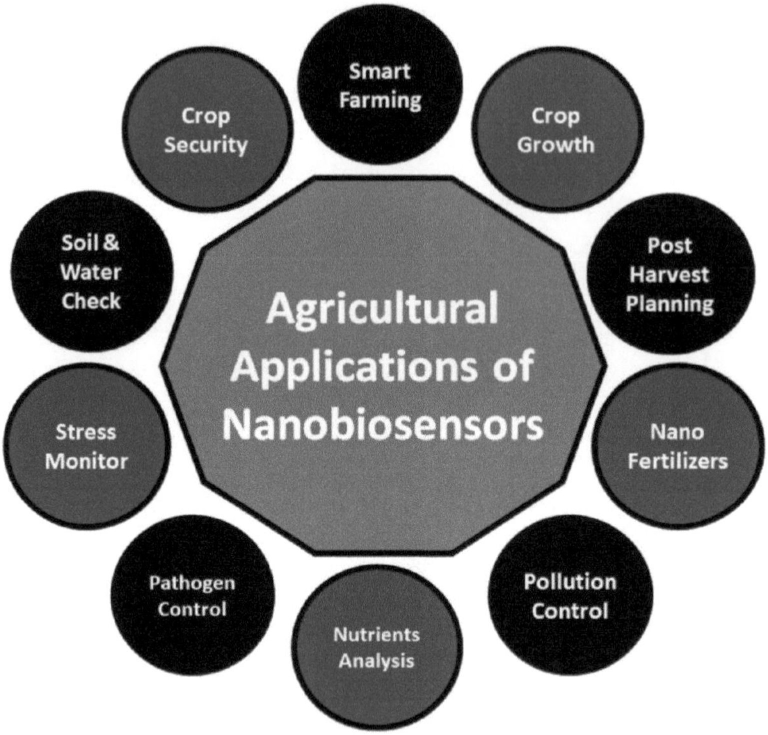

FIGURE 17.4 Applications of nanomaterials-integrated biosensors in the agriculture sector

17.4 APPLICATION OF NANOBIOSENSORS FOR DISEASE DIAGNOSIS

Many recurrent diseases in plants, originated from factors such as deficiencies or infections, significantly impact agricultural efficiency and lead to a decline in crop yields (Mathivanan, 2021; Fang & Ramasamy, 2015). It is a widely stated fact that prevention is always better than cure. Hence, estimation of disease occurrence in the first stage itself can help eliminate the primary factors for its evolution. The current methods for pathogen detection and diagnosis are based on micro-biological, enzyme linked immunosorbent assay (ELISA), or polymerase chain reaction (PCR) techniques (Mathivanan, 2021). These techniques are often sophisticated and accurate; however, they can be time-consuming. Nanobiosensors offers an alternative method of detection for nutrient deficiency, infection or other concerns related to soil and water resources and provides data on crop health to farmers in real-time (Mathivanan, 2021; Fang & Ramasamy, 2015). For example, pathogens such as *Xylella fastidiosa subsp. pauca* strain CoDiRO causes olive quick decline syndrome (OQDS) as well as being capable of infecting other plant species (Raied et al., 2019). Thus, the disease represents a great threat to the economies of agricultural-based countries such as Italy. Consequently, it is highly desirable to develop nanobiosensors for the rapid and early screening of plants for OQDS. Recently, a lab-on-a-chip assay for the determination of *Xylella fastidiosa subsp. pauca* detection in leaf samples was developed and its performance was compared with the standard ELISA and PCR techniques that are conventionally used to detect such pathogens (Chiriacò et al., 2018). The developed technique included a micro-fluidic module, and its performance was found to be competitive with ELISA and PCR in terms of reliability. A further advantage of the developed lab-on-chip technique was its low cost, portability and ease of use. The system was built on a glass substrate and based on a polydimethylsiloxane (PDMS) micro-fluidic module that employed micro-channels and micro-chambers (volume of 20 μL). This allowed for the test sample to be delivered directly to the surface of the electrodes which were designed as an array of interdigitated (IDT) micro-electrodes. The IDT electrodes were functionalized with antibodies that were specific to the *Xylella fastidiosa subsp. pauca*. The system enabled the performance of functionalization steps before measuring the samples in up to quadruplicates by impedance spectroscopy through the addition of hexacyanoferrate (II/III). The multi-measurement capability option improved the reliability of the technique. An earlier study had shown that the lab-on-a-chip technique can be employed for the rapid nucleic acid-based diagnosis of *Phytophthora*, which is a genus of plant-damaging oomycetes (Julich et al., 2011). The single chip allowed for both PCR and hybridization steps to be performed consecutively. The chip consisted of an inflexible and flexible layer with integrated micro-channels and micro-chambers. The inflexible component contained DNA micro-array and integrated half channels placed on it. There were 32 measurement points on the presented setup that allowed for ≥ 5 different DNA fragments to be tested in parallel. The data collection was through electrochemical means where the deposition of silver through enzymatic catalyzation was monitored over a 5–10 minute period. The device was shown to have good potential to be used directly in the field as a diagnostic tool for routine surveys of plant pathogens. In other examples of nanobiosensors, bacterial infectivity by pathogens was successfully distinguished using antibodies-mediated nanostructures (Yao et al., 2009). Fungal infections in wheat and potato crops were detected using gold nanostructures and antibodies-mediated fluorescent silica nanoparticles based on immunosensors working on the principle of optics (Yao et al., 2009; Singh et al., 2010). Witch's broom infection in lemons was conveniently detected using quantum dots by the fluorescence resonance energy transfer (FRET) mechanism (Frasco & Chaniotakis, 2009). Several excellent nanobiosensors have also been shown to display superior sensitivity, and specificity at extremely low concentration ranges (Rad et al., 2012; Safarpour et al., 2012; Kuila et al., 2011; Perez-Lopez & Merkoçi, 2011; Shiddiky & Torriero, 2011). Focussed attention can also be paid by researchers to one of the organic matters produced in higher percentages during the infection stage by crops; that is, methyl salicylate ($C_8H_8O_3$). Early detection of such organic matter can greatly reduce diseases in plants by controlling the infection at the start. Moreover, amperometric nanobiosensors integrated with noble metals and semiconducting metal oxides were reported for fungal, bacterial and viral diseases observed in crops (Cao et al., 2011; Mandler & Kraus-Ophir, 2011; Umasankar & Ramasamy, 2013; Boonham et al., 2008; Chartuprayoon et al., 2013). Recently, zinc oxide (ZnO) and copper oxides (CuO) were prepared using the green chemistry technique with lime peels as reagents for the treatment of black rot infection in citrus fruits (Sardar et al., 2022). Tomato bacterial wilt infection was treated using ZnO, CuO and iron (II) oxide (FeO) nanoparticles (Jiang et al., 2022).

17.5 APPLICATION OF NANOBIOSENSORS FOR FOOD TOXICITY ANALYSIS

Hazardous microbes namely bacteria, parasites, viruses and even toxic chemicals arising from food preservatives, fertilizers and heavy metals from water, soil etc. are communal reasons for the quality deterioration of food items (Kumar et al., 2017). The basic toxins resulting in food toxicity and poisoning are morbific microbes that include *Norovirus, E. coli, Salmonella, Clostridium perfringens, Campylobacter* and *Staphylococcus aureus* (Inbaraj & Chen, 2015). Earlier detection methods had their limitations in terms of being costly, tedious and/or inaccurate. However, with the advent of nanobiosensors, rapid scrutiny and prevention treatment of health hazards due to food contamination and toxicity has been possible (Muthukrishnan, 2022). Nanobiosensors used for food toxicity analysis utilize nanomaterials in various shapes and sizes, for example, particles, tubes, cubes, rods, polygons, fibres, hierarchical structures and superparticles. The standard method for quantification of microbes in food items is a colony forming unit technique utilizing a template made up of agar, which usually uses three daylight hours for preliminary information and as high as seven days for authenticating the presence (Poltronieri et al., 2014). The technique holds risk as food items with a lower shelf life may perish even before test results are out. Alternate techniques such as PCR, ELISA and turbidimetric analysis that are again tedious and costly can be employed instead of colony forming methodology. In such scenarios, nanobiosensors providing advantages such as portability, rapid results, accuracy and cost-effectiveness can be used for various pathogens with scope for on-site testing or real-time monitoring (Srivastava et al., 2018). In the past, *E. coli* and *Staphylococcus aureus* contamination in fresh meat and its freshness over a while were detected using a micro-fluidic immunosensor with dimethylsiloxane, carbon nanotubes and noble metals as a sensitive layer (Abdalhai et al., 2014; Huang et al., 2016). Hybrid colorimetric sensors amplifying β-galactosidase enzymes in an aqueous solution were reported for *E. coli* monitoring using tailored gold nanoparticles (Miranda et al., 2011). Antibodies-mediated $Au@Fe_3O_4$ nanoparticle clusters with lateral flow filters were developed for rapid and accurate investigation of *Salmonella* in dairy products (Hwang et al., 2016). Additionally, zirconia embellished reduced graphene oxide ($nZrO_2@RGO$) heterostructured nanobiosensors were developed for monitoring aflatoxin B1 in food products (Chauhan et al., 2022). Similarly, a contaminant in food that is responsible for major food outbreaks is *Campylobacter jejuni* (Acheson & Allos, 2001). This pathogen is a spiral-shaped motile bacterium that is gram-negative and micro-aerophilic. *Campylobacter jejuni* causes infections in humans which encompass diarrhoea, vomiting, severe abdominal cramps and inflammation. A more recent study reported on a single-stranded DNA aptamer that had a high affinity for *Campylobacter jejuni* (Chen et al., 2020). The DNA aptamer was integrated as a nanobiosensor for the detection of the pathogen through ultraviolet-visible (UV-Vis) responses or colour changes. The group confirmed the accuracy of results by screening milk samples as well as undergoing specificity tests to show the high selectivity toward the pathogen of interest. The limit of detection of the sensor in PBS and milk was reported to be 10 CFU/mL and 13 CFU/mL, respectively. Another interesting application of nanobiosensors in the food industry is the monitoring of mycotoxins which are toxic metabolites secreted by the fungus while undergoing growth and development. Examples of mycotoxins commonly found in fruits are Ochratoxin A., Aflatoxins, Patulin, and Alternaria toxins (Fernández-Cruz et al., 2010). Mycotoxin creates severe health complications even in small concentrations. Given that many food crops are easily affected by different pathogenic fungi, it is appropriate to assume the affected plants are most likely contaminated with mycotoxins. Thus, it is important to develop nanobiosensors that can detect and measure the concentration of mycotoxins. Recently, a study reported on graphene oxide and gold nanocomposites (GO/AuNCs) illustrating high sensitivity toward aflatoxin B1 (AFB1) in peanut samples when used as fluorescent-based nanobiosensor (Li et al., 2018). The authors attributed the high sensitivity of the nanobiosensor to the better quenching ability of the developed nanocomposite. The sensor also showed high selectivity toward AFB1 in the presence of other cross-interferents (including other mycotoxins). Moreover, the nanobiosensor showed a wide linear range and a detection limit of only 0.03 pg/mL toward AFB1. It is evident that nanobiosensors are used for a whole range of areas in the food industry including the detection of heavy metals, pesticide residue, pathogens, toxins and illegal additives as well as the determination of genetically modified organisms.

17.6 CONCLUSIONS

Technological advancement in the fields of materials science, nanotechnology, biology, chemistry and engineering demonstrates the bright future of nanobiosensors in the agriculture sector and food industries. Enhancement in productivity, crop yield, soil quality, nutrient absorption capabilities, deployment of intelligent farming techniques, sustainable food processing, packaging and storage are many of the direct and indirect advantages of nanobiosensors. Integration of tailored nanomaterials synthesized using novel chemical routes improves the sensitivity, selectivity and resolution of the sensor. Although the complete capacity of nanoscience in farming and food industries is still in the nascent stage, the comprehensive understanding reported by the researchers so far through their concerted efforts has been encouraging. In decades to come ahead, innovation in sensor technology is foreseen to have important consequences in the agriculture sector that will result in smarter and sustainable technologies for safeguarding the needs of future generations.

ACKNOWLEDGMENTS

SJ acknowledges the Department of Science and Technology and Biotechnology Industry Research Assistance Council, Ministry of Science and Technology, Government of India for the grants received.

REFERENCES

Abdalhai, M. H., Maximiano Fernandes, A., Bashari, M., Ji, J., He, Q., & Sun, X. (2014). Rapid and sensitive detection of foodborne pathogenic bacteria (*Staphylococcus aureus*) using an electrochemical DNA genomic biosensor and its application in fresh beef. *Journal of Agricultural and Food Chemistry*, *62*(52), 12659–12667.

Acheson, D., & Allos, B. M. (2001). *Campylobacter jejuni* infections: Update on emerging issues and trends. *Clinical Infectious Diseases*, *32*(8), 1201–1206.

Anam, M., Yori, O., Romana, S., Muhammad, A., Waheed, S., & Sadia, Z. (2019). Nanosensors for diagnosis with optical, electric, and mechanical transducers. *RSC Advances*, *9*(12), 6793–6803.

Bhatia, S. (2016). Nanoparticle types, classification, characterization, fabrication methods, and drug delivery applications. In S. Bhatia (Ed.), *Natural polymer drug delivery systems* (pp. 33–93). Springer.

Boonham, N., Glover, R., Tomlinson, J., & Mumford, R. (2008). Exploiting generic platform technologies for the detection and identification of plant pathogens. *European Journal of Plant Pathology*, *121*(3), 355–363.

Cao, X., Ye, Y., & Liu, S. (2011). Gold nanoparticle-based signal amplification for biosensing. *Analytical Biochemistry*, *417*(1), 1–16.

Chartuprayoon, N., Rheem, Y., Ng, J. C. K., Nam, J., Chen, W., & Myung, N. V. (2013). Polypyrrolenanoribbon based chemiresistive immunosensors for viral plant pathogen detection. *Analytical Methods*, *5*(14), 3497–3502.

Chauhan, D., Kumar, Y., Chandra, R., & Kumar, S. (2022). Nanostructured zirconia@reduced graphene oxide based ultraefficient nanobiosensing platform for food toxin detection. *Sensors & Diagnostics*, *1*(3), 550–557.

Chauhan, N., Narang, J., & Jain, U. (2016). Amperometric acetylcholinesterase biosensor for pesticides monitoring utilising iron oxide nanoparticles and poly (indole-5-carboxylic acid). *Journal of Experimental Nanoscience*, *11*(2), 111–122.

Chen, W., Teng, J., Yao, L., Xu, J., & Liu, G. (2020). Selection of specific DNA aptamers for hetero-sandwich-based colorimetric determination of campylobacter jejuni in food. *Journal of Agricultural and Food Chemistry*, *68*(31), 8455–8461.

Chiriacò, M. S., Luvisi, A., Primiceri, E., Sabella, E., De Bellis, L., & Maruccio, G. (2018). Development of a lab-on-a-chip method for rapid assay of *Xylella fastidiosa* subsp. pauca strain CoDiRO. *Scientific Reports*, *8*(1), 7376.

Cui, D., Guo, Q., & Zhang, X. (2015). Recent advances in nanotechnology applied to biosensors. *Sensors*, *15*, 30736–30758.

Dasgupta, N., Ranjan, S., Mundekkad, D., Ramalingam, C., Shanker, R., & Kumar, A. (2015). Nanotechnology in agro-food: From field to plate. *Food Research International*, *69*, 381–400.

Dincer, C., Bruch, R., Costa-Rama, E., Fernández-Abedul, M. T., Merkoçi, A., Manz, A., Urban, G. A., & Güder, F. (2019). Disposable sensors in diagnostics, food, and environmental monitoring. *Advanced Materials*, *31*(30), 1806739.

Doria, G., Conde, J., Veigas, B., Giestas, L., Almeida, C., Assuncao, M., Rosa, J., & Baptista, P. V. (2012). Noble metal nanoparticles for biosensing applications. *Sensors*, *12*(2), 1657–1687.

Ettenauer, J., Zuser, K., Kellner, K., Posnicek, T., & Brandl, M. (2015). Development of an automated biosensor for rapid detection and quantification of *E. coli* in water. *Procedia Engineering*, *120*, 376–379.

Fan, X., White, I. M., Shopova, S. I., Zhu, H., Suter, J. D., & Sun, Y. (2008). Sensitive optical biosensors for unlabeled targets: A review. *Analytica Chimica Acta*, *620*(1–2), 8–26.

Fang, Y., & Ramasamy, R. P. (2015). Current and prospective methods for plant disease detection. *Biosensors*, *5*(3), 537–561.

Farahi, R. H., Passian, A., Tetard, L., & Thundat, T. (2012). Critical issues in sensor science to aid food and water safety. *ACS Nano*, *6*(6), 4548–4556.

Fernández-Cruz, M. L., Mansilla, M. L., & Tadeo, J. L. (2010). Mycotoxins in fruits and their processed products: Analysis, occurrence and health implications. *Journal of Advanced Research*, *1*(2), 113–122.

Frasco, M. F., & Chaniotakis, N. (2009). Semiconductor quantum dots in chemical sensors and biosensors. *Sensors*, *9*(9), 7266–7286.

Gallocchio, F., Belluco, S., & Ricci, A. (2015). Nanotechnology and food: Brief overview of the current scenario. *Procedia Food Science*, *5*, 85–88.

Huang, H., Liu, M., Wang, X., Zhang, W., Yang, D.-P., & Cui, L. (2016). Label-free 3D Ag nanoflower-based electrochemical immunosensor for the detection of *Escherichia coli* O157. H7 pathogens. *Nanoscale Research Letters*, *11*(1), 507.

Hwang, J., Kwon, D., Lee, S., & Jeon, S. (2016). Detection of Salmonella bacteria in milk using gold-coated magnetic nanoparticle clusters and lateral flow filters. *RSC Advances*, *6*(54), 48445–48448.

Inbaraj, B. S., & Chen, B. (2015). Nanomaterial-based sensors for detection of food borne bacterial pathogens and toxins as well as pork adulteration in meat products. *Journal of Food and Drug Analysis*, *24*(1), 15–28.

Jiang, H., Lv, L., Ahmed, T., Jin, S., Shahid, M., Noman, M., Osman, H. E. H., Wang, Y., Sun, G., Li, X., & Li, B. (2022). Nanoparticles mediated amelioration of tomato bacterial wilt disease by modulating the rhizosphere bacterial community. *International Journal of Molecular Sciences*, *23*, 414.

Joshi, S., Ippolito, S. J., Periasamy, S., Sabri, Y. M., & Sunkara, M. V. (2017). Efficient heterostructures of Ag@CuO/$BaTiO_3$ for low-temperature CO_2 gas detection: Assessing the role of nanointerfaces during sensing by operando DRIFTS technique. *ACS Applied Materials & Interfaces*, *9*(32), 27014–27026.

Joshi, S., Tonde, S., Wakhure, U., Bornare, D., Chatterjee, A., Syed, K., & Sunkara, M. V. (2022). Hierarchical $CaTiO_3$ microspheres for acetone sensing. *Sensors and Actuators B: Chemical*, *359*, 131621.

Julich, S., Riedel, M., Kielpinski, M., Urban, M., Kretschmer, R., Wagner, S., Fritzsche, W., Henkel, T., Möller, R., & Werres, S. (2011). Development of a lab-on-a-chip device for diagnosis of plant pathogens. *Biosensors and Bioelectronics*, *26*(10), 4070–4075.

Khalilian, A., Khan, Md. R. R., & Kang, S. W. (2017). Highly sensitive and wide-dynamic-range side-polished fiber-optic taste sensor. *Sensors and Actuators B: Chemical*, *249*, 700–707.

Khanna, V. K. (2008). New generation nano-engineered biosensors, enabling nanotechnologies and nanomaterials. *Sensor Review*, *28*(1), 39–45.

Kuila, T., Bose, S., Khanra, P., Mishra, A. K., Kim, N. H., & Lee, J. H. (2011). Recent advances in graphene-based biosensors. *Biosensors and Bioelectronics*, *26*(12), 4637–4648.

Kulkarni, A., Satbhai, M., Li, W., Bornare, D., Syed, K., & Joshi, S. (2022). Oleic acid induced tailored morphological features and structural defects in CuO for multifunctional applications. *Materials Advances*, *3*, 418–436.

Kumar, V., Guleria, P., & Mehta, S. K. (2017). Nanosensors for food quality and safety assessment. *Environmental Chemistry Letters*, *15*(2), 165–177.

Li, Z., Xue, N., Ma, H., Cheng, Z., & Miao, X. (2018). An ultrasensitive and switch-on platform for aflatoxin b1 detection in peanut based on the fluorescence quenching of graphene oxide-gold nanocomposites. *Talanta*, *181*, 346–351.

Lim, T. C., & Ramakrishna, S. (2006). A conceptual review of nanosensors. *Zeitschrift für Naturforschung – A*, *61*(7–8), 402–412.

Maduraiveeran, G., & Jin, W. (2017). Nanomaterials based electrochemical sensor and biosensor platforms for environmental applications. *Trends in Environmental Analytical Chemistry*, *13*, 10–23.

Malhotra, B. D., Singhal, R., Chaubey, A., Sharma, S. K., & Kumar, A. (2005). Recent trends in biosensors. *Current Applied Physics*, *5*(2), 92–97.

Mandler, D., & Kraus-Ophir, S. (2011). Self-assembled monolayers (SAMs) for electrochemical sensing. *Journal of Solid State Electrochemistry*, *15*(7–8), 1535–1558.

Mathivanan, S. (2021). Perspectives of nanomaterials and nanobiosensors in food safety and agriculture. In K. Krishnamoorthy (Ed.), *Novel nanomaterials* (pp. 1–21). Intech Open.

Miranda, O. R., Li, X., Garcia-Gonzalez, L., Zhu, Z. J., Yan, B., Bunz, U. H., & Rotello, V. M. (2011). Colorimetric bacteria sensing using a supramolecular enzyme nanoparticle biosensor. *Journal of American Chemical Society*, *133*(25), 9650–9653.

Mousavi, S. R., & Rezaei, M. (2011). Nanotechnology in agriculture and food production. *Journal of Applied Environmental and Biological Sciences*, *1*, 414–419.

Mustafa, F., & Andreescu, S. (2018). Chemical and biological sensors for food quality monitoring and smart packaging. *Foods*, *7*(10), 168.

Muthukrishnan, L. (2022). An overview on the nanotechnological expansion, toxicity assessment and remediating approaches in agriculture and food industry. *Environmental Technology & Innovation*, *25*, 102136.

Neethirajan, S., Ragavan, V., Weng, X., & Chand, R. (2018). Biosensors for sustainable food engineering: Challenges and perspectives. *Biosensors*, *8*(1), 23.

Ozoglu, O., Unal, M. A., & Altuntaş, E. G. (2017). Biosensors: Lactate biosensors in food and health field. *Turkish Journal of Life Sciences*, *2*, 180–193.

Perez-Lopez, B., & Merkoçi, A. (2011). Nanoparticles for the development of improved (bio) sensing systems. *Analytical and Bioanalytical Chemistry*, *399*(4), 1577–1590.

Poltronieri, P., Mezzolla, V., Primiceri, E., & Maruccio, G. (2014). Biosensors for the detection of food pathogens. *Foods*, *3*(3), 511–526.

Rackauskas, S., Barbero, N., Barolo, C., & Viscardi, G. (2017). ZnO nanowire application in chemoresistive sensing: A review. *Nanomaterials*, *7*(11), 381.

Rad, F., Mohsenifar, A., Tabatabaei, M., Safarnejad, M. R., Shahryari, F., Safarpour, H., Foroutan, A., Mardi, M., Davoudi, D., & Fotokian, M. (2012). Detection of Candidatus Phytoplasma aurantifolia with a quantum dots fret-based biosensor. *Journal of Plant Pathology*, *94*, 525–534.

Raied, A. K., Giampetruzzi, A., Altamura, G., Saponari, M., & Saldarelli, P. (2019). Infections of the Xylella fastidiosa subsp. pauca strain "De Donno" in lfalfa (*Medicago sativa*) elicits an overactive immune response. *Plants*, *8*(9), 335.

Reddy, M. N. K., & Ratna, N. P. (2013). Nanobiosensors. *Advance in Electronic and Electric Engineering*, *3*, 321–326.

Safarpour, H., Safarnejad, M. R., Tabatabaei, M., Mohsenifar, A., Rad, F., Basirat, M., Shahryari, F., & Hasanzadeh, F. (2012). Development of a quantum dots FRET-based biosensor for efficient detection of *Polymyxa betae*. *Canadian Journal of Plant Pathology*, *34*(4), 507–515.

Sardar, M., Ahmed, W., Al Ayoubi, S., Nisa, S., Bibi, Y., Sabir, M., Khan, M. M., Ahmed, W., & Qayyum, A. (2022). Fungicidal synergistic effect of biogenically synthesized zinc oxide and copper oxide nanoparticles against *Alternaria citri* causing citrus black rot disease. *Saudi Journal of Biological Sciences*, *29*(1), 88–95.

Sekhon, B. S. (2014). Nanotechnology in agri-food production: An overview. *Nanotechnology, Science and Applications*, *7*, 31–53.

Sharma, P., Pandey, V., Sharma, M. M. M., Patra, A., Singh, B., Mehta, S., & Husen, A. (2021). A review on biosensors and nanosensors application in agroecosystems. *Nanoscale Research Letters*, *16*(1), 136.

Shiddiky, M. J., & Torriero, A. A. (2011). Application of ionic liquids in electrochemical sensing systems. *Biosensors and Bioelectronics*, *26*(5), 1775–1787.

Shipway, A. N., Katz, E., & Willner, I. (2000). Nanoparticle arrays on surfaces for electronic, optical, and sensor applications. *ChemPhysChem*, *1*(1), 18–52.

Singh, S., Singh, M., Agrawal, V. V., & Kumar, A. (2010). An attempt to develop surface plasmon resonance based immunosensor for Karnal bunt (*Tilletia indica*) diagnosis based on the experience of nano-gold based lateral flow immuno-dipstick test. *Thin Solid Films*, *519*(3), 1156–1159.

Srivastava, A. K., Dev, A., & Karmakar, S. (2018). Nanosensors and nanobiosensors in food and agriculture. *Environmental Chemistry Letters*, *16*(1), 161–182.

Umasankar, Y., & Ramasamy, R. P. (2013). Highly sensitive electrochemical detection of methyl salicylate using electroactive gold nanoparticles. *Analyst*, *138*(21), 6623–6631.

Velasco, M. N. (2009). Optical biosensors for probing at the cellular level: A review of recent progress and future prospects. *Seminars in Cell & Developmental Biology*, *20*(1), 27–33.

Wang, Y. Z., Qu, K., Tang, L. H., Li, Z. L., Moore, E., Zeng, X. Q., Liu, Y., & Li, J. H. (2014). Nanomaterials in carbohydrate biosensors. *TrAC Trends in Analytical Chemistry*, *58*, 54–70.

Weiss, P. S. (2010). Nanoscience and nanotechnology: Present and future. *ACS Nano*, *4*(4), 1771–1772.

Yao, K. S., Li, S. J., Tzeng, K. C., Cheng, T. C., Chang, C. Y., Chiu, C. Y., Liao, C. Y., Hsu, J. J., & Lin, Z. P. (2009). Fluorescence silica nanoprobe as a biomarker for rapid detection of plant pathogens. *Advanced in Materials Research*, *79*, 513–516.

You, C., Bhagawati, M., Brecht, A., & Piehler, J. (2009). Affinity capturing for targeting proteins into micro and nanostructures. *Analytical and Bioanalytical Chemistry*, *360*(6–7), 1563–1570.